2006 PHYSICS EDUCATION RESEARCH CONFERENCE

Proceedings in the Series of Physics Education Research Conferences

Year	Title	Publisher	ISBN
2006	2006 Physics Education Research Conference	AIP Conf. Proceedings Vol. 883	978-0-7354-0383-3
2005	2005 Physics Education Research Conference	AIP Conf. Proceedings Vol. 818	0-7354-0311-2
2004	2004 Physics Education Research Conference	AIP Conf. Proceedings Vol. 790	0-7354-0281-7
2003	2003 Physics Education Research Conference	AIP Conf. Proceedings Vol. 720	0-7354-0200-0

To learn more about this title, or the AIP Conference Proceedings Series, please visit the webpage **http://proceedings.aip.org**

2006 PHYSICS EDUCATION RESEARCH CONFERENCE

Syracuse, New York *26 – 27 July 2006*

EDITORS

Laura McCullough

University of Wisconsin-Stout, Menomonie, Wisconsin

Leon Hsu

University of Minnesota, Minneapolis, Minnesota

Paula Heron

University of Washington, Seattle, Washington

SPONSORING ORGANIZATION

American Association of Physics Teachers (AAPT)

Melville, New York, 2007

AIP CONFERENCE PROCEEDINGS ■ VOLUME 883

Editors:

Laura McCullough
Department of Physics
University of Wisconsin-Stout
P.O. Box 790
Menomonie, WI 54751
USA

E-mail: mcculloughl@uwstout.edu

Leon Hsu
Department of Postsecondary Teaching and Learning
University of Minnesota
128 Pleasant Street, S.E.
Minneapolis, MN 55455
USA

E-mail: lhsu@umn.edu

Paula Heron
Department of Physics
University of Washington
Box 351560
Seattle, WA 98185-1560
USA

E-mail: pheron@phys.washington.edu

L.C. Catalog Card No. 2006939387
ISBN 978-0-7354-0383-3
ISSN 0094-243X
Printed in the United States of America

CONTENTS

INVITED PAPERS

PEER-REVIEWED PAPERS

Preface

The theme of the 2006 PERC, *Discipline-based education research in other STEM disciplines,* is reflected in a few of the articles, but the bulk continue to represent a very broad spectrum of directions within PER. The Editors would like to thank organizers Steve Kanim, Rebecca Lindell, Michael Loverude and Chandralekha Singh for organizing a stimulating and productive meeting.

This year marks the fourth year of publication of the PERC Proceedings by the AIP and the sixth year that a peer-reviewed volume has been produced. Feedback suggests that the Proceedings continue to play an important role. They provide a snapshot of ongoing research that can help characterize the PER community as a whole. They also provide a mechanism for authors to get feedback, for readers to find ideas, and for both to consider new collaborative efforts. For many members of the community the Proceedings also serve as valuable evidence of scholarly activity. Since the value of the Proceedings is determined as much by the referees as by the authors, it is traditional to thank them in the preface. Most referees are authors themselves. In fact, agreeing to review papers is a condition for submitting one. This system works well up to a point. Many authors, such as junior graduate students, are relative newcomers to the field. Some groups submit a large number of papers that increases the refereeing load but are not able to contribute a corresponding number of referees to the reviewing pool. In order to ensure that every paper submitted receives a fair review in a short period of time it has been the practice of recent editors to ask a few non-authors to review as a service to the community. Without the willingness of these people to devote time and energy to the Proceedings they would not be as good as they are. This year a total of 67 individuals reviewed more than 50 papers. Thanks go out to: Eugenia Etkina, Chandralekha Singh, Charles Henderson, Jennifer Blue, Valerie Otero, Kathleen Harper, Andy Elby, Scott Bonham, Jeffrey Marx, Charles De Leone, David Meltzer, Paula Engelhardt, Karen Cummings, Steven Pollock, Vincent Coletta, Melissa Dancy, Michael Wittmann, Richard Steinberg, Rebecca Lindell, John Thompson, Bob Beichner, Dean Zollman, Alice Churukian, Noah Finkelstein, Sanjay Rebello, Tom Foster, John Clement, Paul Ashcraft, Young-Jin Lee, Anna Karelina, Federica Raia, Katherine Perkins, M. L. Horner, Cathy Mariotti Ezrailson, Gyoungho Lee, Edward Price, Antti Rissanen, Jeffrey Phillips, Gary Gladding, Catherine Ishikawa, Kristin Walker, Zeynep Isvan, Daniel Haileselassie, David Rosengrant, E. Grant Williams, A. Lynn Stephens, Maria Ruibal Villasenor, Danielle Harlow, Jessica Mamais, Noah Podolefsky, C.J. Keller, Eleanor Sayre, Elizabeth Gire, Patrick Kohl, Brandon Bucy, Eugene Torigoe, Ossama Tfeily, Katrina Black, Dedra Demaree, Bijaya Aryal, Edgar Corpuz, Lili Cui, Jeff Saul, and Eric Brewe.

Two recent developments raise some important issues for the future of the Proceedings. One is the launch of *Physical Review Special Topics – PER,* which provides an additional publication venue. The Proceedings were started in the era in which many felt that growth in the community was far outpacing the ability of the *American Journal of Physics* to serve as the *de facto* publication venue. The advent of PRST-PER may change the way in which the PERC Proceedings are used and viewed. The other development is the establishment of the PER Topical Group in the AAPT and the election of its first leadership committee. In the past the three editors of the PERC have invited a new member to join the team each year as the senior editor cycles off. It may be that the PER-TG may provide some input into the editorial succession in the future. In any case it seems probable that the PERC Proceedings will continue to evolve as the community does. I wish the future conference organizers and editors the best of luck.

Paula Heron
Outgoing Editor

Program
2006 Physics Education Research Conference
Syracuse, New York

Wednesday, July 26

3:30 - 5:30: Bridging Session: Invited Talks & Panel Discussion

3:30: Use of Concept Inventories to Identify Misconceptions in Thermal Sciences
Ronald L. Miller, Colorado School of Mines

4:00: Student Thinking About Rate of Change in Differential Equations
Chris Rasmussen, San Diego State University

4:30: Building the Biology Concept Inventory using Ed's Tools, an on-line response analysis system
Michael W. Klymkowsky, Rachel Gheen, Rebecca Koopman, Isidoros Doxas & R. Kathy Garvin-Doxas, University of Colorado, Boulder

5:00: Panel Discussion
Discussant: To be Announced

6:00 - 7:00: Dinner Banquet
Banquet Speaker: Jay Labov, NRC/NAS

7:00 -9:00: Contributed Poster Session

Thursday, July 27

8:15 - 8:30: Orientation

8:15 - 9:45: Workshops , Targeted Poster Sessions & Roundtable Discussions- I

Workshop W-B: PER and Human Subjects: The PER Community and Institutional Review Boards
Gordon Aubrecht, the Ohio State University

Targeted Poster Session TP-A: Issues and Innovations in Concept Inventory Development & Administration
Rebecca Lindell, Southern Illinois University Edwardsville

Targeted Poster Session TP-B: Physics Education Research Across Disciplinary Boundaries
Michael Loverude, California State University Fullerton

Targeted Poster Session TP-C: Investigations of Student Learning in Upper Division Courses that Link Physics, Chemistry and Engineering
Brad Ambrose, Grand Valley State University

Targeted Poster Session TP-D: Bridging the Gap Between Instructors and Curriculum Developers: Facilitating Successful Communication and Promoting Customization
Chandralekha Singh, University of Pittsburgh

9:45 - 10:00: Break

10:00 - 12:00: Invited Talks & Panel Discussion (Session II)
Discussant: Chandralekha Singh

10:00: The Effects of Inquiry-Based Instruction on Elementary Teaching Majors' Chemistry Content Knowledge and Their Views about Teaching Science
Michael Sanger, Middle Tennessee State University

10:30: Rasch Analysis and the Geoscience Concept Inventory
Julie Libarkin, Michigan State University & Steven W. Anderson, Black Hills State University

11:00: Astronomy Concept Inventories: Foundations and Frontiers
Beth Hufnagel, Anne Arundel Community College

11:30: Panel Discussion
Discussant: To be Announced

12:00 - 1:30: Luncheon Banquet
Musician: Peter Haskell

1:45 - 3:15: Workshops , Targeted Poster Sessions & Roundtable Discussions - II

Workshop W-A: Multiple Methods of Analyzing Reasoning About Quantum Tunneling
Sam McKagan, University of Colorado & Michael Wittmann, University of Maine

Targeted Poster Session TP-A: Issues and Innovations in Concept Inventory Development & Administration
Rebecca Lindell, Southern Illinois University Edwardsville

Targeted Poster Session TP-B: Physics Education Research Across Disciplinary Boundaries
Michael Loverude, California State University Fullerton

Targeted Poster Session TP-D: Bridging the Gap Between Instructors and Curriculum Developers: Facilitating Successful Communication and Promoting Customization
Chandralekha Singh, University of Pittsburgh

Roundtable Discussion RT-A: Cross-Pollination in Science Education Research
Eleanor Sayre, University of Maine and Rebecca Lindell, Southern Illinois University Edwardsville

3:15 - 3:45: Break

3:45 - 5:15: Workshops , Targeted Poster Sessions & Roundtable Discussions - III

Workshop W-A: Multiple Methods of Analyzing Reasoning About Quantum Tunneling
Sam McKagan, University of Colorado & Michael Wittmann, University of Maine

Workshop W-B: PER and Human Subjects: The PER Community and Institutional Review Boards
Gordon Aubrecht, the Ohio State University

Targeted Poster Session TP-C: Investigations of Student Learning in Upper Division Courses that Link Physics, Chemistry and Engineering
Brad Ambrose, Grand Valley State University

Roundtable Discussion RT-A: Cross-Pollination in Science Education Research
Eleanor Sayre, University of Maine and Rebecca Lindell, Southern Illinois University Edwardsville

Poster Titles and Authors

Validation Studies of the Colorado Physics Problem Solving Survey
Adams, Wendy University of Colorado, Boulder
Wieman, Carl University of Colorado, Boulder

Use of Physical Models to Facilitate Transfer of Physics Learning to Understand Positron Emission Tomography
Aryal, Bijaya Kansas State University, Department of Physics
Zollman, Dean Kansas State University, Department of Physics
Rebello, N. Sanjay Kansas State University, Department of Physics

Modeling Aspects of Nature of Science to Preservice Elementary Teachers
Ashcraft, Paul Penn State Erie, The Behrend College

Student Perceptions Of A Class's Goals: Interclass Comparisons
Aubrecht, Gordon Ohio State University at Marion
Lin, Yuhfen Ohio State University

Physics By Inquiry: A Model For Effective Education Reform In Mathematics And Science
Barsky, Constance Learning by Redesign, Physics Department, The Ohio State University

Students' Integration Methods For First-Order Differential Equations
Black, Katrina University of Maine
Sayre, Eleanor University of Maine
Wittmann, Michael University of Maine

One Laboratory Course Serving Two Populations: Student Perceptions Of Their Understanding Of Physics And Their Group Roles
Blue, Jennifer Miami University

Reliability, Compliance and Security of Web-based Pre/Post-testing
Bonham, Scott Department of Physics and Astronomy, Western Kentucky University

From Physics To Physiology, A Spread Of Curricular Reform.
Brewe, Eric Hawaii Pacific University
Korsmeyer, Keith Hawaii Pacific University

Student (Mis)Application Of Partial Differentiation To Material Properties
Bucy, Brandon R. Department of Physics and Astronomy, University of Maine, Orono, ME
Thompson, John R. Department of Physics and Astronomy, University of Maine, Orono, ME
Mountcastle, Donald B. Department of Physics and Astronomy, University of Maine, Orono, ME

Impact of a Classroom Interaction System on Student Learning
Bueckman, Joseph Kansas State University, Department of Physics
Rebello, N. Sanjay Kansas State University, Department of Physics
Zollman, Dean Kansas State University, Department of Physics

Comparison of Student Responses to Identical DC Circuit Questions When Presented on Different Diagnostics
Churukian, Alice D. Concordia College
Engelhardt, Paula V. Tennessee Technological University

Thinking in Physics
Coletta, Vincent Loyola Marymount University
Phillips, Jeffrey Loyola Marymount University

Refining Students' Ideas of Microscopic Friction: A Case Study with Two Students
Corpuz, Edgar G. Kansas State University, Department of Physics
Rebello, N. Sanjay Kansas State University, Department of Physics

Using Physics Jeopardy Problems to Assess College Students' Transfer of Learning from Calculus to Physics
Cui, Lili Kansas State University, Department of Physics
Rebello, N. Sanjay Kansas State University, Department of Physics
Bennett, Andrew G. Kansas State University, Department of Mathematics

The Effectiveness of Incorporating Context-Rich Conceptual Writing into Physics Education
Cummings, Karen Southern Connecticut State University
Murphy, Michael Southern Connecticut State University

A Post-Positivist Perspective on Physics Education Reform
Dancy, Melissa University of North Carolina at Charlotte
Henderson, Charles Western Michigan University

Cultivating Problem Solving Skills via a New Problem Categorization Scheme
Harper, Kathleen A. The Ohio State University
Freuler, Richard J. The Ohio State University
Demel, John T. The Ohio State University

Diffusion of Educational Innovations via Co-Teaching
Henderson, Charles Western Michigan University
Beach, Andrea Western Michigan University
Famiano, Michael Western Michigan University

Comparison of Teaching Methods for Energy Conservation
Horner, M. L. Southern Illinois University Edwardsville
Jeng, Momo Southern Illinois University Edwardsville / Syracuse University
Lindell, Rebecca Southern Illinois University Edwardsville

Threatened Because of Gender?
Humpherys, Candice BYU-Idaho Department of Physics
Pyper, Brian A. BYU-Idaho Department of Physics

Tutorials on Coulomb's law and Gauss's law
Isvan, Zeynep University of Pittsburgh
Singh, Chandralekha University of Pittsburgh

Investigating Students' Ideas About X-rays While Developing Teaching Materials for a Medical Physics Course
Kalita, Spartak Kansas State University, Department of Physics
Zolman, Dean Kansas State University, Department of Physics

When And How Do Students Engage In Sense-Making In A Physics Lab?
Karelina, Anna Graduate School of Education, Rutgers University
Etkina, Eugenia Graduate School of Education, Rutgers University

Studying the Use of Computer Simulations in Undergraduate Laboratory Environments
Keller, Christopher University of Colorado at Boulder
Finkelstein, Noah University of Colorado at Boulder
Perkins, Katherine University of Colorado at Boulder
Pollock, Steven University of Colorado at Boulder

Student Estimates of Probability and Uncertainty in Statistical Physics
Mountcastle, Donald B. Department of Physics and Astronomy, University of Maine
Bucy, Brandon R. Department of Physics and Astronomy, University of Maine
Thompson, John R. Department of Physics and Astronomy, University of Maine

Open-Ended Exam Questions In Large-Enrollment Classes: Is The Trouble Worth It?
Murthy, Sahana Massachusetts Institute of Technology (previously, Rutgers University)
Etkina, Eugenia Rutgers University

SWOSing and Its Implications for Learning Physics
Otero, Valerie University of Colorado at Boulder
Jalovec, Stephanie University of Colorado at Boulder
Harlow, Danielle University of Colorado at Boulder

Classification of Common Physics and Astronomy Concept Inventories
Peak, Elizabeth Southern Illinois University Edwardsville
Lindell, Rebecca Southern Illinois University Edwardsville
Foster, Thomas Southern Illinois University Edwardsville

Chemistry Vs. Physics: A Comparison Of How Biology Majors View Each Discipline And Learning In The Discipline
Perkins, Katherine University of Colorado
Barbera, Jack University of Colorado
Adams, Wendy University of Colorado
Wieman, Carl University of Colorado

Reframing Analogy: Framing, Blending, And Representations As Mechanisms Of Learning By Analogy
Podolefsky, Noah S. University of Colorado at Boulder
Finkelstein, Noah D. University of Colorado at Boulder

Sustaining Change: Instructor Effects in Transformed Large Lecture Courses.
Pollock, Steven University of Colorado, Boulder
Finkelstein, Noah University of Colorado, Boulder

Characterization Of Instructor And Student Use Of Ubiquitous Presenter, A Presentation System Enabling Spontaneity And Digital Archiving
Price, Edward California State University, San Marcos
Malani, Roshni University of California, San Diego
Simon, Beth University of California, San Diego

The Concepts of Causality in Complex Dynamic Systems: A Challenge in Science Education
Raia, Federica City College of New York and CUNY Graduate Center
Steinberg, Richard City College of New York and CUNY Graduate Center

Transfer of Learning: Implications for Research, Curriculum Development and Instruction
Rebello, N. Sanjay Kansas State University, Department of Physics

A Project To Make Better Printed Material At Cadet Physics And Engineering Education And To Test Idea Of Reuse The Material Production Work
Rissanen, Antti Teacher of Technology

Free Body Diagrams: How Do We Know if Students Benefit from Using Them?
Rosengrant, David Rutgers, The State University of New Jersey
Etkina, Eugnia Rutgers, The State University of New Jersey
Van Heuvelen, Alan Rutgers, The State University of New Jersey

Reformed Physics Instruction Through The Eyes Of Students
Ruibal Villasenor, Maria Rutgers University
Etkina, Eugenia Rutgers University

A DVD on Basic Neutron Activation and Radioactive Decay Experiments for High Schools and Introductory Physics Courses
Sahyun, Steven C. University of Wisconsin-Whitewater

Resource Plasticity: Detailing a Common Chain of Reasoning in Damped Harmonic Motion
Sayre, Eleanor, University of Maine
Wittmann, Michael, University of Maine

Student Difficulties With Quantum Mechanics Formalism
Singh, Chandralekha University of Pittsburgh

Effects Of Changing Representations In Two-Dimensional Motion
Springuel, R. Padraic University of Maine
Thompson, John R. University of Maine, Center for Science and Mathematics Education Research
Wittmann, Michael C. University of Maine, Center for Science and Mathematics Education Research

Depictive Gestures As Evidence For Dynamic Mental Imagery In Four Types Of Student Reasoning
Stephens, A. Lynn University of Massachusetts Amherst
Clement, John J. University of Massachusetts Amherst

Why do Students Struggle in Freshman Physics?
Tfeily, Ossama University of North Carolina at Charlotte
Dancy, Melissa H. University of North Carolina at Charlotte

Numeric And Symbolic: Equivalent To Us, Different To Them.
Torigoe, Eugene University of Illinois Urbana-Champaign
Gladding, Gary University of Illinois Urbana-Champaign

Analysis Of Student Understanding Of Symmetry In Gauss' Law
Traxler, Adrienne University of California Santa Cruz
Black, Katrina University of Maine
Thompson, John University of Maine
Wittmann, Michael University of Maine

Sustaining Reform: A Qualitative Study of Professors' Beliefs and Classroom Practices
Turpen, Chandra University of Colorado, Boulder
Finkelstein, Noah University of Colorado, Boulder
Pollock, Steve University of Colorado, Boulder

Investigation and Evaluation Of A Physics Tutorial Center
Walker, Kristin N. University of North Carolina at Charlotte
Dancy, Melissa H. University of North Carolina at Charlotte

Item Rearrangement and Its Impact on Student Performance
Wang, Jing The Ohio State University
Bao, Lei The Ohio State University

Strategy Levels for Guiding Discussion to Promote Explanatory Model Construction in Circuit Electricity
Williams, E. Grant University of Massachusetts - Amherst
Clement, John J. University of Massachusetts - Amherst

Probability and probability density in Intuitive Quantum Physics
Wittmann, Michael C. University of Maine

SECTION I

Invited Papers

Avoiding Reflex Responses: Strategies for Revealing Students' Conceptual Understanding in Biology

Michael W. Klymkowsky, Rachel Gheen, and Kathy Garvin-Doxas

The Bioliteracy Project, Molecular, Cellular, and Developmental Biology, University of Colorado, Boulder, Boulder, Colorado 80309-0347, U.S.A.

Abstract. There is widespread concern about the level of scientific literacy in the U. S. An important, although often overlooked, point, is that student learning is generally only a good as the assessments used to measure it. Unfortunately, most assessments measure recall and recognition rather than conceptual understanding, and as a result over-estimate levels of scientific literacy. We have encountered this fact during the construction of the Biology Concept Inventory (BCI). Using the concept of diffusion, which is taught in a wide range of introductory biology, chemistry, and physics courses, as an exemplar, we describe lessons learned and strategies we use to create questions that better probe student understanding.

Keywords: reflexive responses, memory versus conceptual assessments.
PACS: 01.40.-d, 01.40.gf, 01.50.Kw, 01.75.+m

INTRODUCTION

Educators have long realized that authentic assessments drive effective instruction [1]. Without such assessment instruments, instructors do not know whether their teaching has been effective, whether students have learned critical information, or whether student misconceptions remain intact, or have been created. At the same time, students and instructors can be content with ersatz understanding, which is often no more than the ability to recognize terms [2]. Assessments have begun to carry more weight as our country continues to emphasize the importance of developing a scientifically literate population and work force [3], as witness their role in the federal "No Child Left Behind" program.

Good assessments of science education determine students' level of subject mastery; they measure how well students understand fundamental scientific concepts and can use these concepts in problem solving or explanatory situations, that is, whether they can think and express themselves in a scientifically valid manner. Producing authentic tests is often substantially more difficult than it seems. However, given that assessments often determine how students are taught and what they learn, developing assessments that accurately reveal students' conceptual understanding is critical to attaining the goal of a scientifically literate citizenry.

Science educators have begun to develop a range of assessment instruments that focus on conceptual understanding rather than the recall of isolated bits of information. The Bioliteracy Project[1] models the development of a Biology Concept Inventory (BCI) after the Force Concept Inventory (FCI)[4, 5], which was developed for use in introductory college physics classes. Like the FCI, the BCI is a research-based, multiple-choice instrument designed for use in evaluating instructional methods and to assess students' understanding of key concepts. Through the development of the BCI, we are now finding for biology what the FCI found in mechanics: students taught through traditional methods are learning much less than we had assumed. Just as FCI findings served as a wake up call for many physics instructors [e.g., 6, 7], it is our hope that the BCI and similar instruments, such as the Natural Selection Concept Inventory [8] will lead to changes in biology course design and teaching methods.

BUILDING CONCEPT INVENTORIES

The first step in building any concept inventory is to discover what students believe about specific subjects. We began our project by asking

[1] See http://bioliteracy.net

CP883, *2006 Physics Education Research Conference*, edited by L. McCullough, L. Hsu, and P. Heron

students open-ended essay questions on topics often covered in high school and introductory biology classes. To analyze their responses, we developed a web-based Java program (Ed's Tools) that enables us to generate a database of student language that reveals at least part of their thinking in response to a specific question. Very quickly we recognized that certain questions, which appeared likely to be informative, were not. We will look at examples of poor questions and describe the strategies we use to transform poor questions into ones that better reveal students' conceptual assumptions. In describing our question-writing processes we will focus on the concept of diffusion. While this is certainly not the only conceptually difficult subject we have uncovered, it illustrates our question-writing process.

Poor Questions Elicit Reflex Responses

Diffusion is taught in nearly all high school and undergraduate entry-level biology courses, as well as in chemistry and physics courses. Understanding diffusion is critical for understanding how molecules move into and out of cells, as well as the basic mechanisms of molecular interactions. Diffusion can be understood at a number of levels. As recognized by Sutherland and Einstein, it is random molecular motions that drive diffusion. In the context of the biology classroom, diffusion is commonly presented in the context of concentration gradients and the tendency of a given molecule to move from high to low concentrations across a biological membrane. While net flux depends upon diffusion, diffusion often occurs without net flux. We can define those conditions in which a net flux of molecules will occur if we understand the basic random nature of molecular motion.

We began to probe students' understanding of the motion of molecules by asking a simple question: "What is diffusion and why does it occur?" After gathering ninety-seven student essay answers, we found that none of the students understood (or better put - none of the students explicitly stated) that diffusion is based on the random motion of molecules. Instead, we discovered that the question elicited what we call a "reflex response." Reflex responses are automatic answers that rely on recalling the context in which the questions subject was presented, either during the course of lecture or in assigned readings, rather than a response based on understanding the processes involved in actually answering the question. In the case of our initial diffusion question, essentially all of the students described a set of facts related to concentration gradients and membranes, rather than diffusion *per se*. Many students listed the types of molecules that could and couldn't diffuse. While many of these responses were accurate in and of themselves, they failed to answer (and generally ignored) the question posed. All of this points to the key feature of a reflex response: students memorize and supply information as a discrete package with little regard to what the question actually requests.

We have found that many standard science assessments contain a number of questions that elicit reflex responses; in some cases, they actually offer a prompt for what the correct reflex response should be. For example, consider a question from an eighth grade national science assessment (carried out by the Institute of Educational Statistics, US Department of Education and obtained from their website): "*Some people have proposed that ethyl alcohol (ethanol), which can be produced from corn, should be used in automobiles as a substitute for gasoline. Discuss two environmental impacts that could result from substituting ethyl alcohol for gasoline.*" An example of a correct and complete student response is: "*We would need a lot more land, soil, and money to grow enough corn to feed people and to put in cars. We would have to cut down forests, causing to higher CO2 levels and making more animals endangered. We would need more irrigationing (sic), using up our small % drinkable water.*"

The students' answer makes a number of unwarranted assumptions, triggered apparently by the suggestion of using corn, and is a reflex in that it does not explicitly discuss ethanol versus gasoline, which the question asks, but rather corn alone. It fails to note that growing corn leads to the sequestration of CO_2, and so its burning produces a lower net increase in CO_2 levels compared to the combustion of fossil fuels, depending, of course, on the extent to which the generation of ethanol requires the use of fossil-fuels. It also assumes that forested lands are more productive in sequestering CO_2 than agricultural crops and it does not consider other sources for the generation of ethanol (e.g. other plants or crop waste – cellulosic ethanol) that might lead to increased efficiencies. It seems likely that the student is repeating information that he or she heard in class (*e.g.* environmental impacts, often drilled into students beginning with their first science class) as opposed to a dispassionate cost-benefit analysis. There is no evidence that the student is able to build an answer based on the conceptual foundations of the problem (conservation of mass, CO_2 fixation and release upon combustion, mechanisms and costs of generating ethanol versus gasoline). That the test graders considered this students' response one of the best suggests a misplaced value on reflex responses rather than on conceptual understanding. Therefore, it is no surprise that the student answered the way he or she did.

Rather than blame the student, we need to reexamine the question - what, exactly, are the key ideas a student needs to grasp to be able to analyze "the relative cost-benefits associated with using biologically-derived ethanol versus fossil fuels?" Based on our experiences, posing the question in a different manner may well have provoked a more conceptually informative response. As it is, our research suggests that a typical student would be disinclined to answer the ethanol question conceptually since they have already been told the context of the expected answer (*i.e.*, corn as a source of ethanol).

Creating Better Conceptual Questions.

Writing a test with questions that require students to use, and so reveal, their working understanding of a topic can be difficult and often involves multiple rounds of analyzing student responses and revising question. At the same time questions that elicit a reflex response can give the illusion of understanding, are easier to write, are easier for the student to answer, and easier for the instructor to grade. In our diffusion example, we soon realized that students understood that membranes and gradients were associated with diffusive events, but we learned little about whether students understood the root cause of diffusion: the constant motion of molecules. We needed to generate a question that would force students to answer *why* diffusion occurred. Clearly, simply asking "why?" did not work, so we asked a different type of question, a question based on a scenario that students were unlikely to have encountered previously - *"Imagine that you are molecule of ADP inside a cell. Describe how you manage to "find" an ATP synthase, so that you can become an ATP."* The correct answer would be by random motions, or by diffusion. Surprisingly, we found that only approximately five percent of students mentioned the possibility of ADP finding the synthase by chance or diffusion. Most students described ADP as "looking" for the synthase, or hydrogen, concentration, or charge gradient. Thus, the answers to the new diffusion question revealed much more about students' understanding of the target concept than the answer to the first question.

Building Concept Inventory Questions

As we begin to understand how students approach a particular subject, we move on to develop one or more multiple choice questions as part of the BCI. In such questions, the incorrect answers or "distracters," are based upon our database of student responses. These distracters are truly distracting because they represent commonly held alternative responses to the question. Moreover, all parts of the question are written in student language that reduces the chances that students can recognize, rather than know, the correct answer. For example, in the final diffusion question (see below), both the correct response (e) and the distracters (a-d), were based on common alternative conceptions and presented in the students' own words. The final question becomes:

Imagine that you are molecule of ADP inside a cell. Describe how you manage to "find" an ATP synthase, so that you can become an ATP.

a. I follow the hydrogen ion flow
b. ATP synthase grabs me
c. My electronegativity attracts me to the ATP synthase
d. I would be actively pumped into the right area
e. By random motion or diffusion

We next tested this question for accuracy and validity through one to one and group interviews with students (think-alouds). During these interviews, we ask students what they think the question is asking, what they think each possible answer means, which answer they think is correct, and why they chose that answer. From interviews we found that students understand the meaning of the ATP question and the answer choices. However, the majority of students did not think that "e" was correct. Thus, our suspicion regarding most students' understanding of diffusion was confirmed: many students do not understand the fundamental concept behind diffusion.

Why Does Valid Conceptual Assessment Matter?

A cursory analysis of a number of science standards exams suggests that they are heavy on evaluating the recall of terms and facts and weak on the assessment of whether students have a working understanding of conceptual foundations. They tend to be reflexive in their approach, and as such measure whether a student has been paying attention to the material presented, rather than whether they understand how to use that information. Their structure appears to have a pernicious effect on our K-12 educational system. As political demands for higher student performance on these exams increases, teachers are both encouraged and coerced to teach to the test rather than to concept mastery. While it is possible to efficiently drill students on vocabulary (and reflexive responses), leading them to concept mastery is a much more time-consuming process. We would argue that many current testing instruments act to decrease the average students' competence, even as they increase their nominal achievement, because they encourage excessive (reflexive) content at the expense of conceptual confusion. In the absence of valid and confidently held understanding, students tend to generate scenarios that obscure rather than illuminate problem-solving, leading to uncertainty and frustration. If the goal of science education is for students to be able is to apply scientific understanding and analysis to new situations, it is critical that both instructors and evaluators become explicit in what tasks they expect their students to perform. Only then will curriculum and assessments positively reinforce one another, and provide the context for something better, and arguably more engaging, than memorization.

ACKNOWLEDGMENTS

We thank Isidoros Doxas, Valerie Otero, Steve Pollock, Noah Finkelstein and the LA-TEST/PER groups for helpful discussion and continuing insights. We thank Rebecca Klymkowsky for editorial advice. RG was supported by a Noyce fellowship and the BCI project is funded in part by a grant from the National Science Foundation.

REFERENCES

1. L. Shepard, *Educational Leadership* **46,** 4-9 (1989).
2. J.F. McClymer and L.Z. Knowles. J. *Excell. Coll. Teaching* **3**, 33-50 (1992).
3. Committee on Prospering in the Global Economy of the 21st Century, *Rising Above The Gathering Storm: Energizing and Employing America for a Brighter Economic Future*. (National Academies Press 2006).
4. D. Hestenes, M. Wells and G. Swackhamer, *The Physics Teacher* **30**, 141-166 (1992).
5. D. Hestenes and I. Halloun, *The Physics Teacher* **33**, 502-506 (1995).
6. E. Mazur, *Peer Instruction: A User's Manual*. (Prentice Hall, 1997).
7. R.R. Hake. *Am. J. Physics* **66**, 64-74 (1998).
8. D.L. Anderson, K.M. Fisher and G.J. Norman, *J. Res. Sci. Teaching* **39**, 952-978 (2002).

Is Inquiry-Based Instruction Good for Elementary Teaching Majors? The Effects on Chemistry Content Knowledge and Views About Teaching and Learning Science

Michael J. Sanger

Department of Chemistry, Middle Tennessee State University, Murfreesboro, TN 37132, USA

Abstract. Although science educators have advocated that elementary teaching majors learn science concepts using inquiry-based methods, many college professors believe that these courses are merely "watered down" versions of traditional lecture-based courses. This study compared the chemistry content knowledge of elementary teaching majors enrolled in an inquiry-based course and science majors enrolled in traditional lecture-based courses. It also compared the elementary teaching majors' views of how science is taught and learned to the views of secondary science teaching majors. The elementary teaching majors developed chemistry content knowledge comparable to the students enrolled in the traditional lecture-based course, but they developed views regarding how science is taught and learned that were more in line with the constructivist ideals than the secondary science teaching majors. The elementary teaching majors also improved their interest and confidence in teaching science in the elementary school setting. These results suggest that both sets of teaching majors would benefit more from inquiry-based science courses than lecture-based courses.

Keywords: Research in physics education; Teaching methods and strategies; Preservice training; Philosophy of science.
PACS: 01.40.Fk; 01.40.gb; 01.40.jc; 01.70.+w.

INTRODUCTION

The use of inquiry-based instructional methods in the science classroom (both K-12 and beyond) has been widely advocated from a variety of sources including national standards for K-12 teaching [1, 2], national reports on the crisis in K-12 mathematics and science education [3, 4], and editorials and commentaries from science education journals [5-8]. These methods have also been advocated for college chemistry courses for teachers, especially elementary teachers. Reasons cited for this suggestion include the ideas that what students learn and how they view science are greatly influenced by how they are taught [1, 9, 10], that teachers tend to teach using the same methods and in the same ways they were taught [1, 5, 11, 12], and that teachers with poor science content backgrounds tend to feel underprepared and thus avoid teaching science to their students [10, 13-16].

While these recommendations tend to focus on the use of inquiry-based methods as a way to improve students' science content knowledge, they fail to take into account one of the major strengths of inquiry-based methods—their potential to affect students' perceptions of what science is and how it is done, and more importantly for future teachers [17, 18], how science is taught and how it is learned [1, 2, 9, 10]. Piaget's theory of cognitive development [19-21] and the constructivist theory of knowledge [22, 23] support the notion that inquiry-based lessons should improve students' content learning and their views about teaching and learning science. However, many college professors view inquiry-based courses as "watered down" versions of traditional lecture-based courses that will lead to inferior learning.

This study is concerned with two research questions: (1) Do students learning chemistry using different instructional methodologies develop comparable chemistry content knowledge? and (2) How are students' beliefs regarding how science is taught and learned different for students learning chemistry using different instructional methodologies?

METHODOLOGY

Subjects

The elementary teaching majors (N = 16) in this study were junior and senior preservice elementary teaching majors enrolled in a physical science course taught using inquiry-based methods. This course was a one-semester four-credit (five-hour per week)

CP883, *2006 Physics Education Research Conference*, edited by L. McCullough, L. Hsu, and P. Heron

chemistry and physics content course intended for elementary teaching majors based on learning cycles [24] with minimal lecturing and most of the class time spent in the laboratory setting. The experiments were specifically written so that they could be adapted to the elementary science classroom, and involved equipment and chemicals that could be purchased at supermarkets, toy stores, or hardware stores. Prior to this course, these students had completed 2-4 science content courses using similar inquiry methods.

The chemistry content knowledge of the elementary teaching majors was compared to that of students enrolled in general chemistry courses taught by the same instructor. This comparison group was used because elementary teaching majors who did not take the physical science course were allowed to use this chemistry course (or the equivalent general physics course) as a substitute. These students were primarily freshmen majoring in natural science (biology or chemistry). This course was structured in the traditional format, with three hours of lecture and one three-hour laboratory per week.

The elementary teaching majors' views of teaching and learning science were compared to the views of junior and senior secondary science teaching majors enrolled in science methods courses ($N = 24$). Many of these students were planning to teach chemistry in grades 7-12, and all of them had completed several traditional lecture-based college science courses, including the general chemistry courses.

Data Analysis

For the comparison of chemistry content knowledge, both sets of students were asked to answer the same chemistry content question after receiving very different instructional lessons on the same topic. The content questions were conceptual in nature and covered five topics in chemistry: Density; compressibilities of solids, liquids, and gases; the structure of an ionic solid; surface area and evaporation; and the immiscibility of oil and water. For each question, students were given one point when they provided a correct response and when they did not demonstrate a misconception, for a total of two points. Statistical comparisons of the two groups' responses were made using the Student's *t* distribution or a statistical test of proportions [25].

To compare the students' views on teaching and learning science, written reflections from their class assignments were analyzed. The elementary teaching majors were asked to discuss how they would adopt the experiments for elementary students, what worked well in the experiment and what did not, and any questions or concerns they may still have regarding the material covered in these experiments. The secondary science teaching majors provided reflections based on their perceptions of what makes a good science teacher, what the ideal science classroom would look like, the instructional methods that they liked/disliked and with which they learned the best/worst, and their observations and experiences in actual secondary science classrooms. The reflections were analyzed to identify differences in students' beliefs regarding the way chemistry is taught and learned, and according to the instructors who have taught these courses the categories identified from these reflections describe real differences between these two groups.

RESULTS AND DISCUSSION

For four of the five chemistry content questions, there were no significant differences between the responses of the elementary teaching majors and the natural science majors (all $p > .05$), and the percentage of students providing a correct response or demonstrating a particular misconception were similar for the two groups. In all cases, the scores for the elementary teaching majors were slightly higher than those of the science majors. The similarity of the responses between the two groups can be seen in the following responses to the ionic solid question: 'Describe the forces holding a crystal of sodium chloride together'.

> "The Na^+ ions are attracted to the Cl^- ions. The attraction of '+' to '–' is very strong. The '+'/'+' or '–'/'–' repulse each other, but the '+' and '–' hold strong together." [*Natural science major*]

> "The + want to be with the – and as far away from the other + as they can. The – want to be with the + and as far away from the other – as they can. This attraction keeps them together." [*Elementary teaching major*]

For the conceptual density question ('Explain the concept of density at the molecular level without using the words *mass* or *volume*.'), the difference between the students' responses was significantly different ($t_{46} = 2.12$, $p = .04$). The elementary teaching majors were more likely to describe density in terms of the 'crowdedness' or 'closeness' of matter, and were less likely to use alternate words for *mass* (like *weight* or *heaviness*) and *volume* (like *space* or *area*) or provide a mathematical definition.

These results suggest that the inquiry-based physical science course helped the elementary teaching majors learn chemistry content at least as well as (and perhaps a little better than) a traditional lecture-based general chemistry course. This is in direct contradiction to the notion that hands-on chemistry lessons are somehow "watered down" versions of the lessons presented in traditional lecture courses [26, 27], and should provide convincing evidence that inquiry-based instructional methods can be used to develop students' conceptual understanding of science concepts.

Eight topics related to the teaching and learning of science (in which the two groups of students held different beliefs) were identified from the students' written reflections; a summary of these ideas appears in Table 1. The ideas expressed by the elementary teaching majors who studied science using inquiry-based methods were more mature [17, 18] and were more consistent with constructivist ideals [22, 23] than those of the secondary science teaching majors who studied science in traditional lecture-based courses. In particular, the elementary teaching majors viewed chemistry as an explanation of nature based on empirical evidence collected as part of their studies, while the secondary science teaching majors viewed chemistry as a set of unrelated facts that were blindly accepted from the "experts" (teacher or textbook).

In addition to demonstrating more constructivist ideals regarding the teaching and learning of science, the elementary teaching majors in this study also showed remarkable interest and enthusiasm for teaching science, and confidence in being well prepared to teach science lessons to their future students as a result of the inquiry-based science courses they had taken. These students reported that performing these hands-on lessons involving real-world applications had greatly improved their interest and confidence in teaching science.

"When I was in high school I did not enjoy physics or chemistry... However, as it turns out chemistry and physics can be fun—imagine that! Your class really opened my eyes and made me realize that physical science can be fun and interesting—and messy! I no longer hold on to those bad memories of high school chemistry of balancing equation after equation. Now I know that chemistry can be fun and interesting for both my students and I." [*Elementary teaching major*]

"I know that this course has affected me because I have seen my new found knowledge shine through in an actual classroom. I went to Grant Elementary to teach a light lesson right after we had talked about colored light. I was quite nervous worrying that I would 'mess' up or not be able to explain everything correctly. The knowledge that I learned from only a few days of class had made me extremely confident that day in the classroom. My lesson went well and it made me feel that I was accomplishing something in classes—which does not really happen too often." [*Elementary teaching major*]

The lack of meaningful science instruction in elementary schools has been generally attributed to the teacher's lack of interest or enthusiasm for teaching science [10, 13, 15, 16, 26] or the teacher's lack of confidence in their abilities to teach science effectively [10, 13-15]. These results suggest that the use of inquiry-based instructional materials not only improves the elementary teaching majors' conceptions regarding the nature of science, but also improves their interest, enthusiasm, and confidence in teaching science concepts to their future students.

TABLE 1. Topics where the secondary science and elementary teaching majors had different views regarding the teaching and learning of science.

Topic	Secondary Science Teaching Majors' View	Elementary Teaching Majors' View
Role of the teacher in the science classroom	Teacher is the source of information	Teacher guides students to find information on their own
Role of the textbook in the science classroom	Textbook is the source of information	Textbook is a supplement to help students explain their observations
Teacher's role in the science classroom	Teacher transfers knowledge (lectures) to the students	Teacher acts as guide to help students learn on their own
Role of lecturing in the science classroom	Primary (sole) method of knowledge transfer from teacher to students	Minimized (but not eliminated) in favor of hands-on instructional methods
Effect of feeling 'lost' or 'out of control' in the science classroom	Prevents learning because students are uncomfortable	Initial confusion is okay, and will disappear after more experiments
Effect of students making a mistake in the science classroom	Fatal, because students learn the wrong things	An opportunity to discuss experimental techniques, how science is done
Role of conceptual understanding in the science classroom	Unimportant, and gets in the way of learning facts and algorithms for tests	The goal of teaching science, and important
Role of outside experiences / "real world" in the science classroom	Unimportant, and gets in the way of learning facts and algorithms for tests	The goal of teaching science, and important

IMPLICATIONS FOR THE CLASSROOM

This study makes a strong case for the inclusion of inquiry-based methods in science content courses for preservice elementary teaching majors. The elementary teaching majors enrolled in this inquiry-based physical science course learned chemistry content knowledge at a level comparable to (if not slightly better than) students enrolled in traditional lecture-based courses. In addition, these students development more mature views regarding how science is done [17, 18], and how it is taught and learned, compared to secondary science teaching majors who studied their science content in traditional lecture-based courses. The use of inquiry-based methods in the science content courses for the elementary teaching majors also improved their interest and confidence in teaching science, two measures which are also directly tied to their science content knowledge. These results also suggest that we should use inquiry-based instructional methods in the content courses for secondary science teaching majors if we want to improve the way they view how science is taught and learned.

ACKNOWLEDGMENTS

More detailed accounts of the studies described here have been submitted to the *Journal of Chemical Education* as two chemical education research papers. The author would like to thank the editors of this journal for permission to reprint information from these articles.

REFERENCES

1. National Research Council, *National Science Education Standards*, National Academy Press, Washington, DC, 1996.
2. American Association for the Advancement of Science, *Benchmarks for Science Literacy*, Oxford University Press, New York, 1993.
3. National Academy of Sciences, *Rising Above the Gathering Storm: Energizing and Employing America for a Brighter Economic Future*, National Academy Press, Washington, DC, 2005.
4. National Research Council, *America's Lab Report: Investigations in High School Science*, National Academy Press, Washington, DC, 2005.
5. G. A. Crosby, *J. Chem. Educ.* **73**, A200-A201 (1996).
6. S. A. Ware, *J. Chem. Educ.* **73**, A307-A308 (1996).
7. J. W. Moore, *J. Chem. Educ.* **75**, 391 (1998).
8. J. W. Moore, *J. Chem. Educ.* **81**, 775 (2004).
9. J. Z. Ellsworth and A. Buss, *Sch. Sci. Math.* **100**, 355-363 (2000).
10. A. Pell and T. Jarvis, *Int. J. Sci. Educ.* **25**, 1273-1295 (2003).
11. G. H. Roehrig, J. A. Luft, J. P. Kurdziel, and J. A. Turner, *J. Chem. Educ.* **80**, 1206-1210 (2003).
12. P. L. Hammrich, *J. Elem. Sci. Educ.* **10**, 18-38 (1998).
13. P. J. Tilgner, *Sci. Educ.* **74**, 421-431 (1990).
14. J. Mulholland and J. Wallace, *J. Elem. Sci. Educ.* **8**, 17-38 (1996).
15. O. S. Jarrett, *J. Elem. Sci. Educ.* **11**, 47-57 (1999).
16. L. L. Liang and D. L. Gabel, *Int. J. Sci. Educ.* **27**, 1143-1162 (2005).
17. T. S. Kuhn, *The Structure of Scientific Revolutions, 3rd Ed.*, University of Chicago Press, Chicago, 1996.
18. National Science Teachers Association, *NSTA Position Paper: The Nature of Science*, National Science Teachers Association Press, Washington, DC, 2000.
19. J. D. Herron, *J. Chem. Educ.* **52**, 146-150 (1975).
20. J. D. Herron, *J. Chem. Educ.* **55**, 165-170 (1978).
21. D. M. Bunce, *J. Chem. Educ.* **78**, 1107 (2001).
22. G. M. Bodner, *J. Chem. Educ.* **63**, 873-878 (1986).
23. G. Bodner, M. Klobuchar, and D. Geelan, *J. Chem. Educ.* **78**, 1107 (2001).
24. E. A. Marek and A. Cavallo, *The Learning Cycle: Elementary School Science and Beyond, Rev. Ed.*, Heinemann, Portsmouth, NH, 1997.
25. D. E. Hinkle, W. Wiersma, and S. G. Jurs, *Applied Statistics for the Behavioral Sciences, 3rd Ed..*, Houghton Mifflin, Boston, 1994, pp. 228-315.
26. G. A. Crosby, *J. Chem. Educ.* **74**, 271-272 (1997).
27. R. D. Anderson and C. P. Mitchener, "Research on Science Teacher Education" in *Handbook of Research on Science Teaching and Learning*, edited by D. L. Gabel, Macmillan, New York, 1994, pp. 3-44.

Physics Education Research and Human Subjects: The PER Community and Institutional Review Boards

Gordon J. Aubrecht, II

Department of Physics, Ohio State University
Columbus, OH 43210 and Marion, OH 43302, USA

Abstract. This workshop was a discussion among participants about human subjects and Institutional Review Boards (IRBs) and dealt with the following questions: (1) What are the important human subjects issues facing physics education researchers? Do few, many, or most PER projects raise issues of confidentiality, liability, withholding of learning, differences in grading policy, impact of the student lack of informed consent, or other ethical issues? (2) Should PER physicists at each institution create a common IRB form to be used by all PER researchers at that institution? (3) Should the PER community as a group address the IRB issues as a community? If so, what might the outcome be? (4) Should all PER research be exempt from IRB approval, given the extreme unlikelihood of student physical or emotional damage? How could such global exemption be achieved?

Keywords: Institutional Review Boards, human subjects.
PACS: 01.40.Fk, 01.78.+p

I. INTRODUCTION

Because of federal regulations, Physics Education Research (PER) is subject to Institutional Review Board (IRB) oversight over use of human subjects. The regulations addressing human subjects are mainly concerned with medical research issues. Most PER researchers are mainly concerned about students and their learning, which overlaps little with ethical concerns of medical research.

This issue catapulted to public consciousness recently because of both local conditions (for example, a revolt of faculty in the College of Social and Behavioral Sciences at Ohio State) and because of the recent editorial in Science magazine [1]. The authors of Ref. 2 state that "Our IRB system is endangered by excessive paperwork and expanding obligations to oversee work that poses little risk to subjects. The result is that we have simultaneous overregulation and underprotection." They go on to say "IRBs' burdens have grown to include studies involving interviews, journalism, secondary use of public-use data, and similar activities that others conduct regularly without oversight. Most of these activities involve minimal risks—surely less than those faced during a standard physical or psychological examination, the metric for everyday risk in the federal regulations."

Participants in the workshops discussed these issues extensively and shared stories from their respective institutions. Below, we outline the results of the discussions concerning the questions given in the abstract

II. WORKSHOP DISCUSSIONS

What are the important human subjects issues facing physics education researchers? This question was reinforced for me after I was named to a subcommittee of Ohio State University's Research Committee and I discussed these issues with PER colleagues. Various subsequent discussions with other colleagues gave rise to this workshop.

At the end of the workshop, participants shared their conclusions with the entire group. After general discussion, it is fair to say that there was consensus on most issues discussed. Overall, it was clear that most PER projects should be exempt, but that having the possibility of the IRB could be constructive.

What was striking from the discussions was the variability of application of IRB rules (presumed identical) across institutions. A common complaint involved prescribing a human subjects form as necessary for proposing funding of research. One group said that it "takes far too much time to do a report for every grant you write but don't get funded for," and most agreed. It was found that some

CP883, *2006 Physics Education Research Conference*, edited by L. McCullough, L. Hsu, and P. Heron

institutions regarded these proposals as *pro forma* (one participant said he was able to work around things so that none of his research involved an IRB) and others faced the unleashing of a battery of barriers.

(1) What are the Important Human Subjects Issues facing Physics Education Researchers?

Workshop participants generally agreed that PER does not raise issues of confidentiality, liability, or unfairness in comparison to the "standard" instruction. Most don't believe in giving grades as part of PER projects: for example, how equal is it to grade problems phrased differently the same way? As long as a student is not receiving less than "standard" instruction, there was no problem seen in forming treatment and control groups even if the researcher believed PER instruction was better.

Participants thought that pre- and posttests did not involve any serious confidentiality issues as opposed to someone who gathered data for assessing the department's teaching, where it might be good to have the protocol be reviewed.

Informed consent protocols were not seen as really useful; they can induce student and parent anxiety for no reason.

While liability issues cannot be entirely ruled out in a litigious society, most participants did not think that this was a major issue in most research. Confidentiality could become an issue when there are very few students in a class, and it was thought that such a case would benefit from IRB oversight.

(2) Should PER Physicists at each Institution create a Common IRB Form to be used by all PER Researchers at that Institution?

Most participants thought that the PER community should be far more collaborative than at present; it would be good to have IRB boilerplate documents available for everyone at various institutions to use. One problem with this is the "not created here" syndrome. Examples from multiple institutions could still be extremely useful in helping individual PER researchers. One participant has a list of "taboo" words; sharing this sort of information widely can be useful. These could be put on CDs or could be web documents, whatever the community wants.

Graduate students and faculty who are being prepared to do PER definitely need to know about IRB issues before they begin their work. Perhaps the community could develop a research ethics "course" framework.

Although forms might be different at differing institutions, to have an agreed-on mission statement can make a wonderful cover letter. There's no probability of harm in having that information available.

(3) Should the PER Community as a Group address the IRB Issues as a Community?

Bearing in mind that institutions' IRBs are different, participants did not think it would be possible to develop a "magic" form that all would accept. It was thought that there could be a community statement that in general this research is used to improve teaching and learning and did not fall under non-exempt research except under a certain set of categories. Such a community statement to this effect to attach to an individual IRB could be very useful. It is possible that there *is* no consensus: the community hasn't been able to build consensus on other things. Some suggested that Physics Education Research Leadership and Organizing Council (PERLOC) should address this issue by putting up a standard document.

The PER community could advocate for conversion of IRB's image and reality from that of the current watchdog/enforcer to that of resource/aid. If the community uses its leverage, it was suggested that it could get the review boards to pay attention.

Two groups suggested that we should reach out to other disciplinary education research groups. If a researcher can say the human subjects treatment complies with national standards, it will be useful.

(4) Should all PER Research be Exempt from IRB Approval, given the Extreme Unlikelihood of Student Physical or Emotional Damage?

There was a lot of discussion on this question. The groups had varying answers to the question. Here are their thoughts:

"No, but approval should be fast, easy, and almost automatic for research that fits certain standard, common types and obeys accepted standards and protocols."

"No. . . A list of practices or ethical issues would be useful."

"NO! Pragmatism is the reason (it just can't happen);" this group was worried that PER would "attract people who are unscrupulous or whose research goals don't match the community to come into the community; this one bad apple could spoil it for everyone."

"Qualify this to apply to the CATEGORY of exempt research; this is preexisting at some institutions—research is exempted by category."

"Yes, but ... it probably wouldn't fly with the individual institutions."

"For normal class activities, it should be exempt; all others should be 'expedited' with informed consent forms. A defined list of okay things would be very useful."

"Yes, but we didn't have great ideas."

"Yes, but we weren't sure what outcomes would be."

III. WORKSHOP CONCLUSIONS

Human subjects paperwork was not seen in a positive light by workshop participants, but a nuisance (except in rare cases, when they were really needed). To quote once again from Ref. 1, "All this has generated a trend in which researchers increasingly think of IRBs as the 'ethics police.' In fact, all researchers must take primary responsibility for professional, ethical conduct. Our systems should reinforce that, not work against or substitute for it; the IRB should be a resource, not the source, for ethical wisdom. All compliance systems require the buy-in and collaboration of the regulated, and it will be a sad day if scholars come to see human protection in research as the source of frustrating delays and expensive paperwork." This latter view appears to be already quite common within the PER community.

Workshop participants were united that these IRB issues be considered soon by PERLOC and the American Association of Physics Teachers' PER Committee. The possibility of a useful community document setting standards for categories of semi-automatic exemption in education research is very attractive and can be potentially useful.

ACKNOWLEDGMENTS

I thank all the participants of the workshop for their thoughtful input

REFERENCES

1. Gunsalus, C. K., Bruner, E. M., Burbules, N. C., Dash, L., Finkin, M., Goldberg, J. P., Greenough, W. T., Miller, G. A. and Pratt, M. G., "Mission Creep in the IRB World," *Science* **312**, 1441 (2006).

Are They All Created Equal? A Comparison of Different Concept Inventory Development Methodologies

Rebecca S. Lindell, Elizabeth Peak and Thomas M. Foster

Physics, Astronomy, Chemistry, Biology, Earth Science Education Research Group
Southern Illinois University Edwardsville, Edwardsville, IL 62025, USA

Abstract. The creation of the Force Concept Inventory (FCI) was a seminal moment for Physics Education Research. Based on the development of the FCI, many more concept inventories have been developed. The problem with the development of all of these concept inventories is there does not seem to be a concise methodology for developing these inventories, nor is there a concise definition of what these inventories measure. By comparing the development methodologies of many common Physics and Astronomy Concept Inventories we can draw inferences about different types of concept inventories, as well as different valid conclusions that can be drawn from the administration of these inventories. Inventories compared include: Astronomy Diagnostic Test (ADT), Brief Electricity and Magnetism Assessment (BEMA), Conceptual Survey in Electricity and Magnetism (CSEM), Diagnostic Exam Electricity and Magnetism (DEEM), Determining and Interpreting Resistive Electric Circuits Concept Test (DIRECT), Energy and Motion Conceptual Survey (EMCS), Force Concept Inventory (FCI), Force and Motion Conceptual Evaluation (FMCE), Lunar Phases Concept Inventory (LPCI), Test of Understanding Graphs in Kinematics (TUG-K) and Wave Concept Inventory (WCI).

Keywords: Instrument Development, Concept Inventory Development
PACS: 01.40.Fk, 01.40.gf, 01.40.G-, 01.50.Kw

INTRODUCTION

Since the creation of the Force Concept Inventory (FCI) [1] the development of research-based distracter driven multiple-choice instruments has surged. Now nearly every scientific discipline has multiple concept instruments available for their use. A quick Google search yielded concept instruments in Physics [2], Astronomy [3, 4], Engineering [5], Biology [6], Chemistry [7], and Geoscience [8]. Nowhere was the prevalence more evident than at the 2006 Physics Education Research Conference, where 4 out of the 5 invited talks spoke about concept inventory development in their specific scientific fields.

The problem with the development of all of these instruments is that there does not seem to be a concise definition of what exactly a concept inventory actually measures. Many users of these instruments lump all conceptual tests under the classification of "concept inventory," while others argue that we need to differentiate the instruments based on standard definitions of surveys and instruments as defined by the education research community [9].

Not only is there a discrepancy in definitions of the term concept inventory, there also seems to be discrepancies in the methodologies utilized to create these inventories. The purpose of this project is to compare the different methodologies of conceptual instruments in Physics and Astronomy to determine the similarities and differences in the development methodologies.

For the purposes of this project we are adopting the following general definition of a concept inventory.

Concept Inventory: "A multiple-choice instrument designed to evaluate whether a person has an accurate and working knowledge of a concept or concepts."[10]

Results from this study yielded information about the different methodologies utilized to develop concept inventories, as well as providing evidence for the need for a new classification scheme for concept inventories.

CP883, *2006 Physics Education Research Conference*, edited by L. McCullough, L. Hsu, and P. Heron

METHODOLOGY

To begin this study, we selected a sub-population of concept inventories and assessments as listed on the North Carolina State University's Assessment Instrument Information website [2]. To further narrow our focus, the following criteria were used.

- Each instrument needed to focus on student understanding of either Physics or Astronomy concept(s).
- The methodology must have been published and/or communicated to us through private communication and the inventory made available.

This left us with a total of 12 "concept inventory" methodologies to analyze. The inventories are listed in Table 1.

TABLE 1. Concept Inventory Methodologies Analyzed In This Study

Instrument	# of Items	# of Concepts[1]
Astronomy Diagnostic Test (ADT) [3]	21	10
Brief Electricity and Magnetism Assessment (BEMA) [11]	30	3
Conceptual Survey in Electricity and Magnetism (CSEM) [12]	32	2
Diagnostic Exam Electricity and Magnetism (DEEM) [13]	66	
Determining and Interpreting Resistive Electric Circuits Concept Test (DIRECT) [14]	29	1
Energy and Motion Conceptual Survey (EMCS) [15]	25	2
Force Concept Inventory (FCI) [1]	30	1
Force and Motion Conceptual Evaluation (FMCE) [16]	43	2
Lunar Phases Concept Inventory (LPCI) [4]	20	1
Mechanics Baseline Test (MBT)[17]	36	3
Test of Understanding Graphs in Kinematics (TUG-K) [18]	21	1
Wave Concept Inventory (WCI) [19]	20	1

[1]We classified the number of concepts in terms of broad concepts and not specific sub-concepts.

To compare the methodologies, we began by studying the instrument design process as outlined in Crocker and Algina [20]. These steps are listed in Table 2. While there are many different texts and articles on test development and design, many focus on how to develop classroom tests and not educationally valid instruments designed for large-scale use. For the purposes of this study, we will refer to methodologies for educationally valid instruments as instrument design methodologies, as opposed to test development. Similar rubrics can be obtained from other test theory texts. For those unfamiliar with instrument design, Table 3 provides an overview of key concepts [20].

TABLE 2. Steps in Instrument Design

1. Identify purpose
2. Determine the concept domain
3. Prepare test specifications
4. Construct initial pool of items
5. Have items reviewed - revise as necessary
6. Hold preliminary field testing of items - revise as necessary
7. Field test on large sample representation of the examinee population
8. Determine statistical properties of item scores - eliminate inappropriate items
9. Design and conduct reliability and validity studies

TABLE 3. Overview of Key Concepts in Instrument Design

Concept Domain: Refers to the concept/concepts that will be covered on an instrument. It represents the content that will be assessed by the instrument. May contain alternative and scientific understanding as well. Sometimes referred to as Construct Domain.

Test Specifications: Details how items will be constructed. Typically represented as a table or diagram. The test map discussed by Aubrecht and Aubrecht [21] is just a visualization of a test specifications.

Field Testing: The process by which items are tested using a sample population. Item statistics are calculated to determine validity of items. Invalid items are deleted or revised. Field testing is often repeated until all items meet test specifications.

Item Statistics: There are two main statistics calculated in instrument development: difficulty and discrimination. While many test theory texts provide guidelines for excepting or rejecting an item based on these statistics, these guidelines are based on the assumption that responses are randomly distributed among the distracters.

Difficulty: How difficult is the item? Measures the percentage of respondents that answer item correctly.

Discrimination: Refers to how well the item discriminates between the upper quartile and lower quartile.

Validity: Represents how well an instrument measures the construct it is attempting to measure. There are three main types of validity: criterion, construct and content validity.

Criterion Validity: The degree to which scores on inventory predicts another criterion. Typically established through comparison to other standardized instruments or course grades.

Construct Validity: The degree to which scores can be utilized to draw an inference on the content domain. Typically established through open-ended student interviews.

Content Validity: The degree to which inventory measure the content covered in the content domain. Typically established through expert review.

Reliability: The degree to which scores are consistent.

Table 4. Methodology Comparison for Common Astronomy and Physics Concept Inventories[1]

	Concept Domain Determined by			Test Specifications				Item Statistics Reported			Field Testing					Validity Studies			Reliability Statistics Reported		
				Basis of Distracters		Distracter Correspondence to Alternate Models					Size			Location							
Instrument	Qualitative Study	Researcher	Existing Literature	Researcher's Understanding	Student understanding	Corresponds	Does Not Correspond	Difficulty	Discremenation	Concentration Analysis	> 500 students	500 - 1000 students	>1000 students	Local	National	Criterion	Construct	Content	Cronbach Alpha	Kuder - Richardson	Point Biserial
ADT[2]	■	■	■	■			■	■	■	■			■		■		■	■			
BEMA		■		■			■	■	■			■		■				■		■	■
CSEM								■	■				■		■			■		■	
DEEM	■	■						■	■		■			■					■	■	
DIRECT		■	■				■	■	■				■		■		■	■		■	
EMCS		■			■		■	■	■				■		■		■	■	■		
FCI		■		■			■	■					■		■		■	■			
FMCE	■		■		■	■							■		■	■	■		■		
LPCI	■	■	■		■	■		■	■	■			■		■		■	■	■		
MBT		■		■	■		■						■	■			■	■		■	
TUG-K		■	■	■	■		■	■	■			■			■		■	■		■	■
WCI											■			■			■	■			

[1] All analysis based on research reported in original paper or personal communication. Blanks refer to non-reported information.

[2] Questions relating to Lunar Phases and Seasons were based on qualitative investigation conducted by R. Lindell; rest of concept domain determined by researchers.

For the purposes of this study, we focused on five different points in instrument design process: determining the concept domain, preparing the test specifications, determining the item statistics, field-testing of the items and conducting the reliability and validity studies.

RESULTS AND DISCUSSION

Table 4 shows the results of our analysis. As you can see there does not appear to be a consistent methodology being employed, nor do the different methodologies break down according to the accepted definitions of survey and inventory instruments. Below, we will discuss each of the five different points in the instrument design process separately.

Differences in Determining the Concept Domain

Developers utilized three different resources to define the concept domain. These included researcher's understanding, existing literature on students' understanding of the phenomena and performing a qualitative investigation of students' understanding. It must be noted, that we would argue that if the instrument is going to be utilized to diagnose specific alternative conceptions, the concept domain must be grounded in the views of the students and not the researchers. Of the twelve methodologies classified, only four utilized a qualitative investigation to help define the concept domain, and only two others utilized common student difficulties as reported in the literature.

Differences in Test Specifications

When analyzing the differences among the test specifications, we determined that there were two classes of differences: basis of the distracters and the correspondence of the different distracters to different alternate models.

Examining the basis of the distracters, we discovered that there were two main differences. Of the eight methodologies, which reported the basis of their distracters, three based the distracters purely on students' understanding, as discovered through open-ended questions or student interviews. Another three based the distracters purely on the researchers' understanding and two used a combination of the students' and researchers' understandings. It is interesting to note that four out of the twelve methodologies failed to discuss this key feature.

Since many of the concept inventories claim to be able to diagnose specific alternative understanding of the concept, we specifically examined the methodologies to determine if each distracter corresponded to a specific alternative model. In our opinion only two of the instruments meet this criteria and only these two inventories can be reliably used to diagnose alternative understandings.

Differences in Item Statistics

It is interesting to note that three of the different methodologies failed to report any item statistics. Typically these statistics are considered the bare minimum that should be reported in any instrument development. Of the remaining nine methodologies, all but one reported the two standard statistics of difficulty and discrimination. Two instruments utilized the new technique of concentration analysis [22] to help evaluate the items.

Differences in Field Testing

We found that there were two classes of field-testing: size of population and location of the field tests. We found that eight of the twelve instruments were field-tested at the national level, but only one failed to receive at least 1000 data points. Of the four locally field-tested instruments, only one exceeded a 1000 data points.

Differences in Establishing Reliability and Validity

Of the ten methodologies that reported the results of their validity study, only one reported establishing criterion validity and no methodology reported establishing all three types of validity. Finally only two methodologies reported only the lowest content validity results.

Reliability statistics were reported for only nine of the different methodologies. The reliability was typically determined by calculating one or more of the following statistics: Cronbach alpha, Kuder-Richardson 20 or 21, or the Point Biserial coefficient.

CONCLUSIONS

In conclusion, we found that there are many different methodologies being utilized to develop concept inventories. We as a community need to determine guidelines for developing these instruments. We also find that the definition for concept inventories is way too broad and we need to introduce a new classification scheme. Lastly we strongly encourage developers to employ all of the steps in the design process, as well as to publish their methodologies so as the community can determine the appropriateness of utilizing the instruments.

REFERENCES

1. D. Hestenes, M. Wells and G. Swackhamer, *Phys. Teach* **30**, 141-158 (1992).
2. Assessment Instrument Information Page, North Carolina State University, http://www.ncsu.edu/per/TestInfo.html, accessed October 15, 2005.
3. B. Hufnagel, *Astro Ed Rev. http://aer.noao.edu* **1**(1), 47-51 (2002)
4. R. Lindell and J. Olsen, *Proc. 2002 PERC,* New York: PERC Publishing, NY, (2002).
5. Engineering Foundation Coalition, http://www.foundationcoalition.org/home/keycomponents/concept/index.html.
6. D. Anderson, K. Fisher and G. Norman, *J. Res. Sci Teach* **39** (10), 952-978 (2002).
7. D. Mulford and W. Robinson, *J. Chem. Ed.* **79**(6), 739-44 (2002).
8. J. Libarkin and S. Anderson, *J Geosci. Ed.* **53**, 394-401 (2005).
9. Wikepedia definition, http://en.wikipedia.org/wiki/Concept_inventory, accessed October 15, 2005.
10. Education researchers consider a survey to contain 5-7 items per concept, while with inventories you need 20+ items per concept. P. Heller, Personal Communication, 2006.
11. L. Ding, R. Chabay, B. Sherwood and R. Beichner, Personal Communication, (2006).
12. D. Maloney, T. O'Kuma, C. Hieggelke & A. Van Heuvelen, *Am. J. Phys* **69**, S12-S23 (2001).
13. J. Marx, Unpublished Dissertation,: Rensselaer Polytechnic Institute (1998).
14. P. Engelhardt & R. Beichner, *Am. J. Phys* **72**, 98-115 (2004).
15. C. Singh and D. Rosengrant, *Am. J. Phys* **71**, 607-617, (2003).
16. R. Thornton and D. Sokoloff, *Am. J. Phys* **66**, 338-352 (1998) and personal conversation with R. Thornton (2006).
17. I. Halloun and D. Hestenes, *Am. J. Phys* **53**, 1043-1055 (1985).
18. R. Beichner, *Am. J. Phys* **62**, 750-762 (1994).
19. R. Roedel, S. El-Ghazaly, T. Rhoads & E. El-Sharawy, *29th ASEE/IEEE Frontiers in Education Conference* San Juan, Puerto Rico, 10–13 November 1999.
20. L. Crocker and J. Algina, *Introduction to classical and modern test theory.*, Reinhart and Winston, Inc, New York: Holt (1986).
21. G. J. Aubrecht, II and J. D. Aubrecht, *Am. J. Phys* **51**, 613-620 (1983).
22. L. Bao and E. F. Redish, *Am. J. Phys.* **69**, S45 (2001).

Problem Solving Skill Evaluation Instrument – Validation Studies

Wendy K. Adams and Carl E. Wieman

Department of Physics, University of Colorado, Boulder, Boulder, CO 80309, USA

Abstract. Researchers have created several tools for evaluating conceptual understanding as well as students' attitudes and beliefs about physics; however, the field of problem solving is sorely lacking a broad use evaluation tool. This missing tool is an indication of the complexity of the field. The most obvious and largest hurdle to evaluating physics problem solving skills is untangling the skills from the physics content knowledge necessary to solve problems. We are tackling this problem by looking for the physics problem solving skills that are useful in other disciplines as well as physics. We report on the results of a series of interviews comparing physics students' skills when solving physics problems with their anonymous completion of the problem solving instrument. There is an encouragingly good match.

Keywords: Problem Solving, Physics, Undergraduate Education
PACS: 01.40.Fk, 01.50.gf

INTRODUCTION

Physics Education Researchers over the past 20 years have developed many very useful assessment tools including a range of conceptual evaluations such as the Force Concept Inventory [1], Force and Motion Conceptual Evaluation [2], and the Conceptual Survey of Electricity and Magnetism [3] among others. There has also been work in the area of attitude and belief surveys [4]. However, one of the most highly valued assets of physics courses by educators is problem solving; but, very little has been published regarding ways to evaluate students problem solving skills. To date, research on problem solving in physics has been focused predominantly on curriculum improvements and the identification of differences between expert and novice problem solving characteristics [5]. There have been efforts to develop a problem solving evaluation tool, but none have been successful.

In this paper we will discuss the problem solving instrument that is being created at Colorado. First we will describe the direction of our approach and how it avoids problems that have hindered previous attempts. Then we will outline our validation procedures to date, and present the specific set of measurable skills that we have identified.

APPROACH

First we define what we mean by "problem solving", because this term is used very broadly. The fairly specific definition of problem solving used for the purposes of this paper is drawn from the literature: "Problem solving is cognitive processing directed at achieving a goal when no solution method is obvious to the problem solver." [6]. In addition we will break problem solving down into a set of skills, including those motivational aspects relevant to successful completion of solution. Note that this definition of problem solving is based on the solver. If a person is an 'expert' in their field, then it is very likely that tasks that are problems for students will only be an 'exercise' where the solution or path to solution is obvious, for an expert.

The extensive problem solving research in psychology and cognitive science has included transfer of problem solving ability as well as work on characterizing specific skills needed to solve problems. Many researchers have gone as far as to declare that problem solving abilities do not transfer between disciplines [5]. Being an expert problem solver in a particular area is not an indication of that person's problem solving skills in any other area. However, other researchers present evidence that some problem solving skills do transfer between disciplines [6].

We argue that these contradictory results on problem solving transfer can be explained with the same reasons that have made developing a physics problem solving evaluation tool difficult. First, solving physics problems requires content knowledge, both factual and procedural; second, that it is important to view problem solving as a combination of many different specific skills, and the different skills have

CP883, *2006 Physics Education Research Conference*, edited by L. McCullough, L. Hsu, and P. Heron

varying degrees of importance in various types of problems.

The physics content requirement is a fairly straight forward yet seemingly unassailable complication. A particular student may be unsuccessful at solving a problem because they are a poor problem solver or because they simply lack familiarity with the specific bit of physics knowledge needed for the problem. On the other hand, when a student is successful at solving a particular problem, it is very difficult to tell the difference between a strong problem solver or a student who is so familiar with that particular content area that the problem was actually an exercise for that student. In both of these cases, one cannot clearly delineate the student's problem solving ability because the student's content knowledge is inextricably intertwined within these skills.

A great deal of research in problem solving – physics, chess, etc. is result oriented. Researchers are interested only in who can successfully solve problems rather than each solvers' strengths and weaknesses in a variety of specific problem solving skills. This focus greatly limits both the information that can be gained by these studies and the development of assessment tools.

Our approach comes from the hypothesis that a person has a set of skills that vary in strength that they use to tackle problems. This includes problems in any context, the physics classroom or the workplace. If this is the case, students' skills can be analyzed with a problem that does not require physics content knowledge. Although still preliminary, we have growing evidence that our hypothesis is correct.

ASSESSMENT TOOL

We have developed an assessment tool which uses the work of the Cognition and Technology Group at Vanderbilt (CTGV)[7]. CTGV developed the Jasper Woodbury series of problems for 6th – 7th grade math students to solve in small groups. These are long, involved problems that each have no less than 14 steps to solution. CTGV carefully designed and researched each Jasper problem on its effectiveness. Our problem solving instrument was developed through a series of interviews with a wide range of subjects using one of the Jasper problems. The evaluation instrument instructs the student to analyze a script of two individuals working through the solution to the Jasper problem. This format provides several benefits: 1) motivation for students to work through the entire solution, 2) removes the stress of being analyzed 3) scaffolds the problem so that a solution will be reached even if the student has a specific weakness that would have prevented further progress if they were attempting to solve the problem in isolation. There are 57 questions that must be answered as they go through the script. These questions are about planning, procedures, calculations, reflection as well as analysis of the two scripted solvers' skills. This combination of questions evolved from multiple interviews that were used to evaluate students' skills as they solved the scripted problem. In this way, we have identified 34 skills which show up as useful when solving this problem.

VALIDATION

We have performed a series of validation studies that involved several iterations to revise and refine the instrument. The validation process included: face validity – interviews with a wide range of people to confirm and clarify the meaning of questions and the story; anonymous written instrument results compared to a series of physics problem solving interviews; and concurrent validity – comparison of instrument results to professor and employer evaluations. In addition we have two studies which are in process: interviewer observations of physics majors using the instrument will be compared to an independent interviewer's observations of the same students solving quantum mechanics problems; and written instrument responses of undergraduate teaching assistants will be compared to independent instructor evaluation of the teaching assistants using the same rubric.

The development of the survey was intertwined with the first phase of face validation. Interviews were conducted with a variety of subjects including non-science majors, physics majors, professors, elementary education majors, and adult professionals with a variety of backgrounds including a high school drop out. The first few interviewees were asked to think out loud as they solved the Jasper problem. The two person script and the questions that are asked throughout the instrument were created based on these first few interviews. Further interviews were conducted with the scripted instrument. Periodically the script was refined and questions were added based on interviews. A total of 23 interviews were completed, each lasting between one and two hours.

Eventually the instrument and analysis rubric were refined to the point that for the last nine interviewers, the interviewer was able to obtain a complete analysis of the subjects skills without further interaction after initial instructions on think-aloud style. The analysis rubric includes 34 separate problem solving skills that have been identified during the interviews (see Table 1).

The next stage of instrument construction and validation of the rubric involved having the students take a written version (no interview) of the survey. These were then graded using the same rubric. If enough information was not provided in the written responses to rate the student in all 34 categories, then questions were added or adjusted and tested with new students. Currently 16 written responses have been graded and 90% of the skills that are identifiable during interviews can be consistently graded with the latest written version of the survey.

Anonymous Written Results Matched to Physics Problem solving Interviews

The next step in face validation was to test how the set of problem solving skills that are identified while solving this 6th grade math trip-planning problem matched with the skills a student uses to solve a physics problem. For this portion of the validation procedure we gave five science majors who were currently enrolled in introductory physics the written version of the instrument. These students brought the completed instrument to their first interview. The instrument was graded at a later date without knowing the identity of the student. Interviews consisted of having the student sit down with the interviewer whom they had never met and immediately begin solving a slightly modified problem about the Great Pyramid of Giza[7]. In this problem students must determine how many blocks are in the pyramid and how many men were needed to build the pyramid if it took 20 years to build. A few facts about the block size and men's capabilities were included without any other scaffolding. These interviews were mathematically intensive so had to be limited to an hour. Each student required two to three interview sessions to complete the problem. They were evaluated on the skills that

TABLE 1. Problem Solving Skills

Skills	Metaskill	Will	Uncatergorized
Skills that were measured by both the written instrument and the pyramid interview			
◆Math Skills (add/sub/mult/div)	◆Planning What – Question Formation	◆Confidence	◆Real Life vs. Classroom Approach
◆Spatial – Mapping	◆Planning How – Way to get answer.	◆Enjoyed Solving the Problem	◆Overall Success
◆Estimation	◆Planning Big Picture – Visualizes the Problem.	◆Wanted to Succeed on 'Test'	
◆Number Sense	◆Connects Steps and Pieces	◆Attribution (responsible for own mistakes)	
◆Real World (informal) Experience	◆Monitors Own Progress	◆Wanted to Find Best Solution for Themselves.	
	◆Knowledge of Own Strengths and Weaknesses	◆Wanted to find Best Solution to Please Interviewer.	
	◆Creativity		
	◆Judgment of Reasonable Issues		
	◆Ties in Personal Experience		
Skills that were not measured by the pyramid problem interviews and why			
◆Acquires Information 1st Time Through (*pyramid data of a numerical nature)*	◆Keeps Problem Framework in Mind (*Pyramid framework not complicated)*	◆Enjoyed Analyzing Interns (*Specific to instrument scenario)*	
◆Remembers Previously Noted Facts. (*pyramid interviews done in 2 segments w/ a 2 week break.)*	◆Adaptability - Shift Direction (*A forced change of plans was not inserted into Pyramid problem)*	◆Enjoyed Complete Survey (*Same as enjoyed solving the problem for pyramid)*	
◆Outside Factual Knowledge (*Not necessary for pyramid problem or in separate category such as Geometry)*	◆Skepticism – Thinks about Information that is Supplied. (*Specific to instrument scenario)*		
◆Math – Equation Formation (*Difficult to see w/ Pyramid response format)*	◆Checks Scripted Solvers' Calculations (*Specific to instrument scenario)*		
◆Reading Comprehension (*Pyramid presentation makes this difficult to judge*)	◆Aware of How Scripted Solvers Helped (*Specific to instrument scenario)*		
Skills that were not measured by the written problem solving instrument and why			
◆Geometry (*Not necessary for Survey)*	◆Physics/Math They Think They 'Should' Know Blocks Progress (*No outside formal knowledge needed for survey)*		

were exhibited while completing the problem. The interview sessions were separated by two weeks which made some of the grading of skills difficult, as noted in Table 1.

The five graded anonymous written instruments were then compared to each student's set of skill ratings identified during the pyramid interviews. The skills that could be identified with both the anonymous written instrument and the pyramid interviews are listed in Table 1. There were a few skills that only worked in one situation or the other. These are listed in Table 1 with explanations. Many of the skills identified by the instrument that could not be graded by the pyramid interviews have been recognized as important skills when solving physics problems, they simply could not be rated with the structure of the pyramid interviews.

The five students' written instruments were easily matched up with their interview results. This was done without any identifying information about the students other than the rating of problem solving skills from each problem scenario. The same set of problem solving strengths and weaknesses were evident in how each student solved these two different problems, After the students written results were matched with their interview results, the specific ratings in each of the skills matched up for 75% of the 22 skills listed in Table 1. Of the 25% that did not match exactly, most were a neutral response versus (strong/weak). Only two of the students had a single skill that appeared strong for one problem and weak on the other.

Concurrent Validation

Our concurrent validation studies include comparison between the instrument results and evaluation of problem solving skills in various environments. We asked three different instructors (class sizes of 40 or less) to evaluate particular students' strengths and weaknesses over the course of a semester. These same students were interviewed or took the written version of the problem solving evaluation tool. The instructor's assessment of the students strengths and weaknesses in problem solving have matched the skills identified by instrument results in all 15 cases. In addition we had an employer to do the same for five different employees. Again, all five instrument results matched the employer's evaluations.

We are in the process of conducting an additional validation study involving the comparison of instrument results with skills identified while students solve quantum mechanics problems. Six students underwent a series of seven interviews where they solved quantum mechanics problems. The interviewer is in the process of scoring these students quantum mechanics problem solving skills on the 34 areas identified in Table 1. A separate interviewer is interviewing these same students as they solve the problem solving evaluation instrument. The results of these separate interviews will be compared.

CONCLUSION

We have created an evaluation instrument that identifies 34 specific problem solving skills. We have evidence to support our hypothesis that a particular student has the same strengths and weaknesses when solving a complicated trip planning problem as they do when solving a physics problem or performing in the work place. There are a few specific skills that may be necessary for one type of problem that are not required to solve another; however, the strengths and weaknesses of a particular person are the same, regardless of the environment. Consistent strengths and weaknesses match intuition – certain people are stronger problem solvers – however; on the surface this may seem to disagree with research on context and transfer. A closer look at problem solving shows that our results are in fact consistent with other research on problem solving as well. The difference is that most research is complicated with the need for specific content knowledge to solve a particular problem and many researchers focus on the end result rather than looking at the specific skills used by the solver during the problem solving process.

ACKNOWLEDGMENTS

Supported in part by the NFS Distinguished Teaching Scholar.

REFERENCES

1. D. Hestenes, M. Wells and G. Swackhamer, *The Physics Teacher* **30**, 141-158 (1992).
2. R. K. Thornton and D. R. Sokoloff, *American Journal of Physics* **66**, 338-352 (1998).
3. D. P. Maloney, T. L. O'Kuma, C. J. Hieggelke and A. Van Heuvelen, *Am. Jrnl. of Phys.* **69**, S12-S23 (2001).
4. W. K. Adams, K. K. Perkins, M. Dubson, N. D. Finkelstein and C. E. Wieman, *Phys. Rev. Special Topics – PER* **2** (2006).
5. D. P. Maloney, "Research on Problem solving: Physics," in *Handbook of Research on Science Teaching and Learning* edited by D. L. Gabel, Macmillan, Toronto 1993, pp 327-354.
6. R.E. Meyer, *Thinking, problem solving, cognition (2nd ed)* Freeman, New York, 1992.
7. The Cognition & Technology Group at Vanderbilt, *Ed. Tech. Rsrch. and Dev.* **40**, 65-80 (1992).
8. K. Vick, E. Redish, and P. Cooney, Retrieved 8/18/06 from Activity-Based Physics (ABP) Alternative Homework Assignments (AHAs) Problem site: http://www.physics.umd.edu/perg/abp/aha/pyramid.htm
9. A copy of the problem solving instrument can be viewed at: http://cosmos.colorado.edu/CPPSS/

Physics Education Research in an Engineering Context

Christian H. Kautz

Institute of Mechanics and Ocean Engineering, Hamburg University of Technology
Schwarzenbergstr. 95, D–21073 Hamburg, Germany

Abstract. We report on an ongoing investigation of student understanding in several introductory engineering courses at Hamburg University of Technology. Preliminary results from a first-year electrical engineering course indicate that many students did not gain a conceptual understanding of the material. Some students had difficulty interpreting graphical representations of information or displayed a lack of understanding of basic principles. Specific examples concerning load lines and three-phase systems are used to illustrate how general findings from physics education research can guide investigations of student understanding and the development of curriculum in an introductory engineering context.

Keywords: Physics education research, engineering, electric circuits, load lines, three-phase systems.
PACS: 1.40.Fk

INTRODUCTION

In recent decades, Physics Education Research (PER) has helped identify many student difficulties with specific concepts in introductory physics. Results from many individual studies that focus on specific topics combine to support a number of general conclusions about the teaching and learning of physics that by now are widely accepted [1].

Many introductory engineering courses cover topics in which basic principles from physics are applied or extended. It therefore seems plausible that methods from physics education research could be successfully applied to investigate student understanding in introductory engineering.

At Hamburg University of Technology (TUHH), we have begun to follow this approach in several engineering disciplines. In this paper, we report on our study and illustrate how many of our findings are in good agreement with various general results from physics education research.

ENGINEERING CONTEXT

So far, our work has mostly focused on two courses: *Fundamentals of Electrical Engineering,* which covers circuits and fields, and *Engineering Thermodynamics,* which covers the laws of thermodynamics and their applications in industrial processes. More recently we have also turned our attention to a three-semester course in *Engineering Mechanics* that includes topics from *Statics, Mechanics of Materials,* and *Dynamics*. For the purpose of this paper, we focus on work done in the context of the electrical engineering (EE) course. Below, we briefly describe this course and then present some sample tasks used to probe student understanding.

Description of EE Course

Fundamentals of Electrical Engineering is a two-semester course taken by general and electrical engineering majors. In the first semester, it covers DC circuits, static electric fields and stationary currents, and magnetic fields and induction. Since *Physics* is only a co-requisite for this course (and for these students does not include electricity and magnetism) instruction for all three topics begins at a level comparable to a typical calculus-based introductory physics course but includes methods and examples usually not seen in the context of physics (such as Thevenin's and Norton's theorems, and the concept of a magnetic circuit). In the second semester, the course covers AC circuits as well as nonlinear and active circuit elements. While building on basic concepts that are taught early in the first semester, coverage extends substantially beyond typical physics content, including such topics as three-phase systems, transistor circuits, and operational amplifiers (op amps).

CP883, *2006 Physics Education Research Conference*, edited by L. McCullough, L. Hsu, and P. Heron

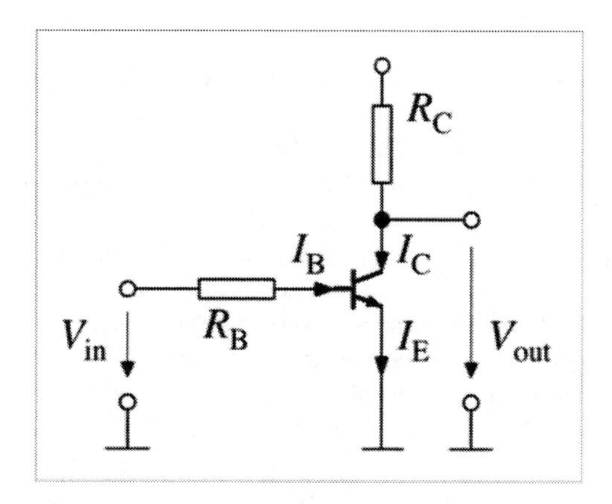

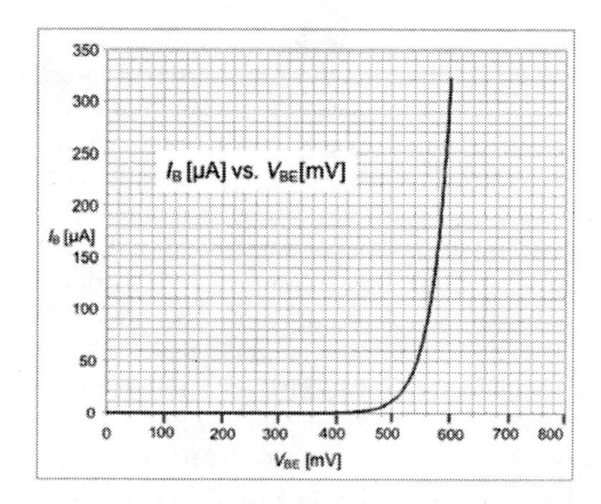

FIGURE 1. Circuit diagram and base-current characteristic included in the examination problem on transistor circuits.

The course has an enrolment of about 150 students and follows a format that is frequently found at German engineering institutions. Three hours of lectures each week are supplemented by two hours of small-group recitation sections. Although the audience includes a considerable number of foreign students, the language of instruction (and of all the questions posed as part of this study) is German. Examinations take place during lecture-free time periods at the end of each semester. Since students who do not pass may re-take the exam two more times after subsequent semesters, the cohort of students taking a particular exam may include students who attended classes one or two years earlier. This situation severely limits the changes that can be made to the course at one time. Moreover, it complicates the task of determining whether a particular modification to the course has had any effect on student learning.

Tasks for Probing Student Understanding

To probe student conceptual understanding, we use non-graded quizzes, examination tasks (including standard quantitative problems and qualitative questions) and multiple-choice questions asked during lecture ("clicker questions"). Below we describe two sample examination questions from the second semester of *Fundamentals of Electrical Engineering* and briefly outline a solution in each case.

Sample examination questions
from second-semester electrical engineering course

One of the questions presented students with a circuit diagram for a simple transistor circuit and a graph of the base-current characteristic (I_B *vs.* V_{BE}) as shown in Fig. 1. Among other tasks, students were asked to determine graphically the operating point of the transistor. This task required them to use the circuit data in the problem to draw a load line and then find its intersection with the given characteristic curve. The load line, which is a graphical representation of the current-voltage relationship in the base-emitter junction imposed by the circuit external to the transistor, can be found by applying Kirchhoff's voltage law (and Ohm's law for R_B). Although there is an algorithm that can be used, a conceptual understanding of simple DC circuits is helpful to arrive at the answer.

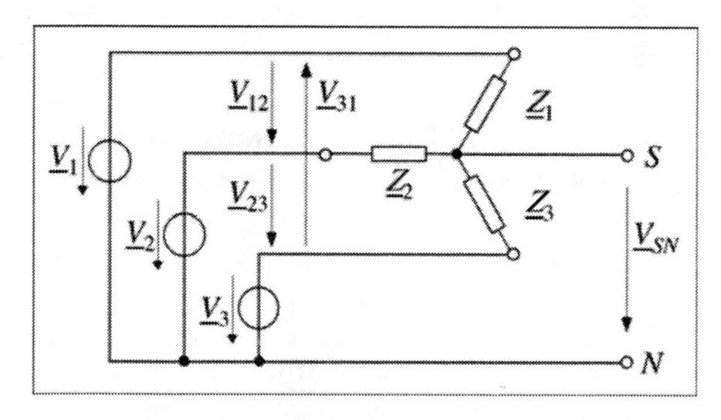

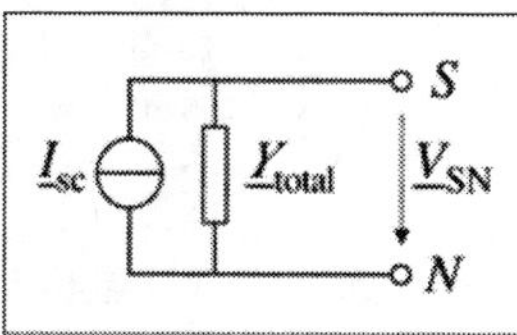

FIGURE 2. Circuit diagram for three-phase network and blueprint for Norton equivalent as given in the problem.

Another question that we used in our study concerned the three-phase circuit shown in Fig. 2. The three impedances ($\underline{Z}_1$, $\underline{Z}_2$, $\underline{Z}_3$) were of equal magnitude but different phase angle (i.e., one was purely resistive while the others had inductive or capacitive parts). Students were first asked to draw the phasors (vectors in the complex plane used to represent sinusoidally time-dependent quantities) for the three source voltages ($\underline{V}_1$, $\underline{V}_2$, $\underline{V}_3$) in a diagram containing the line voltages ($\underline{V}_{12}$, $\underline{V}_{23}$, $\underline{V}_{31}$). Next, the Thevenin and

Norton equivalents were to be found (i.e., the total impedance or admittance, the short-circuit current, and the open-circuit voltage needed to be determined for a given set of terminals). Finally, students were asked if the currents through the three load impedances would change if the terminals *SN* were short-circuited.

A correct answer required students to perform the following steps: draw three phasors at 120° angles that would correctly combine to form the given line voltages ($\underline{V}_{12} = \underline{V}_1 - \underline{V}_2$), combine impedances in parallel, graphically or algebraically add (complex) currents, and apply the AC equivalent of Ohm's law. To answer the final task, students did not have to use any of their answers to the preceding parts. Instead, they only needed to recognize that the three load impedances are not identical, and that as a result their currents would not add to zero if *SN* were short-circuited.

RESULTS

Many of the patterns of reasoning and student difficulties that we have observed in these courses so far mirror well established results in PER. Similarly, common strategies for helping students overcome typical difficulties in physics seem to have a good chance at being successful in engineering as well.

General Findings from *PER* Relevant to Engineering Contexts

Research in physics education has led to a number of general conclusions about student understanding for which evidence can be found in student responses to questions from various areas of physics. While these generalizations are now common knowledge in physics education (and are therefore stated here without reference to specific sources), it is interesting to note that many of them seem to be valid for introductory engineering as well. Moreover, these ideas seem to have affected instruction in engineering to a much lesser degree than in physics.

Failure To Form Conceptual Models

When the examination question on the transistor circuit in Fig. 1 was first administered to about 150 students, about 60% of them did not even attempt to draw a load line (and, in most cases, then failed to answer any of the subsequent parts of this question). On a similar question given after modified instruction in the following year, about 10% of the students drew "load lines" with a positive slope. We interpret this as resulting from the students' failure to develop a qualitative understanding of the relationship represented by the load line, such as "Increased load current results in greater voltage drop across internal resistance, thus leading to smaller terminal voltage."

This conceptual model had not been very strongly emphasized in the course. After deriving the load line equation and pointing out the relevance of the open-circuit voltage and short-circuit current, the lecturer presented three examples in some detail but did not make any additional effort to help students develop the kind of qualitative understanding stipulated above. The results suggest, however, that for many students this development is unlikely to happen spontaneously.

Difficulty Interpreting and Applying Graphical Representations of Information

In their responses to the examination question on the three-phase circuit in Fig. 2 about 15% of the students drew circuit symbols for sources or loads on their arrows in the phasor diagram. This observation suggests that a number of students in the course still fail to distinguish between two different (but frequently used) graphical representations at the time of the final exam. Moreover, virtually none of the students (less than 2%) attempted a graphical solution for finding the total admittance or the short circuit current. Students also did not use diagrams to plan or check their algebraic solutions.

As has been shown in various contexts in physics, students often have difficulty interpreting graphical representations of information. Students tend to avoid graphical means of solving problems even when these may be simpler or faster than alternative methods. Finally, relating graphical and algebraic representations of the same information to each other (as would be necessary in order to use one method for checking a solution obtained by the other) seems to present a particular challenge to many students.

Lack of Understanding of Basic Ideas and Principles

About 15% of the students taking the examination that included the three-phase circuit question arrived at incorrect results for the total admittance of the circuit due to difficulties with basic ideas in circuit analysis. These included confusion between series and parallel connections, a failure to distinguish between impedance and its inverse (admittance), and incorrect application of Ohm's law. Another 10% of the students inappropriately added the three source voltages (which always add to zero due to their phase relationship) and then concluded that the short-circuit current would be zero. It is likely that these students had not yet mastered the distinction between current and voltage, which would include an understanding of when either type of quantity can be added.

While this result, like the others above, will not be surprising to physics education researchers, it does not seem to be widely recognized in the engineering disciplines that many student difficulties with more advanced topics (and engineering applications) often have their roots in a lack of understanding of basic ideas and principles.

Conceptual Difficulties Associated with Vector Quantities

In their responses to the last part of the three-phase circuit question, a number of students claimed that the currents would not change since the "impedances are symmetric except for their phase angles." Apparently, these students had difficulty recognizing that vector quantities of the same magnitude but different direction are not equal. It is interesting to note that similar difficulties have been observed in the context of vector quantities in mechanics (e.g., velocity and momentum). More generally, we conclude that when using vector quantities in engineering contexts, students are likely to encounter the same difficulties as in introductory physics.

Use of *PER*-based Strategies for Instructional Materials in Engineering

For more than a decade, results of physics education research have strongly affected curriculum development in physics. While there exists now a variety of research-based materials, many of these use, implicitly or explicitly, similar instructional strategies. Below, we illustrate how we have used some of these in designing collaborative-group worksheets for *Fundamentals of Electrical Engineering.*

"Making Qualitative Predictions before Carrying Out Calculations"

In an activity on load lines, we ask students to consider a non-ideal voltage source that can be connected to different loads. Before calculating specific values, students predict the general shape of the graph of I_{load} *vs.* $V_{terminal}$. Later, students predict how the load line would change if the source voltage were increased or the internal resistance decreased.

"Relating Graphical, Verbal, and Algebraic Representations of the Same Relationships"

After drawing a load line for the given circuit, students are asked to interpret the *I*- and *V*-axis intercepts of the graph (beyond stating the terms "short-circuit current" and "open-circuit voltage"). They are also asked to spell out the general circuit law that is expressed in the equation of the load line.

"Relating Different Ways of Reasoning to Obtain the Same Answer"

In an activity on three-phase systems, students consider Y- and Δ-connected loads made up of identical resistances. After finding the magnitudes and phases of the various voltages and currents, students are asked to compare the power dissipated in the two configurations. Subsequent questions help students recognize that this question can be answered by considering either the source or the load side of the circuit.

Preliminary Assessment of Instructional Strategies

Responses to two different examination questions on transistor circuits that were given after the activities had been implemented indicate that these strategies can help improve student learning. While the fraction of blank responses decreased from about 60% to below 20%, there was a corresponding increase of correct answers from below 30% to about 55%. There is evidence that the frequency of some typical errors was also reduced. After the second iteration of the load-line worksheet, only 2 out of about 150 students drew a line that had a positive slope.

CONCLUSIONS

Preliminary results from our study of student learning in introductory engineering courses indicate that methods from physics education research can be successfully applied to probe student understanding in these courses. Furthermore, our results suggest that research-based strategies for instruction in physics also have some merit for teaching introductory engineering.

ACKNOWLEDGMENTS

The work described in this paper was carried out in several implementations of a course taught by Prof. M. Kasper, who also contributed to our interpretation of student ideas presented here. Some of the questions used in the study were written by various members of his research and teaching staff. The contributions of all those involved are greatly appreciated.

REFERENCES

1. L. C. McDermott, *Am. J. Phys.* **61**, 295-298 (1993).

The Cognitive Blending of Mathematics and Physics Knowledge

Thomas J. Bing and Edward F. Redish

Department of Physics, University of Maryland, College Park, MD 20742, USA

Abstract. Numbers, variables, and equations are used differently in a physics class than in a pure mathematics class. In physics, these symbols not only obey formal mathematical rules but also carry physical ideas and relations. This paper focuses on modeling how this combination of physical and mathematical knowledge is constructed. The cognitive blending framework highlights both the different ways this combination can occur and the emergence of new insights and meaning that follows such a combination. After an introduction to the blending framework itself, several examples from undergraduate physics students' work are analyzed.

Keywords: cognitive modeling, blending, mathematics
PACS: 01.30.lb, 01.40.Fk, 01.40.Ha

INTRODUCTION

An important sign of physics students' progress is their combining the symbols and structures of mathematics with their physical knowledge and intuition, enhancing both. The numbers, variables, and equations of the mathematics come to represent physical ideas and relations. Likewise, physical intuitions become encoded in a precise way that readily lends itself to the complex manipulations often required of a physicist. New ideas and inferences emerge after this combination.

The language of cognitive blending provides a framework for analyzing students' combination of mathematics and physics. This framework emphasizes both the emergence of new relations and the different ways the combination itself can be constructed. Following an overview of the theory of cognitive blending, the framework will be applied to examples of physics students' work.

THEORETICAL OVERVIEW

The knowledge we have in long-term memory that we use to interpret our perceptions is organized into associational patterns, knowledge elements or resources that tend to be primed or activated together. See [1] and the references therein for a detailed discussion. Cognitive blending theory refers to these as mental spaces. Fauconnier and Turner [2,3] describe how the mind combines two or more mental spaces to make sense of linguistic input in new, emergent ways. Blending usually occurs at a subconscious level, although the explicit thought required in classroom activity causes many of its details to become explicitly apparent. Like the learning process itself, blending is nonlinear and nondeterministic. The precise way a person blends two input mental spaces together depends strongly on cues in the linguistic input and on physical and mental context.

Example of a Blend

Consider an example from outside of physics. If someone hears "Bill Gates knocked out Steve Jobs," that person will construct a meaning for the statement by blending two mental spaces. Mental spaces typically contain both elements and an organizing frame of relationships, processes, and transformations. A Boxing mental space containing elements such as opposing fighters, punches, injuries, and so on would be blended with a Business mental space containing elements like Gates, Jobs, price cuts, advertising campaigns, and profits. In the blended space, Boxing CEOs, statements like "Bill Gates knocked out Steve Jobs", "Jobs hurt Gates with a new advertising campaign", and "Gates landed a punch with his price cuts" make sense. Such statements can't occur in the Business space because no one is literally knocked out in business. Nor can such statements occur in the Boxing mental space; Gates and Jobs are not boxers.

CP883, *2006 Physics Education Research Conference*, edited by L. McCullough, L. Hsu, and P. Heron

These statements are all examples of emergent meaning, ideas and relations that could not exist in either input mental space alone.

Single-Scope And Double-Scope Blends

Blends can be constructed in different ways. Fauconnier and Turner describe two types of blending topologies that are especially useful in considering students' combinations of physical and mathematical reasoning. These two types of blends are called single-scope and double-scope. Single-scope blends essentially only import elements from one input mental space into the organizing frame of the other. Double-scope blends display a blending of the organizing frames of the input mental spaces.

The earlier Boxing CEOs example can be diagrammed.

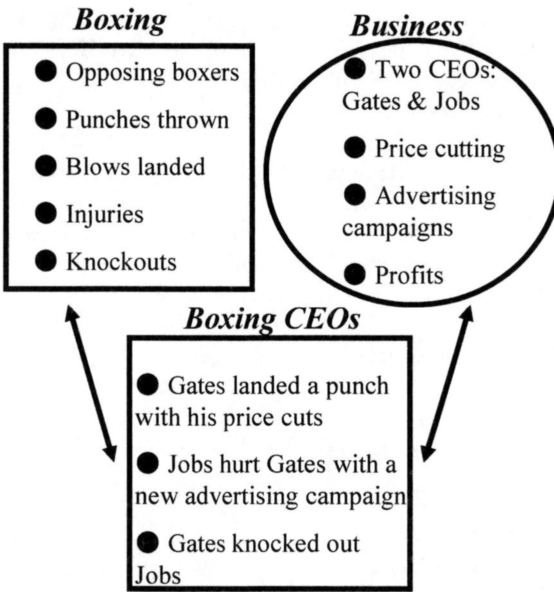

FIGURE 1. Single-Scope Boxing CEOs Blend

This blend is single-scope. The blended space's organizing structure is almost entirely that of the Boxing input mental space. To illustrate this point, consider the verbs used in the blended space. Landing a punch, hurting, and knocking out all come from the Boxing input space. The timescale of the blend is short and abrupt, like a boxing match, not drawn out over fiscal years. Combat in the blend is one-on-one and not corporation against corporation.

Just because a blend is single scope does not imply that no new insights are gained in the input space that only contributes elements to the blended space. On the contrary, most elements in a single-scope blended space inform the deemphasized input space upon projection back up through the blending network. All three of the elements in the Boxing CEOs space in Figure 1 ultimately convey a new idea or relation about business. The single-scope distinction arises from the structure of the blended space itself. That most of the emergent meaning of this blending network pertains to business is a consequence of the single-scope topology.

We claim that this sample blend is a common, reasonable way a person could construct meaning for a statement like "Bill Gates knocked out Steve Jobs" and not necessarily the only possible way. With the blending framework now illustrated, we continue to several examples from physics students' work. Here mathematical and physical mental spaces will be blended to produce emergent meaning.

STUDENT BLENDING EXAMPLES

The examples of student work analyzed here come from video tapes of physics students working outside of class on their homework. In addition to the video, the data library contains copies of the written work the students ultimately submitted.

Double-Scope Air Drag Example

This first example comes from a group of three physics majors enrolled in a Mathematical Methods in Physics class taught by the physics department, usually to students in their second or third year. They are working on a homework problem where an object is thrown straight up and falls back down under the forces of gravity and air resistance. In the following excerpts, they are working to understand the expression for the viscous force, $F_V = -bv$, as given in the homework problem. Their conversation includes

> S1 Because the negative means that friction operates in the opposite direction of whatever v is.

and

> S2 Well, but let's do the first one first, if you're going down … what is v gonna be? Is it gonna be negative or positive?
> S3 It's gonna be negative.
> S2 OK, so a negative times a negative
> S3 Is gonna be positive.
> S2 Right, and a positive points up.

and

> S2 We need to leave this negative in so that it cancels out that one

S1 Right, because v has that negative built into it, and so we need another negative out here to make sure that the two negatives cancel out and you end up with a positive, which is up, which is the direction of friction because it's going down—it's falling down, being dragged in the upward direction.

All of these excerpts come from the same two-minute clip. As before, a blending diagram can be drawn.

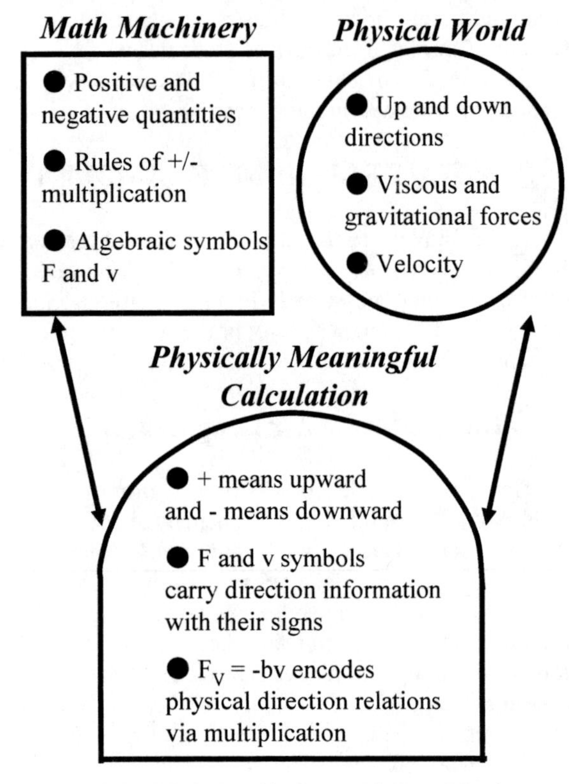

FIGURE 2. Double-Scope Air Drag Blend

This blend is double-scope. It represents a combination of organizing frames, not an importation of elements into the other input space's frame. Unlike the uni-directionality seen in the Boxing CEO example, the transcript examples above exhibit a bi-directionality. Sometimes the students start with a mathematical statement and translate it to a physical statement, as in the first excerpt. In the second excerpt, a physical statement is translated into a mathematical idea. Examining the role of positive and negative signs in this blend also indicates its double-scope nature. In the second and third excerpts these signs behave according to algebraic rules, but at the same time they encode physical information on the direction of the drag force that is carried along and expressed through these algebraic manipulations.

Whereas most of the emergent meaning that arose in the single-scope Boxing CEOs network pertained largely only to one input space, the situation is more balanced in the double-scope example. The third blended space element refines and codifies the physical observation that the air drag force must act opposite the direction of motion. At the same time, this compact expression will be used within a larger piece of mathematical machinery when the students insert it into Newton's Second Law and proceed to solve the resulting equation of motion.

The next example illustrates a single-scope blend of mathematics and physical reasoning. It shares the structural features of the Boxing CEOs example.

Single-Scope Travel Time Example

This example comes from a student enrolled in an algebra-based Introductory Mechanics course for biological science majors. The student is working with a teaching assistant on a homework problem asking how much time it would take a car traveling 95 feet per second to go 500 feet.

S4 So, I was trying to do a proportion, but that doesn't work. I was like 95 feet per second, oh wait, yeah in 500 feet, like, x would be the time—that doesn't, I get like this huge number and that doesn't make any sense.

Having constructed the equation $95/1 = x/500$, the student has arrived at $x = 47{,}500$ seconds.

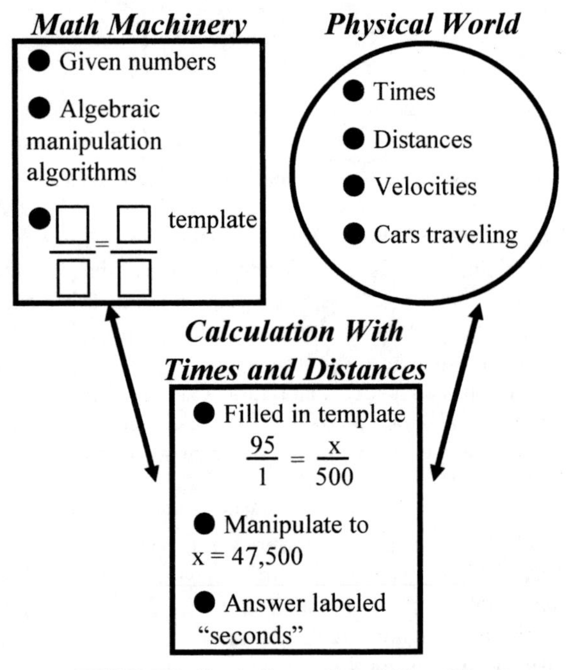

FIGURE 3. Single-Scope Travel Time Blend

The student's work in the blended space is organized almost entirely by the Math Machinery input space's frame. Only elements like the label "seconds" that is tacked onto the final numeric answer come from the other input mental space, leading to the characterization of this blend as single-scope. Just as in the Boxing CEOs example, the emergent meaning of this network largely informs the deemphasized input space, in this case the Physical World mental space. The student interprets the final answer of 47,500 seconds as newly constructed information about how long the car requires to travel 500 feet. It is the unreasonableness encountered in this projection back up through the single-scope network that signals a problem to the student.

It is important to note that this student was quite capable of constructing a rich Physical mental space for the situation at hand. Immediately following the excerpt given above, the teaching assistant probes the student's physical intuitions in more detail.

TA That doesn't make any sense. So what if I said something like if you're traveling two feet per second and you go four feet. How long would that take you?
S4 Two seconds
TA Or, if you tried different numbers, if I was traveling eight feet per second and you traveled sixteen feet, how long would that take you?
S4 Two seconds

Both times the student answers quickly without formal, explicit calculation. The simple numbers the teaching assistant offered allowed the student to operate within a rich Physical mental space with excellent intuitions about rates of travel, distances, and times.

DISCUSSION

Use of the cognitive blending framework emphasizes the demands students face concerning the integration of their mathematical and physical knowledge. This integration process is complex and can be achieved in different ways, as the single-scope and double-scope distinction indicates. Difficulties experienced by the latter student stem not from a lack of prerequisite knowledge but from an inappropriate integration of two well established mental spaces.

In and of themselves, single-scope blends are not more or less appropriate than double-scope blends. The utility of a particular blending topology depends on the specific context. Double-scope blends of mathematical and physical mental spaces are more common in areas like mechanics and electricity and magnetism where a student's physical experiences and intuitions are more robust and available. It is quite possible that a more mathematically dominated topic like quantum mechanics would be marked by a heavier reliance on single-scope networks, even among experts.

Fauconnier and Turner note that the terms single-scope and double-scope are merely two convenient markers along a continuum of blending topologies. Observation of students' work with mathematics and physics, however, indicates that both types of blends have inertia. Students who initially approach a physics question with a heavily single-scope blend tend to continue working within that topology. Transition to a more double-scope blend of physical and mathematical knowledge often requires a strong perturbation to the system, such as a student suddenly considering a new piece of evidence or an instructor asking questions aimed at activating the deemphasized mental space. Awareness of this cognitive blending framework can help instructors more readily understand how students are thinking and offer appropriate guidance for the specific situation at hand. It can help researchers in providing a theoretical framework for description of student thought and perhaps even a structure for understanding what cues prompt students for blending in particular ways. Such a framework could potentially provide predictive power for development of effective instructional materials.

ACKNOWLEDGMENTS

The authors wish to thank David Hammer, Rachel Scherr, and the other members of the Physics Education Research Group at the University of Maryland for their insightful discussion. This work is supported by National Science Foundation grants DUE 05-24987 and REC 04-40113 and a Graduate Research Fellowship.

REFERENCES

1. E. F. Redish, "A Theoretical Framework for Physics Education Research: Modeling Student Thinking" in *Proceedings of the Varenna Summer School, "Enrico Fermi" Course CLVI*, Italian Physical Society, Bologna, Italy, 2004, pp. 1-63.
2. G. Fauconnier and M. Turner, *Cognitive Science* **22**(2), 133-187 (1998).
3. G. Fauconnier and M. Turner, *The Way We Think: Conceptual Blending and the Mind's Hidden Complexities*, Perseus Books Group, New York, 2002.

Probing Student Reasoning and Intuitions in Intermediate Mechanics: An Example With Linear Oscillations

Bradley S. Ambrose

Department of Physics, Grand Valley State University, Allendale, MI 49401, USA

Abstract. The study of linear oscillations—including simple harmonic, damped, and driven oscillations—is not only fundamental in classical mechanics but lies at the heart of numerous applications in the engineering sciences. Results from research conducted in the context of junior-level mechanics courses suggest the presence of specific conceptual and reasoning difficulties, many of which seem to be based on fundamental concepts. Evidence from pretests (ungraded quizzes) will be presented to illustrate critical difficulties in understanding conceptual underpinnings, relating concepts to graphical representations (*e.g.*, motion graphs), and connecting the physics to the relevant differential equations of motion. Preliminary results from the development of a tutorial approach to instruction, modeled after *Tutorials in Introductory Physics* by McDermott, *et al.*, [1] suggest that such an approach can be effective in both physics and engineering courses. (Supported by NSF grants DUE-0441426 and DUE-0442388.).

Keywords: Mechanics, upper division, simple harmonic motion, damped harmonic motion, 2-D oscillations.
PACS: 01.30.Cc, 01.40.Fk, 45.20.D-, 45.30.+s

INTRODUCTION

As part of an ongoing investigation of student learning in intermediate mechanics, we are probing how advanced undergraduate majors in physics, math, and engineering think about oscillations in one and two dimensions. Instructors often expect their students to extend what they have learned at the introductory level about oscillatory motion (*e.g.*, simple harmonic motion) to situations that are physically and mathematically more sophisticated. However, evidence from this study corroborates previous studies that demonstrate how difficulties with basic concepts can hinder meaningful learning in upper level courses [2]. Furthermore, analysis of student responses to numerous research tasks, including written qualitative questions that require explanations of reasoning, often suggests that students need guidance in organizing their knowledge.

This report will focus specifically on the following research questions: (a) How well do students understand the factors that affect the *frequency* of different types of linear oscillations? (b) How well do students interpret and understand *formal representations* of oscillatory motion, such as *x vs. t* graphs of 1-D oscillators and *x-y* trajectories of 2-D oscillators? (c) To what extent do students answer qualitative questions by bringing to bear their knowledge of *general principles* relevant to the physical situation at hand?

CONTEXT OF INVESTIGATION

The student populations discussed here come from junior-level intermediate mechanics courses at Grand Valley State University (GVSU), the University of Maine (UME), and Seattle Pacific University (SPU). Although the details of the courses vary somewhat, all courses cover linear oscillations (simple harmonic, damped, driven) and other topics that require the synthesis of Newton's laws, work and energy, and differential equations. In addition, the classes discussed here were taught either by the author or Michael Wittmann (UME), with whom the author is collaborating on *Intermediate Mechanics Tutorials (IMT),* a set of research-based curricular materials modeled after *Tutorials in Introductory Physics* [1].

The results presented in this paper were taken primarily from the analysis of responses to written pretests (ungraded quizzes). In all cases the pretests were given after lecture instruction but before the tutorial (from *IMT*) on the relevant topic. At GVSU and SPU each pretest was administered during class for 10 min; at UME students were instructed to complete each pretest outside class for 10-15 min. All pretest questions asked for explanations of reasoning.

CP883, *2006 Physics Education Research Conference*, edited by L. McCullough, L. Hsu, and P. Heron

PROBING STUDENT THINKING OF SIMPLE HARMONIC MOTION IN ONE AND TWO DIMENSIONS

In this section we describe results from pretests that probe the ability of students to apply (in 1-D) and extend (to 2-D) the idea that the frequency [$\omega_o = (k/m)^{1/2}$] of a simple harmonic oscillator is determined solely by the spring constant and mass. Students in all classes discussed here covered 1-D simple harmonic oscillators at the introductory level.

Simple Harmonic Motion

The first pretest on oscillations includes questions that elicit student ideas about the factors that affect the frequency of simple harmonic oscillations. For this report we describe the results from 4 classes ($N = 35$) at GVSU and 1 class ($N = 11$) at SPU.

Students are shown a strobe picture illustrating a block connected to an ideal spring that is released from rest on a level, horizontal surface. They are asked how the period would be affected by: (i) changing the release point of the block from 0.5 m to 0.7 m from equilibrium, (ii) replacing the original spring with one that is stiffer, and (iii) replacing the original block with one having four times the mass. The students were expected to recognize that the period will not change in case (i), decrease in case (ii), and increase (double) in case (iii).

Incorrect Intuitions Relating Period and Amplitude

Although most students gave correct responses (ignoring reasoning) for each case, case (i) yielded the lowest percentage of complete and correct explanations (~10%). Many correct responses were supported by "compensation arguments" [3] relating amplitude, average speed, and period. As one student explained, "It may seem that the block is moving faster, but it is also moving farther to compensate." While such justifications make it *plausible* that the period is unaffected by changing the amplitude, they show no evidence of understanding that *only* the spring constant and mass affect the period. Even more telling, the most common *incorrect* explanation (from ~25% of the students) was based on the incorrect intuition that the greater initial displacement from equilibrium (and hence the larger amplitude) would cause the period to increase because, for example, "the block travels farther during each period."

The above results are interesting because they suggest persistent, incorrect intuitions that may lead to confusion in the context of 2-D oscillations. Even though students completed a tutorial (not discussed here) on 1-D harmonic oscillators, the above pretest results suggested the need to explore how students proceed from 1-D to the 2-D case.

Harmonic Motion in Two Dimensions

Students often were introduced to 2-D oscillations as an application of conservative forces, several weeks after covering 1-D oscillations. The following pretest, given after relevant lectures to 4 classes ($N = 31$) at GVSU and 1 class ($N = 17$) at UME, was designed to probe student understanding of the relative frequencies along the x- and y-axes of a 2-D oscillator.

Students are asked to consider an undamped 2-D oscillator with $U(x, y) = \frac{1}{2}k_1x^2 + \frac{1}{2}k_2y^2$. (They are also reminded about the relationship $k_i = m\omega_i^2$ for each force constant.) For each x-y trajectory shown in Fig. 1, students are asked whether that trajectory is possible for such an oscillator and, if so, whether ω_1 is greater than, less than, or equal to ω_2. (The original version of the pretest asked for a comparison of the force constants k_1 and k_2 instead of the frequencies. The change in wording, however, did not significantly alter overall student performance.)

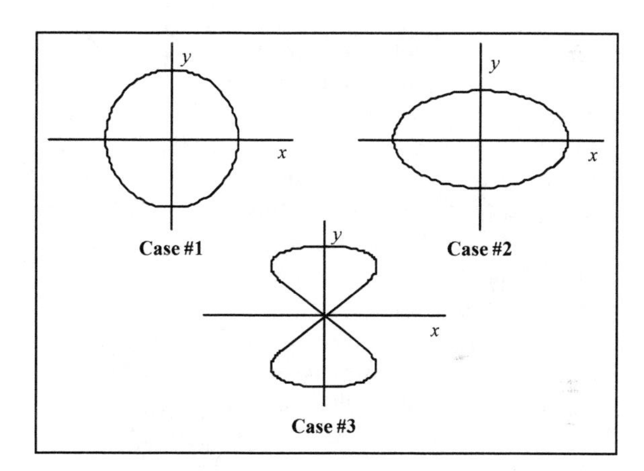

FIGURE 1. Three x-y trajectories shown on the pretest for 2-D oscillators. Students were asked for each case how the frequencies along the x- and y-axes compared.

Students were expected to infer from each trajectory how many cycles occurred along one axis for each cycle along the other. Two isotropic cases (#1 and #2) were included, and showing different x- and y- amplitudes for Case #2 was intended to elicit incorrect intuitions about frequency and amplitude.

Use of "Compensation Arguments" Relating Frequencies (or Force Constants) and Amplitudes

In each class very few students (between 0% and 15%) gave correct responses for all cases, even when explanations were ignored. Most students incorrectly compared the relative frequencies (or relative force constants) by using inappropriate "compensation arguments" involving the relative amplitudes along the *x*- and *y*-axes. For example, for case #2 most incorrectly predicted that $\omega_1 < \omega_2$ (or $k_1 < k_2$) for reasons such as: "the spring goes farther in the *x*-direction, so [the] spring must be less stiff in that direction," or "since we now have an oval curve with the *x*-axis longer, ω_2 must be greater to compensate."

The prevalence of this type of "compensation" reasoning is striking for two reasons. First, it strongly suggests that most students fail to recognize that *x*-*y* trajectories like those from the pretest yield frequency information about the 2-D oscillator. Second, the tendency for students to link amplitudes with frequencies (or force constants) appears to be analogous to the most common incorrect mode of reasoning used on the 1-D oscillator pretest. This result suggests the recurrence of conceptual difficulty with fundamental ideas.

PROBING STUDENT THINKING OF DAMPED HARMONIC MOTION

Students encountering damped oscillations for the first time usually do so at the intermediate level, after simple harmonic motion but before 2-D oscillations. Typically the lecturer demonstrates shows how to set up and solve the differential equation. For the underdamped form of the solution the students are shown that amplitude decreases exponentially with time and that the frequency is smaller than that for the undamped oscillator: $\omega_d = (\omega_o^2 - \gamma^2)^{1/2}$, where γ is the damping constant. Given this typical treatment of damped oscillators it was desired to study how well students understood qualitatively how the presence of damping affects the motion of an oscillator.

Underdamped Motion

After lecture instruction on damped oscillators, students in 4 GVSU classes ($N = 35$) and 1 SPU class ($N = 11$) were given the following two-part pretest. The pretest began by showing students the *x vs. t* graph of a simple harmonic oscillator (no damping) released from rest (see solid curve in Fig. 2). They were then told to assume that a linear damping force is applied, causing the oscillator to become underdamped. (Students were reminded of the meaning of the term.) In part A of the pretest, students were asked to sketch a qualitatively correct graph of the underdamped oscillator having the same initial conditions as the original (undamped) one. In part B, they were asked to consider the instant it first passes $x = 0$: at that instant is the oscillator speeding up, slowing down, or moving with maximum speed?

(*Note:* Part B was not included on the pretest for one GVSU class. A slightly different version of part B was given to two of the GVSU classes: Does the [underdamped] oscillator first attain a maximum speed before, after, or exactly at the same instant when it passes through $x = 0$? The change in wording did not seem to affect the overall performance of the students.)

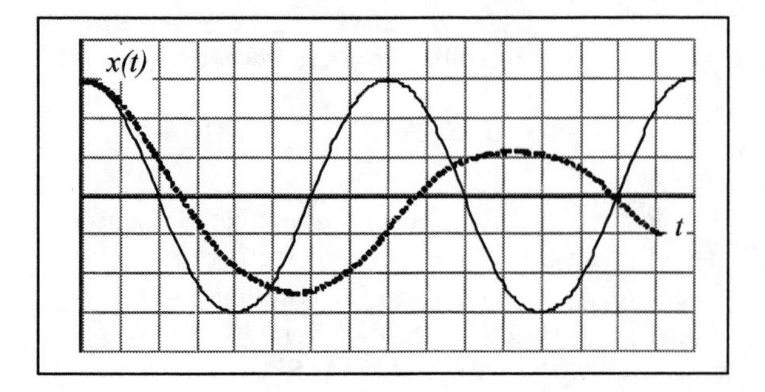

FIGURE 2. Motion graph from part A of the pretest on underdamped oscillators. The solid curve represents the motion of a simple harmonic oscillator. The dashed curve (not shown to students) illustrates a qualitatively correct graph for an underdamped oscillator.

Inappropriate Generalizations from the Case of Simple Harmonic Motion

Few students (~25% or fewer) in each class answered part A correctly. Any curve like the dashed curve shown in Fig. 2 would have been acceptable. However, most students (60% to 70%) drew graphs like the one shown in Fig. 3, showing a gradually decreasing amplitude (which is correct) but a frequency that is equal to that for the undamped case. Most explanations—for example, "the amplitude shrinks in time but the period shouldn't change since they are independent of each other"—suggest an overgeneralization from simple harmonic motion.

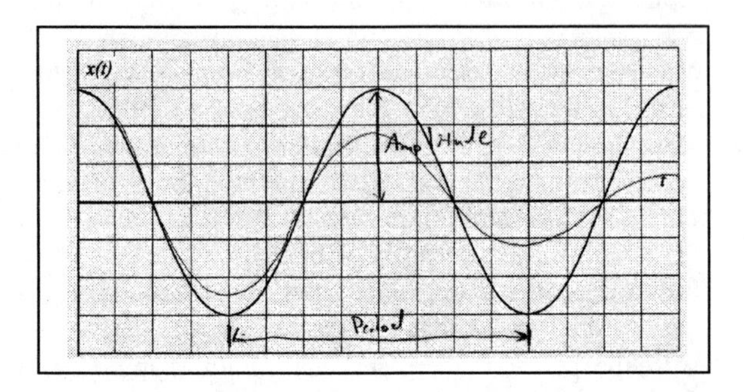

FIGURE 3. Example of a typical incorrect student graph elicited by the pretest on underdamped oscillations. Most students drew graphs like this one, showing equal frequencies for the undamped and underdamped cases.

Other errors arose on part A, including the tendency to show *both* the amplitude *and* period as gradually decreasing. These responses could be interpreted as recurrences of the belief that amplitude and frequency are connected, a belief that was detected on each of the two pretests described previously. More research is needed, though, to tell for certain.

Part B of the pretest was equally difficult for students. Students could answer part B by taking the differential equation of motion, $ma = -cv - kx$, setting $x = 0$, recognizing that acceleration and velocity must be opposite in direction, and concluding that the oscillator must be slowing down at $x = 0$. Students could get the same answer by drawing a free-body diagram and finding that the net force opposes the velocity.

Only 20% to 30% of the students in each class gave correct responses. The most common incorrect answer was to state that the oscillator experienced its maximum speed upon passing $x = 0$. Some did not seem to take the damping into account, saying that there was no acceleration because the spring was neither pushing nor pulling. Others did not at all invoke forces or Newton's laws, saying simply that the slope of the x *vs.* t graph would be at a maximum at $x = 0$. Both modes of reasoning strongly suggest a tendency to overgeneralize from the case of simple harmonic motion rather than to bring to bear one's knowledge of Newton's laws.

IMPLICATIONS FOR INSTRUCTION AND FUTURE RESEARCH

Although the pretests discussed here do not necessarily measure depth of student understanding, they show what physics majors often cannot do after traditional lectures. They need guidance in recognizing which factors affect the frequency of various types of oscillations, including simple harmonic motion (covered at the introductory level) and underdamped motion (covered usually for the first time at the sophomore or junior level). Students also have difficulty interpreting representations of oscillatory motion, including x-y trajectories of 2-D oscillators. As with many investigations conducted at the introductory level and beyond, traditional lecture instruction has found to do little to promote conceptual development in students, even students who are physics majors [4].

The prevalence of incorrect or incomplete explanations suggests that many students entering the intermediate mechanics class lack a strong conceptual framework upon which to build. Rather than recognize the relevance and utility of physics principles (*e.g.*, Newton's second law), many students tend to make inappropriate generalizations from special cases (*e.g.*, to incorrectly infer the behavior of underdamped oscillators from results that are valid only when damping is absent).

Although some types of incorrect responses, including "compensation arguments" linking amplitude to frequency (or period) are prevalent, it is possible that they may not indicate hard-and-fast conceptual difficulties as much as the tendency for students to proceed from incorrect assumptions or even inadvertent "triggers" from the research task. For instance, modifications are being considered for the 2-D oscillator pretest in which the students will not be shown a set of possible x-y trajectories. In the event that presenting both circular and elliptical trajectories "triggered" a high percentage of amplitude-frequency explanations, students will instead be given the initial conditions of motion and a specified k_2/k_1 (or ω_2/ω_1) ratio. They will then be asked to sketch a possible x-y trajectory for the oscillator. It is hoped that analysis of student responses on the revised pretest will allow a measure of the robustness of amplitude-frequency explanations. Such results would be used to guide refinements to existing *IMT* materials, so that they will become even more effective in helping students make the qualitative and quantitative extensions from introductory to intermediate level mechanics.

ACKNOWLEDGMENTS

Special thanks are due to Michael Wittmann (U. Maine) and his colleagues for their collaborations on the *IMT* project, as well as to the National Science Foundation for their support (DUE-0441426 and DUE-0442388). Thanks are also due to Stamatis Vokos and John Lindberg (Seattle Pacific U.); and Lillian McDermott, Peter Shaffer, and Paula Heron (U. Washington). Finally, the author thanks Chandralekha Singh for co-organizing the 2006 PERC targeted poster session in which this report was first presented.

REFERENCES

1. L. C. McDermott, P. S. Shaffer, and the Phys. Ed. Group at the U. of Washington, *Tutorials in Introductory Physics*, Prentice-Hall, Upper Saddle River, NJ, 2002.
2. B. S. Ambrose, *Am. J. Phys.* **72**, 453-459 (2004).
3. T. O'Brien Pride, S. Vokos, and L. C. McDermott, *Am. J. Phys.* **66**, 147-156 (1998).
4. L. C. McDermott and E. F. Redish, *Am. J. Phys.* **67**, 755-767 (1999).

Reforming a large lecture modern physics course for engineering majors using a PER-based design

S. B. McKagan*, K. K. Perkins† and C. E. Wieman*,†

*JILA, University of Colorado, Boulder, CO 80309, USA
†Department of Physics, University of Colorado, Boulder, CO 80309, USA

Abstract. We have reformed a large lecture modern physics course for engineering majors by radically changing both the content and the learning techniques implemented in lecture and homework. Traditionally this course has been taught in a manner similar to the equivalent course for physics majors, focusing on mathematical solutions of abstract problems. Based on interviews with physics and engineering professors, we developed a syllabus and learning goals focused on content that was more useful to our actual student population: engineering majors. The content of this course emphasized reasoning development, model building, and connections to real world applications. In addition we implemented a variety of PER-based learning techniques, including peer instruction, collaborative homework sessions, and interactive simulations. We have assessed the effectiveness of reforms in this course using pre/post surveys on both content and beliefs. We have found significant improvements in both content knowledge and beliefs compared with the same course before implementing these reforms and a corresponding course for physics majors.

Keywords: physics education research, course reform, modern physics, quantum mechanics
PACS: 01.40.Di,01.40.Fk,01.40.G-,01.40.gb

INTRODUCTION

It is well-documented that PER-based interactive engagement techniques improve student learning in introductory physics courses. However, the use and study of these techniques has been much less common in upper-level physics courses. Some physics professors who accept the use of interactive engagement techniques in introductory physics courses claim that these techniques are inappropriate for more advanced courses. At the University of Colorado, we are working to systematically reform both introductory and advanced courses in physics and other sciences, and to document the results of these reforms. In the fall of 2005 and spring of 2006, the authors taught and reformed Physics 2130, a modern physics course for engineering majors [1]. This course is the third semester of physics, and is typically taken by sophomore mechanical engineering majors and senior electrical engineering majors. This is the last physics course that most of these students will take.

This course has been reformed using a research-based design, based on the following principles learned from Physics Education Research:

1. Using interactive engagement techniques can lead to higher learning gains than traditional lecture. [2]
2. Directly addressing common misconceptions can lead to higher learning gains. [3]
3. Unless physics content is presented in a way that explicitly addresses student beliefs about science, these beliefs tend to become more novice-like. [4, 5]
4. People have a limited short term memory, so material should be presented in a manner that reduces cognitive load by focusing on the important points, having a coherent structure, and eliminating nonessential details. [6]
5. For students to gain a conceptual understanding of the material, all aspects of the course, including homework and exams, must address conceptual understanding, not just numerical problem-solving.

PROCESS OF REFORM

In order to develop a clear set of learning goals for the course, we interviewed seven physics faculty members about the most important concepts they thought students should learn from this course. These interviews elucidated an important issue in reforming more advanced courses: unlike introductory physics, in which there is a well-established set of topics on which most experts agree, there is no general consensus about what should be taught in more advanced classes. This issue is particularly acute in this course, which by default has often closely resembled the corresponding course for physics majors, who will see the topics in several later courses.

To determine how to make our course most relevant to our target audience, we met with a group of engineering professors, to whom we posed the question, "What do your students need to know about modern physics?" The general consensus was that engineering students need to know about applications of quantum mechanics such

CP883, *2006 Physics Education Research Conference*, edited by L. McCullough, L. Hsu, and P. Heron

as electron devices, lasers, STMs, and MRIs; they need to know about the quantum origin of molecular bonding and material structure; and they need some experience solving differential equations describing physical systems. The engineering professors said that their students do *not* need to know about special relativity or a lot of abstract mathematical formalism, topics that had typically been emphasized in this course.

To address principle 2, we reviewed the existing PER literature on student difficulties in modern physics and quantum mechanics. In addition, one of us (SBM) hosted a weekly problem-solving session for students in the course the semester before our reform. Field notes from this session provided insights into common problems for our student population.

Concurrently with our reform effort, we have been developing the Quantum Mechanics Conceptual Survey (QMCS), a multiple choice survey designed to test student understanding of the fundamental concepts of quantum mechanics [7]. The questions in the QMCS are based on faculty interviews, studies of textbooks and syllabi, existing conceptual tests of quantum mechanics [8, 9], research studies of student misconceptions [10–12], informal observations of students in problem-solving sessions and class, and formal interviews with students. The student interviews conducted to validate this survey were also useful in gaining a better understanding of student misconceptions.

The content of the course was chosen to reflect the concepts most commonly cited in faculty interviews (fundamental principles of quantum mechanics), the priorities of the engineering faculty (real world applications and the relationship of microscopic principles to macroscopic properties of materials), and expert beliefs about the relevance and coherence of physics (real world applications, grounding in experiment, conceptual understanding, and reasoning development).

INTERACTIVE ENGAGEMENT TECHNIQUES

We encouraged interactive engagement in class by assigning students to 3 person consensus groups for peer instruction. We asked an average of about 5 questions per 50 minute class, to which students submitted their answers using clickers. Most of the questions included a period of group discussion. We used several different kinds of clicker questions, including interactive lecture demos in which we asked students to predict the results of an experiment, usually demonstrated with a simulation; eliciting misconceptions in order to address them; polling students to find out more about their background or what they wanted us to address; asking students to work through difficult multi-step problems; and quizzes on the assigned reading. During clicker questions, the three instructors as well as three undergraduate learning assistants circulated through the room in order to facilitate group discussions and to listen and report back on what students were thinking. We occasionally used other interactive engagement techniques in class, for example working through a tutorial on quantum tunneling.

Outside of class, we encouraged interactive engagement by hosting collaborative problem solving sessions where students could work together on homework. These sessions were staffed by instructors, undergraduate learning assistants, and graduate teaching assistants, all of whom were trained to facilitate discussion and help students figure out answers on their own, rather than telling students the answers. These sessions were voluntary, but attendance was encouraged by advertising their value and making the homework sufficiently difficult that students could seldom complete it on their own. According to the end of term survey, one third of the students attended the problem-solving sessions at least 80% of the time, and another third attended 20-60% of the time.

We have developed a suite of interactive computer simulations on quantum mechanics specifically for this course as part of the Physics Education Technology (PhET) Project [13]. These simulations follow research-based design principles and are extensively tested through student interviews and classroom studies. In the course, we incorporated simulations into lecture, clicker questions, and homework. The homework included a large number of guided inquiry activities designed to help students explore and learn from the simulations. By providing visual representations of abstract concepts and microscopic processes that cannot be directly observed, these simulations help students to build mental models of phenomena that are often difficult to understand.

The simulations incorporate many of the principles listed in the introduction, such as reducing cognitive load by focusing student attention only on essential features. For example, many students have difficulty understanding the circuit diagram for the variable voltage supply in the photoelectric effect experiment, which distracts them from seeing the main point of the experiment. By illustrating the variable voltage supply as a battery with a slider, our Photoelectric Effect simulation eliminates this distraction.

CONCEPTUAL UNDERSTANDING AND REASONING DEVELOPMENT

Throughout the course, we focused on helping students develop conceptual understanding and reasoning skills, such as making inferences from observations and understanding why we believe the ideas of quantum mechanics. This was emphasized in all aspects of the course. We wrote all our own homework, which was online and composed of computer-graded multiple choice and numeric questions, and TA-graded essay questions. The homework was designed to be extremely difficult conceptually, though only moderately difficult mathematically. Thus students were required to write essays explaining a conceptual model or to determine the underlying reasons for a complex physical phenomenon.

For example, students worked through a series of homework questions using the Lasers simulation to build up an understanding of how a laser works, at the end of which they had to write essays on questions such as why a population inversion is necessary to build a laser and why this requires atoms with three levels instead of two.

REAL WORLD APPLICATIONS

We incorporated applications into every aspect of the course, presenting at least one application of each major concept discussed. We presented photomultiplier tubes as an application of the photoelectric effect; discharge lamps, fluorescent lights, and lasers as applications of atomic structure and transitions, alpha decay and STMs and applications of quantum tunneling, LEDs and CCDs as applications of the quantum theory of conductivity, and MRIs as an application of spin. A lecture on Bose Einstein Condensation tied together many of the concepts introduced throughout the course.

TEXTBOOK

Finding a textbook appropriate for this course was difficult, given the focus on conceptual understanding and applications, which are not suitably covered in standard texts. The first semester we used Tipler and Llewellyn [14], a popular modern physics textbook that was consistent with our level of math and contained most topics we covered. Students complained about the text on a weekly basis, both verbally and in feedback forms, and our top students reported that they stopped reading the textbook because they couldn't understand it. Many students used our power point lecture notes, which were posted online, as an alternative to the textbook. The second semester we switched to portions of volumes 3 and 5 of Knight's introductory physics textbook based on physics education research [15]. This textbook is at a lower level than our course mathematically, and it does not include many of our topics such as the time-independent Schrodinger equation. However, the pedagogical focus for the topics it does include is much more consistent with our with our approach. There were almost no complaints about the textbook second semester, and the average student ranking of the usefulness of the textbook for their learning on a scale of 1 (not useful) to 5 (a great deal) went up from 2.1 to 3.2. The usefulness rankings for other aspects of the course did not change significantly between the two semesters, and were between 3.5 and 4.3, with the posted lecture notes receiving the highest ranking.

ASSESSMENT OF COURSE

We used several methods for assessing the effectiveness of instruction in this course, including giving the QMCS pre/post as measure of conceptual learning and the Colorado Learning and Attitudes about Science Survey (CLASS) [16] as a measure of the change in student beliefs about science. We have also done several studies to assess learning in particular content areas of the course, the results of which will be reported elsewhere.

We gave the QMCS as a pretest and posttest during the two semesters in which we taught this course (Fa05 and Sp06) both to our students (engineering majors) and to the students in the corresponding course for physics majors. We also gave it as posttest to both the engineering and physics majors' courses the semester before our reforms (Sp05). We use the other four modern physics as baseline "traditional" courses.[1] It is worth noting that our class size was approximately 180 students both semesters, more than double the typical size of this course in previous semesters, about 80. The physics majors' course ranges in size from about 30 to 80.

Table 1 shows the QMCS results. We calculate the average normalized gain (<g>) for each course, which measures how much students learned as a fraction of how much they could have learned.[2] The values of <g> for the reformed courses are consistent with typical normalized gains on the Force Concept Inventory (FCI) in reformed introductory physics courses [2]. There are wide variations in <g> for the traditional courses, but all are consistently lower than in the reformed courses. It is interesting to note that the physics majors started consistently higher than the engineering majors, but ended up lower than the engineering majors in the reformed courses.

It should be noted that the QMCS covers only the fundamental concepts of quantum mechanics, and not any of the applications that constituted a substantial fraction of our course. However, all six courses spent a comparable amount of time on the material covered by the QMCS, since the engineering majors' course in Sp05 covered statistical mechanics and the physics majors' courses covered special relativity, neither of which were covered

[1] Two of these courses used clickers, but in a quite limited fashion.

[2] <g> and its uncertainty Δ<g> are computed as in Ref. [2].

TABLE 1. Average percent correct responses and normalized gains on 12 common questions on QMCS for six modern physics courses. We do not have pretest data for Spring 2005, so in the analysis of these courses, we have assumed that the average pretest scores would have been the same as the following spring. Because the survey is still under development, different versions were used different semesters. This analysis includes only the 12 common questions that were asked all three semesters.

Course	Pre	Post	<g>	Δ<g>	N
Ref. Eng. Sp06	30	65	0.49	0.04	156
Ref. Eng. Fa05	32	69	0.54	0.05	162
Trad. Eng. Sp05	(30)	51	0.30	0.05	68
Trad. Phys. Sp06	44	64	0.37	0.15	23
Trad. Phys. Fa05	40	52	0.21	0.06	54
Trad. Phys. Sp05	(44)	63	0.34	0.08	64

TABLE 2. Average percent favorable (expert-like) responses on CLASS for the same six modern physics courses shown in Table 1. The shifts for the reformed courses are not statistically significant, unlike the traditional courses, which all have large statistically significant shifts down.

Course	Pre	Post	shift	Δshift	N
Ref. Eng. Sp06	66.1	67.1	+1.0	1.1	135
Ref. Eng. Fa05	70.2	68.0	-2.1	1.1	150
Trad. Eng. Sp05	68.5	60.5	-8.0	1.9	55
Trad. Phys. Sp06	72.1	67.2	-4.9	2.4	25
Trad. Phys. Fa05	78.6	72.9	-5.7	1.5	47
Trad. Phys. Sp05	78.5	74.8	-3.7	1.5	61

in the reformed courses or tested in the QMCS.

It is difficult to evaluate the relative success of our treatment of the real world applications that constituted a major part of our course, because this material is simply not covered in other courses. However, this is likely to impact students beliefs about science, and this can be compared with other classes.

We gave the CLASS to assess student beliefs. It is a well known result [4, 5] that in a typical physics course, these beliefs tend to shift towards novice-like. In other words, students leave most physics courses believing that physics is less coherent, less logical, and less relevant to their everyday lives than when they started the course. There is some evidence that, because the subject is so abstract and counterintuitive, teaching modern physics can have a negative impact even in courses where special efforts are taken to address beliefs [17].

Table 2 shows that while the traditional modern physics courses had large shifts towards novice-like beliefs, there were no statistically significant shifts in the overall beliefs of students in the reformed courses. While it is difficult to pinpoint a single cause of this difference, it seems reasonable that the emphasis on real world applications and reasoning development helped students to see the subject as more relevant and coherent.

CONCLUSIONS AND NEXT STEPS

We have reformed a large lecture modern physics course for engineering majors by implementing peer instruction, collaborative homework sessions, and interactive simulations, and by emphasizing real world applications, conceptual understanding, and reasoning development. These reforms have been successful in producing increased learning gains and eliminating the substantial decline in student beliefs. We are now working on the next step, archiving and sustaining these reforms. This course will be taught next semester by a different professor in the PER group, who will use our materials. We will continue to work to improve the course and package it in a way that is easy to pass on to other instructors.

ACKNOWLEDGMENTS

We would like to thank the undergraduate learning assistants and graduate teaching assistants who helped with the course, as well as the PhET team and the Physics Education Research Group at the University of Colorado. This work was supported by the NSF.

REFERENCES

1. For course materials, see: http://jilawww.colorado.edu/%7Emckagan/2130archive/ (2006)
2. R. Hake, *Am. J. Phys.* **66**, 64-74 (1998).
3. L. C. McDermott, *Am. J. Phys.* **69**, 1127-1137 (2001).
4. E. Redish, J. Saul, and R. Steinberg, *Am. J. Phys.* **66**, 212-224 (1998).
5. K. K. Perkins et al., *2004 PERC Proceedings*, AIP Conf. Proc. No. 790 (AIP, Melville, New York, 2005).
6. R. E. Mayer, *Learning and Instruction*, Upper Saddle River, NJ: Merrill, 2003.
7. http://cosmos.colorado.edu/phet/survey/qmcs/ (2006)
8. E. Cataloglu and R. W. Robinett, *Am. J. Phys.* **70**, 238-251 (2002).
9. C. Singh, *Am. J. Phys.* **69**, 885-895 (2001).
10. B. Ambrose, Ph.D. thesis, University of Washington (1999).
11. D. Styer, *Am. J. Phys.* **64**, 31-34 (1996).
12. P. R. Fletcher, Ph.D. thesis, University of Sydney (2004).
13. PhET simulations can be downloaded for free from http://phet.colorado.edu/ (2006)
14. P. A. Tipler and R. A. Llewellyn, *Modern Physics*, 4 edn, W.H. Freeman and Company, 2002.
15. R. Knight, *Physics for Scientists and Engineers*, 1 edn, Pearson, 2004.
16. W. K. Adams et al., *Phys. Rev ST: Phys. Educ. Res.* **2**, 010101-1-14 (2006).
17. T. McCaskey and A. Elby, Contributed poster at Foundations and Frontiers of PER Conference, Bar Harbor, ME, Aug. 15-19, 2005.

Investigation Of Student Learning In Thermodynamics And Implications For Instruction In Chemistry And Engineering

David E. Meltzer

Department of Physics, University of Washington, Seattle, WA 98195, USA

Abstract. As part of an investigation into student learning of thermodynamics, we have probed the reasoning of students enrolled in introductory and advanced courses in both physics and chemistry. A particular focus of this work has been put on the learning difficulties encountered by physics, chemistry, and engineering students enrolled in an upper-level thermal physics course that included many topics also covered in physical chemistry courses. We have explored the evolution of students' understanding as they progressed from the introductory course through more advanced courses. Through this investigation we have gained insights into students' learning difficulties in thermodynamics at various levels. Our experience in addressing these learning difficulties may provide insights into analogous pedagogical issues in upper-level courses in both engineering and chemistry which focus on the theory and applications of thermodynamics.

Keywords: physics education, chemical education, engineering education, thermodynamics.
PACS: 01.30.Cc; 01.40.Fk

INTRODUCTION

For the past seven years, my group has been investigating student learning in thermodynamics in physics courses at both the introductory and advanced levels. Through this investigation we have probed students' learning difficulties at different points in their undergraduate training. Our experience in addressing these learning difficulties may provide insights into analogous pedagogical issues in upper-level courses in both engineering and chemistry which focus on the theory and applications of thermodynamics.

A particular focus of our work has been to examine the learning difficulties encountered by physics, chemistry, and engineering students enrolled in a junior/senior-level thermal physics course that included many topics also covered in physical chemistry courses. In this paper I will compare the initial knowledge (before instruction) of students enrolled in this course with the post-instruction knowledge of students finishing the introductory calculus-based general physics course. (Both courses were taught at Iowa State University, where the introductory course is populated primarily by engineering majors). I will also compare these results with data from a physical chemistry course taught at the University of Maine.

ASSESSMENT DATA

We [1] and others [2] have recently reported results which indicate that students finishing introductory university physics courses emerge with significant learning difficulties related to fundamental concepts in thermodynamics, such as heat, work, cyclic processes, and the first and second laws of thermodynamics. Some of these data, previously reported, will be repeated here in order to compare with student response data from the upper-level thermal physics and physical chemistry courses.

We administered a short set of written questions, after instruction, to 653 students in the introductory general physics course over a three-year period. In the fourth year, we carried out individual interviews (also after instruction) with 32 volunteers drawn from the students enrolled in that same course. This group, referred to below as the "Interview Sample," had course grades far above the class average; half of the Interview Sample had grades above the 81st percentile of the class as a whole. The students in the Interview Sample responded to the same set of written questions, in addition to other related questions.

The set of written questions (along with other questions) was administered on the first day of class to a total of 33 students enrolled in the upper-level thermal physics course during 2003 and 2004. Some of the questions were also administered in a physical

CP883, *2006 Physics Education Research Conference*, edited by L. McCullough, L. Hsu, and P. Heron

chemistry course during 2005 at the University of Maine; in that case, students responded to the questions after they had already received instruction on those topics in the physical chemistry course.

Two of the questions that were administered to all of the students are shown in Figure 1. The first question (the "Work" question) may be answered by examining the area under the curves representing Process #1 and Process #2, respectively. Since the area under the curve representing Process #1 is the larger, the work done by the system during Process #1 is greater than that done during Process #2. Interpretation of curves drawn on *P-V* diagrams in similar problems is often a focus of study in introductory physics courses; most instructors would probably consider this question to be relatively simple. Nonetheless, we found that a significant proportion of students in all samples responded by claiming that the work done by the system in both Processes #1 and #2 was the same. This response was given by 30% of the 653 students who responded to the written questions, and also by 22% of the students in the Interview Sample. Similarly, 21% of the students in the thermal physics course gave this answer. In the physical chemistry course at the University of Maine, six out of the eight students enrolled also gave this response, even after the relevant material had been discussed in that course.

The explanations offered by the students indicated that many of them believed that work done by a system during a thermodynamic process either is, or behaves as, a state function. Some of the explanations stated that idea explicitly, while others used words and phrases that carried the same implication. Examples of such explanations are these: "Equal, path independent; "Equal, the work is the same regardless of path taken."

A correct response to the second question (the "Heat" question) required some understanding of the first law of thermodynamics. Since the change in internal energy is the same in both processes but more work is done by the system in Process #1, the system must absorb more energy in the form of heat in Process #1 in order to reach the same final state (so we will have $Q_1 > Q_2$).

This *P-V* diagram represents a system consisting of a fixed amount of ideal gas that undergoes two ***different*** processes in going from state A to state B:

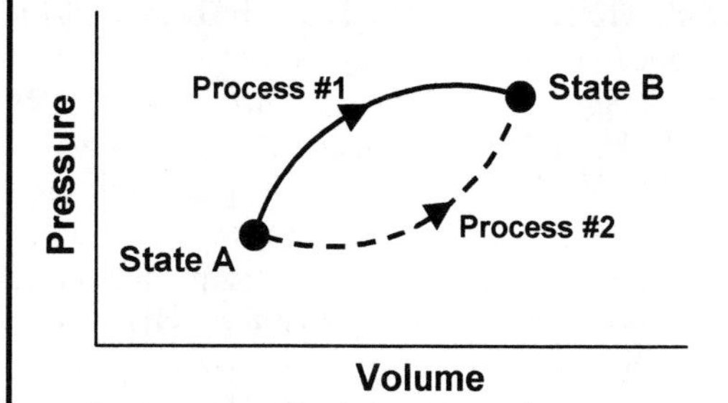

[In these questions, ***W*** represents the work done ***by*** the system during a process; ***Q*** represents the heat ***absorbed*** by the system during a process.]

1. Is *W* for Process #1 ***greater than***, ***less than***, or ***equal to*** that for Process #2? Explain.
2. Is *Q* for Process #1 ***greater than***, ***less than***, or ***equal to*** that for Process #2? Please explain your answer.

FIGURE 1. Two of the questions posed to students in both introductory and upper-level physics courses. Answers: (1) *greater than*; (2) *greater than*.

The results we obtained on this question are shown in Table 1; 2003 and 2004 results are combined. If one neglects consideration of students' explanations, it seems that both the students in the high-performing interview sample and the upper-level thermal physics students performed more poorly than did the introductory students who gave written responses. However, when considering only correct answers that

TABLE 1. Responses to Diagnostic Question #2: Heat Question. The Physical Chemistry course was given at the University of Maine; all other courses were given at Iowa State University. (Figures in the last column are rounded to the nearest 5%.)

	1999-2001 Introductory Physics Written Sample (Post-test) *N* = 653	**2002 Introductory Physics Interview Sample (Post-test) *N* = 32**	**2003-2004 Thermal Physics (Pretest) *N* = 33**	**2005 Physical Chemistry (Post-test) *N* = 8**
$Q_1 > Q_2$	45%	34%	33%	15%
Adequate explanation *(Correct or partially correct)*	11%	19%	30%	15%

are accompanied by correct or partially correct explanations, it becomes clear that the thermal physics students actually had superior results compared to the broader sample of introductory students. (Results in 2004 were similar to those in 2003.) Nonetheless, a correct-response rate of less than one-third would probably not be considered adequate by most instructors for a group that is supposed to be beginning study of statistical mechanics.

In analogy to the explanations offered for the Work question, the most popular incorrect response was that $Q_1 = Q_2$, and the most popular explanation for that answer was that the heat transfer to a system during a process was independent of the path taken by the system during the process. Since the initial and final states were the same, many students argued, the heat absorbed also had to be the same. Thus we found that many students at both the introductory and advanced level, when referring to work and to heat, use words and phrases that are only used by textbooks and instructors when referring to state functions.

Students' responses on related problems [3] corroborated the finding that a majority of the upper-level students were unable apply the first law of thermodynamics effectively in problem solving. Similarly, among students enrolled in the physical chemistry course at the University of Maine, only one out of eight was able to respond correctly to the Heat question even *after* having studied the first law of thermodynamics and related topics in that course.

The fact that we obtained consistent results in three separate upper-level courses at two different universities suggests that a significant proportion of upper-division students beginning advanced study of thermodynamics, in both physics and chemistry, are still struggling with fundamental concepts of heat, work, and the first law of thermodynamics that are normally presumed to have been mastered in their introductory courses. This appears consistent with the report by Towns and Grant [4] that portrays students in an advanced physical chemistry course finding a challenge in working similar problems based on *P-V* diagrams.

We also explored students' understanding of the second law of thermodynamics and the principle of entropy increase during spontaneous processes. We illustrate our findings with the problem shown in Figure 2; the results we obtained are given in Table 2.

The results show several similarities and some differences between the introductory and upper-level students. The introductory students have a tendency to argue that the "system entropy" must always increase, even in cases where inadequate information is available to make such a determination. At the same time, these students are slow to accept the idea that the *total* entropy of system and surroundings must increase during naturally occurring ("spontaneous") processes. In contrast to the introductory students, the students in the thermal physics course readily accept the principle that entropy increases in naturally occurring processes. However, they share with the introductory students the tendency to assume that "system entropy" must always increase regardless of process and regardless of how the "system" is defined. This finding is consistent with results reported by Thomas and Schwenz in 1998 for students enrolled in a physical chemistry course [5].

For each of the following questions consider a system undergoing a naturally occurring ("spontaneous") process. The system can exchange energy with its surroundings.

A. During this process, does the entropy of the **system** [S_{system}] *increase, decrease,* or *remain the same*, or is this *not determinable* with the given information? ***Explain your answer.***

B. During this process, does the entropy of the **surroundings** [$S_{surroundings}$] *increase, decrease,* or *remain the same*, or is this *not determinable* with the given information? ***Explain your answer.***

C. During this process, does the entropy of the system *plus* the entropy of the surroundings [$S_{system} + S_{surroundings}$] *increase, decrease,* or *remain the same*, or is this *not determinable* with the given information? ***Explain your answer.***

FIGURE 2. "Spontaneous Process" question posed to students in both introductory and upper-level physics courses. Answers: (a) *not determinable*; (b) *not determinable*; (c) *increase*.

TABLE 2. Proportion of correct responses to Spontaneous-Process question.[a,b,c,d]

Question	Course	Pretest	Post-test
A (S_{system})	Introductory	42%	40%
	Thermal Physics	50%	65%
	Physical Chemistry	--	75%
B ($S_{surroundings}$)	Introductory	42%	39%
	Thermal Physics	50%	75%
	Physical Chemistry	--	50%
C (S_{total})	Introductory	19%	30%
	Thermal Physics	90%	100%
	Physical Chemistry	--	25%

[a]Introductory: *N* (pretest) = 1184; *N* (post-test) = 255
[b]Thermal Physics: *N* = 12 , matched sample
[c]Physical Chemistry: *N* = 8
[d]Most popular incorrect response on question *A* was *increase* (Introductory: 26% on Pretest, 34% on Post-test; Thermal Physics: 50% on Pretest, 25% on Post-test.

In our own small sample, most of the physical chemistry students asserted that *total* entropy would not change, similar to assertions that were made by the introductory physics students.

ISSUES ASSOCIATED WITH ENGINEERING STUDENTS

Although there were some notable exceptions, it seemed that the majority of the engineering students (and physics-engineering double majors) were relatively unfamiliar and uncomfortable with the need to provide explanations for reasoning in problem-solving, in comparison to the majority of the physics majors. There seemed to be a greater tendency to favor "plug-and-chug" methods, and even to defend them as being the more appropriate method for an upper-level science course. Some students demonstrated a persistent tendency to employ notations and formulations learned in engineering courses, even when they conflicted with those used in the thermal physics course. In general, overt *expressions* of dissatisfaction with the course and the interactive-engagement instructional methods (though not necessarily dissatisfaction itself) seemed more common among the engineering students.

IMPLICATIONS FOR TEACHING CHEMISTRY

Based on the similar course preparation for chemistry and physics majors at the introductory level, it seems probable that students beginning upper-level physical chemistry courses would have the same or similar difficulties regarding fundamental concepts in thermodynamics as was noted among the physics students. This would be consistent with the findings reported by researchers in Chemical Education [4, 5], and with the results from our small sample of physical chemistry students. One implication that could be drawn from this is that there is a need for a strong focus on fundamental concepts—including qualitative reasoning—at the beginning (at least) of the standard physical chemistry course. We also noted that unfamiliarity with standard physics notations and conventions caused difficulties not only for some of the engineering majors, but for a chemistry major enrolled in the course. These difficulties were more persistent than anticipated. This suggests a need for additional attention to addressing confusions related to diverse notations and conventions when students from varied backgrounds are enrolled in an upper-level physics course.

METHODOLOGICAL ISSUES

In physics education research, a vital role is often played by researchers' interpretations of students' explanations as presented in both written and verbal form. Our experience in this course emphasized a need to take into account the different backgrounds and notational conventions of engineering students when analyzing, interpreting, and categorizing their responses to diagnostic questions. Our difficulty in following students' chains of reasoning was often increased by their adherence to non-standard (from the physics standpoint) notations and lines of argument.

Another potentially significant issue for researchers arises when a class under investigation—particularly an upper-level course—includes students from a diversity of majors. When a significant sub-group of a class has a background substantially different from the majority (e.g., engineering vs. physics), data that represent the "class average" can easily tend to obscure patterns that may correlate strongly with sub-group membership. This problem is compounded by the small sample sizes that typify research investigations in upper-level courses. It is likely that patterns in the data that correlate with sub-group membership, even if they do actually exist, may fail to show up as statistically significant with the small numbers of students typical in upper-level courses.

ACKNOWLEDGMENTS

This work has been supported in part by NSF DUE-9981140 [T. J. Greenbowe, Co-Principal Investigator], PHY-0406724, and PHY-0604703 [M. McDermott, Principal Investigator]. Warren Christensen and Ngoc-Loan Nguyen contributed significantly to the research reported here. I am grateful to John Thompson for the University of Maine data.

REFERENCES

1. D. E. Meltzer, *Am. J. Phys.* **72**, 1432-1446 (2004).
2. M. E. Loverude, C. H. Kautz, and P. R. L. Heron, *Am. J. Phys*. **70**, 137-148 (2002); M. J. Cochran and P. R. L. Heron, *Am. J. Phys.* **74**, 734-741 (2006).
3. D. E. Meltzer, "Student Learning in Upper-Level Thermal Physics: Comparisons and Contrasts with Students in Introductory Courses" in *2004 Physics Education Research Conference [Sacramento, California, 4-5 August 2004]*, edited by J. Marx, P. R. L. Heron, and S. Franklin, AIP Conference Proceedings 790, American Institute of Physics, Melville, NY, 2004, pp. 31-34.
4. M. H. Towns and E. R. Grant, *J. Res. Sci. Teach.* **34**, 819-835 (1997).
5. P. L. Thomas and R. W. Schwenz, *J. Res. Sci. Teach.* **35**, 1151-1160 (1998).

Helping Students Learn Quantum Mechanics for Quantum Computing

Chandralekha Singh

Department of Physics and Astronomy, University of Pittsburgh, Pittsburgh, PA, 15260, USA

Abstract. Quantum information science and technology is a rapidly growing interdisciplinary field drawing researchers from science and engineering fields. Traditional instruction in quantum mechanics is insufficient to prepare students for research in quantum computing because there is a lack of emphasis in the current curriculum on quantum formalism and dynamics. We are investigating the difficulties students have with quantum mechanics and are developing and evaluating quantum interactive learning tutorials (QuILTs) to reduce the difficulties. Our investigation includes interviews with individual students and the development and administration of free-response and multiple-choice tests. We discuss the implications of our research and development project on helping students learn quantum mechanics relevant for quantum computing.

BACKGROUND

Quantum computing is a rapidly growing interdisciplinary area of research involving researchers from physics, chemistry, electrical engineering, computer science and engineering, and material science disciplines [1]. Feynman was the first to realize that quantum systems performed computations that might be exploitable for large scale computing [2]. In 1994, Shor [3] developed a powerful and efficient algorithm to factor prime numbers on a quantum computer which is exponentially faster than the classical algorithms. The importance of Shor's algorithm to national security instantly started a race to develop a "real" scalable quantum computer because the difficulty in factoring large prime numbers is at the heart of the protocols used for encoding/decoding secret information. The encoding and decoding protocols that rely on the inability of the hackers to factor large prime numbers are also responsible for most secure communication, *e.g.*, credit card transactions over the internet. Feynman's vision from more than 25 years ago that quantum mechanics should be exploited to perform fast computing has come alive with government agencies investing large resources into quantum computing technologies. Unfortunately, the quantum mechanics curriculum in various departments is not suited to prepare students for research in quantum computing. For those interested in quantum algorithms, learning the quantum mechanics formalism is easier than for those involved in the experimental realization of quantum computers. While the first group must have a good grasp of the quantum formalism for a two level system and product states, the latter group must also consider practical issues involved in experimentation.

Quantum infomation is stored in quantum bits (qubits). Unlike a classical bit which can only take two values (0 and 1), a qubit can be in a quantum superposition of $|0\rangle$ and $|1\rangle$: $|\Psi\rangle = \alpha_0|0\rangle + \alpha_1|1\rangle$ where the only constraints on the complex coefficient is that $|\alpha_0|^2 + |\alpha_1|^2 = 1$. For n-qubit system, 2^n complex numbers are required. For example, for two qubits, $|\Psi\rangle = \alpha_0|00\rangle + \alpha_1|01\rangle + \alpha_2|10\rangle + \alpha_3|11\rangle$. A state with $n = 100$ qubits is specified by $2^{100} \sim 10^{30}$ coefficients! A quantum program is specified by $(2^{100})^2 = 10^{60}$ coefficients but the final answer is a string of $n = 100$ classical bits.

DiVincenzo [4] has put forward these five criteria for solid state implementation of a quantum computer:

- Scalable physical system with well-defined qubits
- Be initializable to a simple state such as $|000...>$
- Have much longer decoherence times than computation time
- Have a universal set of quantum gates
- Permit high quantum efficiency, qubit-specific measurements

As can be seen from these criteria, the practical issues in building a "scalable" quantum computer include challenges in making an actual qubit considering most quantum systems will have more than two levels, issues related to state preparation (*e.g.*, for initializing the register at the start of a computation), making real quantum gates (which involves practical issues related to the time evolution of a quantum state), minimizing decoherence in the system, and performing efficient measurements to read the output of the computation. It can be shown that two qubit gates are universal for quantum computation [4].

CP883, *2006 Physics Education Research Conference*, edited by L. McCullough, L. Hsu, and P. Heron

RESEARCH OBJECTIVES AND METHODOLOGY

We have been carrying out research on the types of difficulties students have with the formalism of quantum mechanics and developing Quantum Interactive Learning Tutorials (QuILTs) to reduce the difficulties [5]. Issues related to state preparation (*e.g.*, for initializing a quantum computer), time development (for making quantum gates for performing the actual computation), measurement (for reading the output of a computation), and basics of two level systems (*e.g.*, spin one-half) and product space are some of the topics we are targeting. In the following section, we briefly describe some of the findings.

The research methodology involves administering written surveys to advanced undergraduate students and beginning graduate students. In these written surveys, students were asked to explain their reasoning. In addition, we also conducted individual interviews with students using a think-aloud protocol. In these interviews, we initially allowed students to answer the questions posed to the best of their ability without interruption and then probed them further about issues they did not otherwise make clear. Many of the probing questions were developed ahead of time, but some were generated on-the-spot in light of student responses.

DIFFICULTY WITH QUANTUM MEASUREMENT

Students were posed the following question: "Consider the following conversation between Andy, Caroline, and John about the measurement of an observable A:

• Andy: When an operator $\hat{A}$ corresponding to a physical observable A acts on a wave function Ψ, it corresponds to a measurement of that observable. Therefore, $\hat{A}\Psi = \lambda_a\Psi$.

• Caroline: I disagree. The measurement collapses the wave function so $\hat{A}\Psi = \lambda_a\Psi_a$ where Ψ_a is an eigenfunction of $\hat{A}$.

• John: I disagree with both of you. You cannot represent the instantaneous collapse of a wave function upon the measurement of A by either equation. Rather, you can write the wave function right before the measurement as a linear superposition of the eigenfunctions of $\hat{A}$, *i.e.*, $\Psi = \sum_a \beta_a\Psi_a$. Then, the absolute square of the coefficients $|\beta_a|^2$ give the probability of collapsing into Ψ_a and measuring λ_a.

• Andy: Then, what is $\hat{A}\Psi = ?$

• John: $\hat{A}$ acting on Ψ is not a statement about the measurement of A. Rather, $\hat{A}\Psi = \hat{A}\sum_a \beta_a\Psi_a = \sum_a \lambda_a\beta_a\Psi_a$.

With whom do you agree? Explain why the other two are not correct."

John's statement is correct. Surprisingly, many interviewed students incorrectly stated that both Caroline and John are actually saying the same thing and they are both correct despite the fact that John explicitly says that he disagrees with the other two. Then, students were explicitly asked to explain how a linear combination of the eigenfunctions of $\hat{A}$ that John proposes can be the same as only one term in the sum proposed by Caroline in $\hat{A}\Psi = \lambda_a\Psi_a$. Most of these students explained their reasoning by claiming that the Hamiltonian operator acting on the wave function corresponds to the measurement of A as Caroline proposes. They incorrectly added that John's equation $\hat{A}\Psi = \hat{A}\sum_a \beta_a\Psi_a = \sum_a \lambda_a\beta_a\Psi_a$ is true only before the measurement of A has actually taken place and Caroline's statement $\hat{A}\Psi = \lambda_a\Psi_a$ is true right after the measurement of A has taken place and lead to the collapse of the wave function. Many students explicitly stated that right at the instant the measurement takes place both Caroline and John are correct because the wave function undergoes an instantaneous collapse and the right-hand-side (RHS) of the equation changes.

When the interviewed students were explicitly asked how the RHS of an equation can change when the left-hand-side (LHS) remains the same, many students appeared not to be concerned about such an anomalous situation in linear algebra where depending upon the context, the same LHS yields different RHS. Students were often very focused on the context. They were convinced that the collapse of a wave function upon the measurement of an observable in quantum mechanics must be represented by an equation and Caroline's equation must correspond to the equation after the collapse of the wave function has occured. They often reiterated that such changes occur only to the RHS (and the LHS is the same for both John and Caroline) because RHS corresponds to the "output" and the LHS corresponds to the "input". According to their reasoning, it is only the output that is affected by the measurement process (and not the input) so the LHS for John and Caroline are the same. When students were asked to explicitly choose the observable to be energy so that the operator is the Hamiltonian operator, their qualitative responses were unchanged even in that concrete case.

The above example shows how difficult the quantum measurement postulate based upon the Copenhagen interpretation is and how students have built a locally coherent knowledge structure (inconsistent with the quantum postulate) to represent the measurement process with equations. It is also interesting to note that since students were often convinced about the physical process of the wave function collapse as represented by the equations that John and Caroline wrote (related to $\hat{A}$ acting on Ψ), they blurred out the linear algebra involved and did not question the anomaly regarding the same LHS yielding different RHS. We plan to administer a modified version of the question to students who have taken linear algebra but not quantum mechanics. Students can be asked to explain why they agree or disagree with a person who says that a physical process can change the equation written by John to that written by Caroline (*i.e.*, the LHS of the equation remains the same but the RHS changes). Our hypothesis is that in the absence of the knowledge of the "collapse" hypothesis and an attempt to represent the collapse by an equation, students who know linear algebra will agree with John and argue that Caroline's equation does not make sense.

Written tests and interviews suggest that students have difficulty figuring out what the wave function will be at time t after the measurement of a physical observable. Many students believe that after the measurement of *any* observable, the system gets "stuck" in the eigenstate of the corresponding operator forever unless an external perturbation is applied. For example, many students believe that the wave function continues to be a position eigenfunction after the measurement of position of a quantum mechanical particle because an eigenfunction cannot change with time. Of course, the statement is true only for observables whose operators commute with the Hamiltonian, but students seem to have overgeneralized this property to include all observables.

Incidentally, when asked to plot an example, many students do not know what a position eigenfunction may look like. During the interviews, students were asked to plot a position eigenfunction on a $\Psi(x)$ vs. x graph but such explicit instruction also did not help. Written tests and interviews suggest that many students do not understand the meaning of "an eigenfunction of an operator corresponding to a physical observable" and believe that the eigenfunctions of all observables are the same as the energy eigenfunctions. In interviews, many students explicitly stated that eigenfunctions do not evolve in time. When they were asked if a delta function in position is an eigenfunction of any physical observable, some students said that it cannot be an eigenfunction because it evolves in time and does not remain a delta function forever. The following response from a student is a typical response: "Energy eigenfunctions must be related to momentum and position eigenfunctions...they are all eigenfunctions after all...shouldn't they at least be proportional to each other?" One reason for such misconception is that energy eigenfunctions which are emphasized in the courses are often simply called "eigenfunctions".

When students were asked to write an eigenvalue equation for the position operator, many students had great difficulty. In the interivews, if students had difficulty writing an eigenvalue equation for the position operator, they were then asked to write an eigenvalue equation for any operator. Roughly half of the students wrote the Time-Independent Schroedinger Equation (TISE) which is an eigenvalue equation for the Hamiltonian operator but the other half could not come up with anything reasonable. Even when prodded to recall a general eigenvalue equation for a generic operator from a math course, they often only recalled that there was a λ involved (perhaps because they were asked to write an "eigenvalue" equation). Some claimed that TISE is not an eigenvalue equation when explicitly asked about it.

While many students believe that a quantum system gets stuck in an eigenfunction after a measurement, a large number of students believe the opposite, *i.e.*, if one waits long enough, the time-evolution will guarantee that the wavefunction after the measurement will go back to the "original" wavefunction (right before the measurement took place). As one student put it: "...well it may not happen immediately but if you wait for a sufficiently long time, it has to go back to the wavefunction before measurement." Incidentally, this notion of going back to the original state was somewhat more prevelant if the wavefunction before the measurement was the ground state wavefunction but it was also quite common when the wavefunction right before the measurement was a linear superposition of the ground state and first excited state wavefunctions. In the case of the ground state wavefunction, students often provided the justification that since the ground state is the equilibrium state, the system must go back to it eventually. In the interviews, the interviewer told students to consider the system to be completely isolated from the environment but very few students felt the need to re-evaluate their claims that the system will go back to the "original" state if one waits long enough. One student described the original state as the "home" state and said that after the collapse, the wavefunction has to find its home state eventually. When asked to show how the wavefunction will evolve from the collapsed state to the "home" state, the student added: "I do not remember the calculation but the wavefunction's goal is to somehow get to the home state." Some students with this belief felt that the collapse of the wavefunction upon measurement is a mathematical construct and they only half-heartedly believed that the collapse can actually change the wavefunction permanently.

On further prodding, responses of the interviewed students about their views on what should happen to the wavefunction a time t after the measurement is also intriguing. When the interviewed students who believed that the wavefunction must go back to the "original one" were told that they should reconsider their response because their initial response is not correct, many quickly switched to the notion that the wavefunction must then get stuck in the collapsed state. When they were told that neither of these possibilities is correct, many students responded in a manner similar to that of the following student: "aren't you contradicting yourself?" A similar situation occured when students who initially said that the system will get stuck in the collapsed state were told that they should reconsider their response. After this hint, many students promptly said that the system must go back to the original state. Stating that neither of these responses is correct and asking students to reconsider their responses again made many students feel that the interviewers were contradicting themselves.

Thus, many of the advanced undergraduates and graduate students interviewed believed that there are only two possibilities for the wavefunction: being stuck in the collapsed state or going back to the original state before the measurement took place. They just could not contemplate the actual situation in which the wavefunction evolves according to the Time-Dependent Schroedinger Equation (TDSE) and may neither be "stuck" (unless the state in which the wavefunction collapsed is an eigenfunction of an operator that commutes with the Hamiltonian) nor ever go back to the "original" state. When students were told that after the position measurement the wavefunction is a delta function in position about a particular position and they were explicitly asked to calculate the wavefunction after a time t, none of the students were able to perform the calculation. The calculation involves expanding the delta function in terms of a linear superposition of eigenfunctions and then computing the wavefunction at time t by introducing appropriate phase factor $e^{-iE_nt/\hbar}$ to each term. After the interviews, some of the students were very surprised to learn that in a majority of cases, the wavefunction will never go back to the "original" wavefunction if allowed to evolve according to TDSE. Studying the real part of the wave function $\Psi(t)$ at different times t via a suitable simulation can convince students that the wavefunction is neither "stuck" in position eigenfunction nor must it go back to the original state. Such simulations can help students learn that the time evolution operator can change the position eigenfunction significantly because the phase factors $e^{-iE_nt/\hbar}$ of each term when the delta function is expanded in terms of energy eigenfunctions will evolve differently.

DIFFICULTY WITH PRODUCT SPACE

When students were given two spin one-half particles and asked to choose a basis and write down a Hamiltonian $\hat{H}$ proportional to $\vec{S}_1 \cdot \vec{S}_2$ describing this system in a matrix form, a majority of the advanced undergraduates who had learned about product space had great difficulty. More than 85% of students tried to construct a 2×2 matrix because they did not realize that they should consider a product space which is four dimensional. In the interviews, when students were specifically told that the product space of two spin one-half particles cannot be two dimensional, many of them remembered that the vector space should be four dimensional. Despite this realization, none of these students could actually choose a basis set and construct the Hamiltonian correctly. Some claimed that $\hat{H}$ must always be diagonal regardless of the basis because it is the "unperturbed" Hamiltonian (survey did not mention it was the unperturbed Hamiltonian). Some students had a "cancellation" model in mind and they felt that if $\vec{S}_1$ and $\vec{S}_2$ are operators for two spins, they have to conspire together to make the total spin of the system zero. For example, one student said: "If the contribution of S_1 is positive then the contribution of S_2 will be negative because their contributions must cancel." They were often unable to articulate their reasoning.

Written tests and interviews suggest that many students have difficulty understanding that if two operators are in different Hilbert spaces, they will always commute and can be treated independently of each other, *e.g.*, spin and position of an electron or spins of two different electrons or positions of two different electrons. Because of this difficulty, students often have difficulty figuring out how the operators in different Hilbert spaces act on states in a product space. For example, for the above Hamiltonian, several students felt that $\hat{S}_{1x}$ and $\hat{S}_{2y}$ will not commute with each other.

Our research also shows that determining the dimensionality of a product space is challenging for students. For example, for a two spin system, students have a tendency to add (instead of multiply) the dimensionality of the individual Hilbert spaces of each spin to obtain the dimensionality of the product space. This process will not introduce an error for two spin one-half particles because $2 + 2 = 2 \times 2$ but it will introduce error for other cases. For example, we have found that for two spin-one particles, many students believe that the product space is 6-dimensional as opposed to 9-dimensional.

CONCLUSION

Research on student understanding of aspects of quantum mechanics relevant for quantum computing is necessary. We have been investigating the difficulties students have in learning quantum mechanics and developing Quantum Interactive Learning Tutorials (QuILTs) to reduce the difficulties. The tutorials can provide scaffolding support to students in science and engineering pursuing quantum computing research.

ACKNOWLEDGMENTS

We are grateful to the NSF for award PHY-0244708.

REFERENCES

1. M. Nielsen and I. Chuang, *Quantum Computation and Quantum Information* (Cambridge University Press, Cambridge 2000).
2. R. P. Feynman, *International Journal of Theoretical Physics* **21**, 467-488 (1982).
3. P. W. Shor in *Proc. 35th Annual Symposium on the Foundations of Computer Science*, edited by S. Goldwasser (IEEE Computer Society Press, Los Alamitos, CA, 1994), 124-134; E. Gerjuoy, *Am. J. Phys.* **73**(6), 521-540 (2005).
4. See *http://www.research.ibm.com/quantuminfo/*; D. DiVincenzo, *Phys. Rev. A* **51**(2), 1015-1022 (1995).
5. C. Singh, Am. J. Phys. **69**(8), 885-895 (2001).

Enabling Informed Adaptation of Reformed Instructional Materials

Rachel E. Scherr and Andrew Elby

Physics Education Research Group, Department of Physics, University of Maryland, College Park, Maryland 20742 USA

Abstract. Instructors inevitably need to adapt even the best reform materials to suit their local circumstances. We offer a package of research-based, open-source, epistemologically-focused mechanics tutorials, along with the detailed information instructors need to make effective modifications and offer professional development to teaching assistants. In particular, our tutorials are hyperlinked to instructor's guides that include the rationale behind the various questions, advice from experienced instructors, and video clips of students working on the materials. Our materials thus facilitate their own implementation and develop instructor expertise with PER-based instructional materials.

Keywords: curriculum adaptation, curriculum development, professional development.
PACS: 01.40.Fk, 01.40.G-, 01.40.jh

MOTIVATION

Instructors who adapt reformed instructional materials to suit their local circumstances may be limited by institutional constraints, inflexible materials, or their own lack of expertise. Nonetheless, they make adaptations. In order to assist instructors in making informed modifications and implementing materials effectively, we offer a package of open-source physics worksheets integrated with implementation resources. In what follows we describe the resource package and three different ways instructors have used these materials.

OPEN-SOURCE MATERIALS INTEGRATED WITH IMPLEMENTATION RESOURCES

The package that we provide includes a sequence of tutorial and interactive lecture worksheets, homework and solutions, an instructor's guide, video of students working on the tutorials, and commentary and discussion questions to accompany the video.

Tutorials and Interactive Lecture Worksheets

A *tutorial* is an active-learning worksheet intended for use by small groups of students, optimally 3 or 4 students per group, in a small-class setting (typically 20 students or so). If experiments are involved, students usually work on them within their small groups. The instructor or instructors float around interacting with individual groups. The tutorial worksheets are typically not graded; instead, students get feedback from TAs during the tutorial sessions, and on homework and quizzes. We find that grading the tutorial worksheets tends to result in students giving the answers they think we want to hear, rather than saying what they really think.

An *interactive lecture demonstration* (ILD) is much the same, except it is intended for use in large lectures. Students work with whomever they happen to be sitting. Instead of interacting with individual groups, the instructor leads a full-class discussion at designated points in the worksheet. Any relevant experiments are set up at the front of the class. As with tutorials (and for the same reasons), we do not typically grade ILD worksheets.

The tutorials and interactive lecture worksheets that we offer are developed by the Physics Education Research Group at the University of Maryland (UM) for the algebra-based introductory physics course, based on the model developed at the University of Washington (UW).[1] The materials cover core concepts in 1st-semester introductory physics (kinematics, forces, momentum, energy, and hydrostatic pressure). UM tutorials and ILDs are developed to promote students' epistemological development along with

CP883, *2006 Physics Education Research Conference*, edited by L. McCullough, L. Hsu, and P. Heron

their conceptual understanding, as part of our response to research indicating that even the best reform materials don't typically improve students' views about the nature of physics knowledge and learning.[2]

With each tutorial, we provide homework questions that reinforce and in some cases build upon the tutorial. Like the UW tutorial developers, we find that to maximize the effects of tutorials and ILDs, it is essential to use at least some associated homework items. Similarly, we find it is important to use some of the included exam questions, which are designed to reward students for gaining the kind of conceptual understanding that the tutorials emphasize.

All the text materials that we provide are fully editable Microsoft Word documents. Instructors can easily make any changes desired with no restriction. They may add or delete material, change wording, divide worksheets into smaller segments, or change the form of the worksheet (perhaps turning a tutorial into an interactive lecture, or even homework).

Each tutorial and interactive lecture worksheet is hyperlinked in multiple places to an instructor's guide containing an overview of the worksheet and the rationale behind it, in addition to section-by-section discussions of common student responses, teaching tips, and so on (see below).

Instructor's Guide

The instructor's guide includes information that we would hope would be helpful for someone implementing the tutorials without other expert assistance. In addition to detailing the necessary equipment and the flow of the lesson, we provide an overview of the purpose and method of the lesson; the curriculum developers' reasons for writing the lesson in a particular way; and references to the physics education literature relevant to the lesson topic. For specific questions within the worksheet, we also relate common student responses; expert instructors' experiences in helping students make progress; and questions for instructors to ask students at particular points in the lesson. Our goal is for the text of the instructor's guide to contain information similar to what we would tell a new tutorial instructor at our own institution.

In addition to the text, the instructor's guide has a video component: video clips of students answering specific questions on the worksheet, accessed through hyperlinks that appear in the guide and accompanied by full transcripts. These videos have the potential to give instructors a vivid sense of how tutorials work and what they are really like for the students who experience them – what difficulties they have, what skills they bring to bear, and how they interact with one another. Many of the video clips also show interactions between students and instructors, providing new instructors with diverse models of tutorial teaching. We originally envisioned these video clips as supplementing the in-person observations that a new instructor might make while learning to teach tutorials. We find, however, that the video clips in some ways go beyond what in-person experiences can provide: they can be played over and over again for detailed observation and analysis, and – perhaps most importantly – they show what the students do when no instructor is present. (Abundant and sometimes incriminating evidence from the videotapes assures us that the students are not inhibited by the presence of the camera.) Text accompanying the video clips includes researcher observations intended to direct instructors' attention to features of interest.

TA Video Workshops

The video clips have the potential not only to help instructors in the ways described above, but also to serve as resources for professional development of teaching assistants. In order to facilitate the use of the video clips for this purpose, we provide "TA Video Workshops" that integrate tutorial excerpts, references to video clips of students working on that part of the tutorial, line-numbered transcripts, and discussion questions. Shorter workshops, with just one video clip, might structure a half-hour's discussion of a particular teaching issue; longer ones, with three or four video clips, show the development of students' thinking over the course of an hour-long lesson. We have found these video workshops to be useful at UM in weekly tutorial preparation sessions for physics graduate teaching assistants.

THREE IMPLEMENTATIONS

We provide our materials for free to anyone who requests them. We are interested to learn what uses people identify for the materials as well as how they implement the tutorials at their institutions. Three examples of such implementations are described below. These are not intended to represent "ideal" use; they are constrained by issues of class size, student motivation and preparation, instructor time and experience, curriculum, and so on.

Public Research University

A lecturer at a public research university in the Northeast, "Jim," used our tutorials in his introductory algebra-based physics course. Jim had a number of colleagues and graduate students in his department who were already experienced with *Tutorials in Introductory Physics,*[1] and thus was able to offer our tutorials in small recitation sections with adequate staffing, as at the University of Maryland. He used the materials exactly as provided (no modifications to the worksheets), but did not use the complete sequence of tutorials in the package; instead, he alternated our tutorials with *Tutorials in Introductory Physics.* Jim used ILDs occasionally throughout the semester, integrating them with his lecture room's infrared personal response system. He used the video workshops in weekly teaching assistant preparation sessions.

Jim felt that the tutorials worked fairly well in his course and were easy to integrate with the recitation sections already in place. Both standardized and course-specific assessments showed good conceptual gains by the students. Jim and his TAs, however, had mixed feelings about using our materials. They felt that the combination of instructional materials that they used (UM tutorials, UW tutorials, and other instructional strategies in other parts of the course) resulted in confusion for students about the course's priorities. Specifically, our tutorials' informal tone and emphasis on intuition building and epistemological questions were not reflected elsewhere in the course, and were sometimes not taken seriously by the students.

Public Comprehensive University

A professor at a public comprehensive university in the Midwest, "Kate," uses our tutorials in her introductory algebra-based physics course. Kate had prior experience with reform instruction through Context-Rich Problems,[3] but had not used any tutorials before using ours. She does not have sufficient staffing for small recitation sessions; she teaches the class of 60-70 students either alone or with one undergraduate peer instructor. She therefore conducts the tutorials in a large class setting, alternating them week by week with context-rich problem-solving sessions. Her students work in small groups, but get only minimal feedback on their work (if any), since Kate and her one TA do not have time for sustained discussions with every student group. She does not conduct whole-class discussions because the student groups work at different paces. Kate's modifications to the tutorial worksheets have included removing the "checkpoints" (points where students are required to consult with an instructor before proceeding), since she was unable to enforce them. She also removed or reduced the equipment required for some tutorials. In another significant departure from the practice at UM and UW, Kate collects and grades the tutorial worksheets, perceiving that her students will not participate in work that does not count directly towards their grade. She does not use the instructor's guide or video clips.

Kate is very happy with the tutorials and finds several of them to be ideally matched to her teaching. Like Jim, she has observed that the epistemological questions are unpopular with her students, and she finds it difficult to get her students out of a "right answer" mindset so that she can hear what they really think. However, she has no wish to remove the epistemological questions from the worksheets; she greatly values that feature of the materials and wants her students to engage in those questions. Kate wants more expertise in facilitating frank discussions of physics concepts and epistemological issues. At the same time, she feels that her expertise as a tutorial instructor has already increased. She has also been inspired to create new materials in a similar format (including a tutorial on coordinate systems).

Community College

A professor at a community college on the West Coast, "Pam," uses only one tutorial from our package (on Newton's second law) in her introductory calculus-based course. However, she has used it several times, making new modifications each round, and intends to retain it in her future teaching. Pam is experienced with reform instruction and integrates our single tutorial into a comprehensive program of other reforms to her course, mostly conceptual labs in the style of *RealTime Physics*[4] and worksheets combining conceptual discussion questions with quantitative problems. Pam is the sole instructor in a class of 24 students; she has her students work in small groups, and leads a class discussion at checkpoints. Her modifications to the single tutorial that she uses include removing the epistemological questions, adding kinematic graphing exercises, and modifying language to promote clearer distinctions between acceleration and velocity and to improve readability for her many students who speak English as a second language. Pam does not use the instructor's guide or video clips.

Pam sees herself as having gained significant expertise as a result of her use of our materials, feeling that they have helped her come to recognize common student difficulties. She now sees "the whole thing about [students] being able to state different arguments" as a key step in conceptual change, and uses that insight elsewhere in her teaching.

Pam particularly appreciates the tone set by the modifiability of our materials and the fact that the developers welcome feedback and encourage modification. She says that "a lot of PER stuff is perceived as being exclusive...looking down their noses," and most PER materials are hard to adopt because they are "presented as just a done deal. So often PER stuff is presented as, you have to do it this way...it's a big turn-off."

DISCUSSION

The three implementations discussed above illustrate several implementation issues that, we hypothesize, are fairly typical. First, few users implement a curricular package as a complete set; they typically mix and match worksheets and other curricular materials from multiple sources. The modifiability of our materials makes it possible for instructors to try to minimize "epistemological" mismatches of the sort Jim encountered, or mismatches between tone, style, level, and so on.

Second, even when an instructor isn't modifying the substantive questions posed by a tutorial, she may still want to make minor modifications to adjust to local circumstances, such as Kate's eliminating checkpoints, or Pam's simplifying the language. Electronically-supplied worksheets make it easy for instructors to do this.

Third, as previous research indicates,[5] instructors often make modifications even when curriculum developers try their best to enforce "faithful" implementation. For this reason, providing resources to help instructors make productive modifications — partly by helping them understand why the developers wrote the worksheet as they did — may be more productive than striving for "faithful" implementations. In our judgment, the latest iteration of Pam's version of our Newton's second law material retains much of the *spirit* of the original version despite not following the "letter of the law," so to speak.

Fourth, as Kate and Pam's experiences illustrate, using tutorials and other reform-oriented instruction in a reflective way can itself serve as professional development. Pam's iterations of the Newton's second law tutorial reflect her evolving view of what her students need and how they learn, and it's not clear if her thinking could have evolved similarly were she unable to iteratively modify the tutorial. It's possible that such modifications could result in a disastrous experience for the students; but this doesn't appear to have happened in her case, and our instructor resources are designed to help instructors avoid lethal mutations.

In conclusion, we want to emphasize that we are not trying to promote our particular materials. As the three implementations show, in many classes, our explicit emphasis on epistemological development is either ill-suited or very difficult to implement. Instead, we hope to promote our *approach* to curriculum development, dissemination, and professional development. In this approach, users are encouraged to make modifications — to become, in a sense, co-developers; and the resources needed to facilitate implementation and to help guide effective modifications are integrated with the tutorials and interactive lecture worksheets themselves.

ACKNOWLEDGMENTS

We are grateful to Tim McCaskey, Raymond A. Hodges, and Thomas J. Bing for major contributions to the work of this project. We are also grateful to David Hammer, E. F. Redish, and the other members of the Physics Education Research Group at the University of Maryland for substantive discussions of this research. This work was supported in part by the National Science Foundation (DUE 0341447).

REFERENCES

1. L. C. McDermott, P. S. Shaffer, and the Physics Education Group at the University of Washington, *Tutorials in Introductory Physics* (Prentice-Hall, Upper Saddle River, NJ, 1998). For a description of the use of tutorials, see L. C. McDermott, P. S. Shaffer, and M. Somers, *Am. J. Phys.* **62**, 46-55 (1994); E. F. Redish, J. Saul, and R. N. Steinberg, *Am. J. Phys.* **65**, 45-54 (1997); and E. F. Redish and R. N. Steinberg, *Physics Today* **52**, 24-30 (1997).
2. E. F. Redish, *Am. J. Phys.* **66**, 212-224 (1998)
3. P. Heller, R. Keith, and S. Anderson, *Am. J. Phys.* **60**, 627-636 (1992); P. Heller and M. Hollabaugh, *Am. J. Phys.* **60**, 637-644 (1992)
4. D. R. Sokoloff, R. K. Thornton and P. W. Laws, *RealTime Physics Module 1: Mechanics* (John Wiley and Sons, Hoboken, NJ, 2004)
5. M. A. Ruiz-Primo, R. J. Shavelson, L. Hamilton, and S. Klein, *J. Res. Sci. Teach.* **39**, 369-393 (2002); K. Tobin and C. McRobbie, *Sci. & Educ.* **6**, 355-371 (1997)

SECTION II

Peer-Reviewed Papers

Chemistry vs. Physics: A Comparison of How Biology Majors View Each Discipline

K.K. Perkins, J. Barbera, W.K. Adams, and C.E. Wieman

Departments of Physics and Chemistry, University of Colorado, Boulder, CO 80309

Abstract. A student's beliefs about science and learning science may be more or less sophisticated depending on the specific science discipline. In this study, we used the physics and chemistry versions of the Colorado Learning Attitudes about Science Survey (CLASS) to measure student beliefs in the large, introductory physics and chemistry courses, respectively. We compare how biology majors – generally required to take both of the courses – view these two disciplines. We find that these students' beliefs are more sophisticated about physics (more like the experts in that discipline) than they are about chemistry. At the start of the term, the average % Overall Favorable score on the CLASS is 59% in physics and 53% in chemistry. The students' responses are statistically more expert-like in physics than in chemistry on 10 statements ($P \leq 0.01$), indicating that these students think chemistry is more about memorizing disconnected pieces of information and sample problems, and has less to do with the real world. In addition, these students' view of chemistry degraded over the course of the term. Their favorable scores shifted -5.7% and -13.5% in 'Overall' and the 'Real World Connection' category, respectively; in the physics course, which used a variety of research-based teaching practices, these scores shifted 0.0% and +0.3%, respectively. The chemistry shifts are comparable to those previously observed in traditional introductory physics courses.

Keywords: Beliefs, Interest, Learning, CLASS, Undergraduate education, Chemistry, Physics.
PACS: 01.40.-d, 01.40.Fk

INTRODUCTION

Over the last decade, students' beliefs about physics and learning physics has become an active area of research within the physics education research community. It is well established that, in introductory college physics, the majority of students start the term with relatively novice-like beliefs about physics – seeing it as isolated pieces of information that have little connection to the real world but must be memorized.[1,2,3] These studies show that students' enrolled in algebra-based physics courses have more novice-like beliefs than those enrolled in calculus-based physics. In both courses, students' beliefs typically degrade – that is become more novice-like – over the course of most introductory physics classes.[1,2,3] In addition, our prior work found correlations between students' beliefs and other important educational outcomes, such as content learning gain, choice of major, and level of interest in physics.[1,4]

These findings have led to efforts to identify teaching practices that explicitly target and support the development of expert-like beliefs within these introductory courses. Some courses with modest efforts to implement these practices have succeeded in avoiding the typical regression, while others with more extensive interventions have shown improvements. [1,5]

In this paper, we compare students' beliefs about physics and chemistry – two disciplines where experts' beliefs about their respective disciplines are quite similar. While we are aware of some efforts in the chemistry education community to investigate beliefs [6], we are not aware of any direct comparisons between physics and chemistry. We ask several questions: Do introductory students have more expert-like beliefs about chemistry or physics? In which categories of beliefs are students' views different between the two disciplines? Do chemistry and physics courses have similar impacts upon students' beliefs about the disciplines? Do students' beliefs in chemistry and physics show similar correlations with other educational outcomes?

STUDY DESIGN

Over the past year, we used the physics and chemistry versions of the Colorado Learning Attitudes about Science Survey (CLASS-Physics and CLASS-Chemistry) [3] to measure student beliefs both at the start (pre) and end (post) of introductory courses within these disciplines. The CLASS-Physics and

CP883, *2006 Physics Education Research Conference*, edited by L. McCullough, L. Hsu, and P. Heron

CLASS-Chemistry surveys consist of 42 and 50 statements, respectively, to which students respond using a 5-point Likert scale. Thirty-four of these statements meet the criteria for the comparative 'Overall' score – the statements are common between the two surveys (with the word "physics" replaced with the word "chemistry") and have a consistent expert response. These 34 statements are used to determine each student's 'Overall' % favorable belief score – the percentage of statements for which his/her response agrees with that of an expert. Eight belief categories (e.g. 'Real World Connection') are scored using groupings of 4 to 8 statements. The details of the design, categorization, and validation of the CLASS-Physics are reported by Adams et al. [3]. All of the CLASS-Chemistry statements have been validated with student interviews and faculty responses. A paper describing this work is in preparation. [7]

In addition to these statements, we included two supplementary questions on the survey to monitor students' level of interest in physics (chemistry):

Currently, what is your level of interest in [discipline]? (very low, low, moderate, high, very high)

During the semester, my interest in [discipline]...

(increased, decreased, stayed the same)

We purposely chose to use vague questions as opposed to questions that are more specific measures of interest, such as whether students would like to learn more physics. This approach was taken in an effort to measure students' composite affective response towards physics or chemistry. The student's answer naturally depends upon the range of factors relevant to how she personally identifies what makes something interesting.

We have collected CLASS-Physics responses in a first-term, algebra-based Physics I course (Phys I) and CLASS-Chemistry responses in a first-term, introductory general chemistry course (Chem I). In prior research, we have found differences in students beliefs correlating with choice of major [1,3]; thus, in this study, we focus on just the biology majors[1]. We choose first-term courses because college physics courses have been shown to alter student beliefs. [1,2,3] With these data, we are able to compare two large and similar populations of students because there are a large number of biology majors and the students are required to take both courses to fulfill their majors.

From Table 1, we see that both courses were large lecture courses (over 500 enrolled) with a large number of biology majors, the majority of which are women. The results presented here are for the 156

TABLE 1. Introductory courses surveyed

		Chem I	Phys I – Alg
Total # of students		812	553
Total # of bio students		362	330
Bio with pre/post surveys	Total #	156	212
	% women	70.3%	72.2%
	% men	28.4%	27.4%

(Chem I) and 212 (Phys I) biology-majors for which we collected matched pre- and post- surveys. The instructor for the Phys I course is very familiar with research findings in PER and incorporated many research-based practices into the course, including: in-class concept questions where student-student discussions are highly encouraged and reasoning/sense-making is emphasized; interactive feedback in lecture where students use H-ITT [8] "clickers" to vote; conceptual questions on homeworks and exams; and labs revised to incorporate more discovery. In implementing these practices, the instructor explicitly worked to promote expert beliefs. Although the Chem I course used concept tests and clickers, the course was comparatively traditional, with less emphasis on peer discussion, reasoning, and conceptual learning.

RESULTS AND DISCUSSION

Incoming students' beliefs. Students' responses on the pre-course surveys were compared to identify difference between biology majors' view of physics and chemistry prior to any college instruction in the discipline. Statistically significant differences ($p<0.01$) between their responses to the CLASS-Physics and CLASS-Chemistry surveys were measured for the 'Overall' score and for three categories – 'Real World Connection', 'Conceptual Connections' and 'Applied Conceptual Understanding'. In Figure 1, we see that biology majors have consistently more expert-like beliefs about physics and learning physics than about

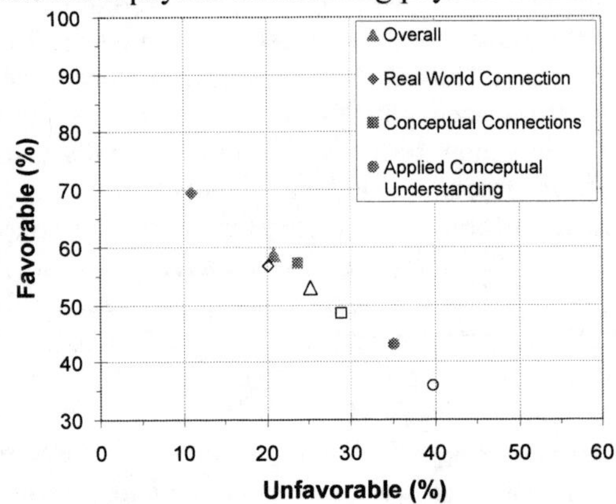

FIGURE 1. CLASS survey scores for biology majors at the start (pre) of Phys I (solid) and Chem I (hollow). The students were consistently more expert-like in their view of physics, with the difference being statistically significant ($p<0.01$) for both 'Overall' and the 3 categories shown.

[1] including students majoring in Molecular, Cellular, and Developmental Biology, Ecology and Evolutionary Biology, and Integrative Physiology as well as the now-discontinued majors of Kinesiology and Environmental, Population, and Organismic Biology

TABLE 2. Comparison of *pre* responses on individual CLASS-Physics and CLASS-Chemistry statements

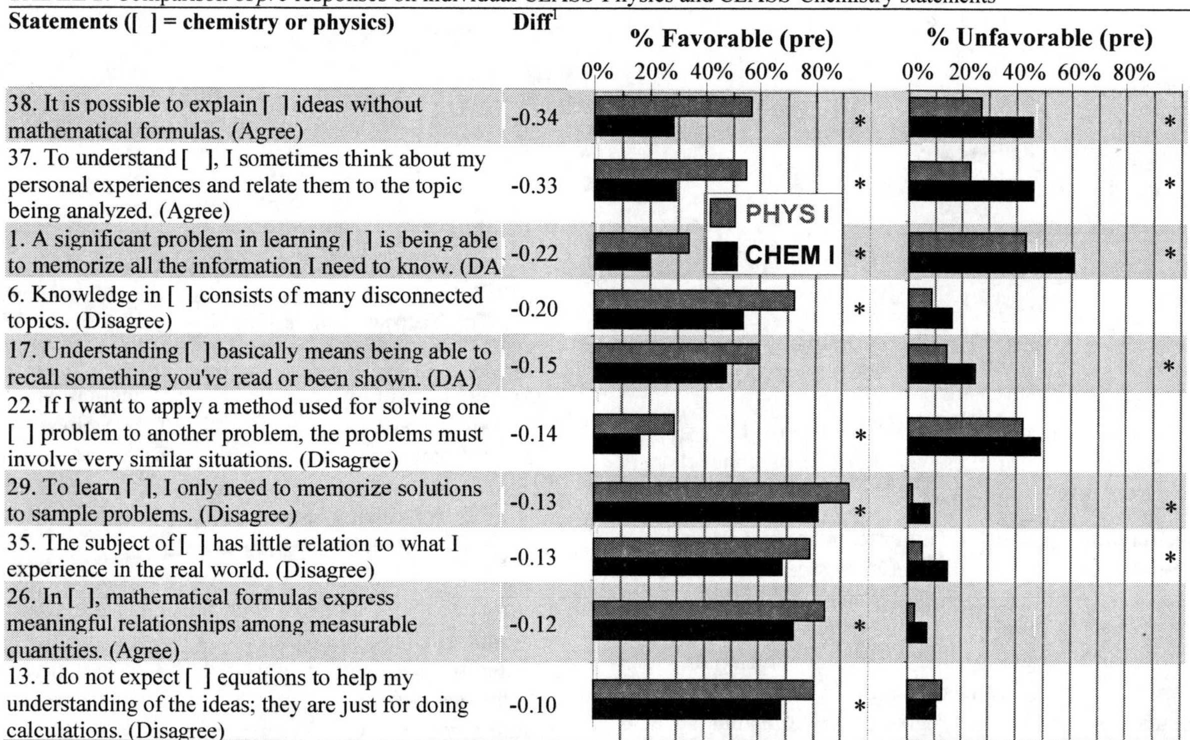

Statements ([ ] = chemistry or physics)	Diff[1]	% Favorable (pre)	% Unfavorable (pre)
38. It is possible to explain [ ] ideas without mathematical formulas. (Agree)	-0.34	*	*
37. To understand [ ], I sometimes think about my personal experiences and relate them to the topic being analyzed. (Agree)	-0.33	*	*
1. A significant problem in learning [ ] is being able to memorize all the information I need to know. (DA	-0.22	*	*
6. Knowledge in [ ] consists of many disconnected topics. (Disagree)	-0.20	*	
17. Understanding [ ] basically means being able to recall something you've read or been shown. (DA)	-0.15		*
22. If I want to apply a method used for solving one [ ] problem to another problem, the problems must involve very similar situations. (Disagree)	-0.14	*	
29. To learn [ ], I only need to memorize solutions to sample problems. (Disagree)	-0.13	*	*
35. The subject of [ ] has little relation to what I experience in the real world. (Disagree)	-0.13		*
26. In [ ], mathematical formulas express meaningful relationships among measurable quantities. (Agree)	-0.12	*	
13. I do not expect [ ] equations to help my understanding of the ideas; they are just for doing calculations. (Disagree)	-0.10	*	

1. Diff is measured by calculating the "linear distance from expertness", or SQRT((1-fraction favorable)^2 + (fraction unfavorable)^2) for the Phys I and Chem I responses and taking the difference of these two values (Phys-Chem). *These responses are statistically different PHYS I vs. CHEM I using two-tailed z-tests p<0.01.

chemistry across these categories. The most dramatic difference is observed in the 'Real World Connection' category where the % favorable scores are 69.3% for physics and only 56.8% for chemistry, indicating that these biology majors see physics as substantially more connected to the real world – as both describing and being useful to understand real world experiences.

Comparing students' responses to individual statements provides more insight into how their beliefs about physics and chemistry differ. We used "distance from expertness" (as described in the table) to compare the two groups and to quantify the difference. This measure allows one to account for both the % favorable and % unfavorable responses. In Table 2, we list the 10 statements for which the biology majors had statistically different ($p<0.01$) responses for physics and chemistry. In all cases, the students responded more expert-like in physics than in chemistry. Several themes are apparent. In comparison to physics, biology majors see chemistry as being *more* about memorizing disconnected pieces of information and sample problems, as having *less* to do with the real world, and as being *less* conceptual, needing math to explain chemistry but not making sense of the math. There are some differences between the responses for men and women in chemistry and physics, but there appears to be no strong discipline-specific aspect to these differences.

We find it especially interesting that the biggest difference is on the statement, "It is possible to explain physics (chemistry) ideas without mathematical formulas." With physics being more mathematically intensive than chemistry, we would have predicted that students would have thought physics was more about math than was chemistry.

The cause of these differences in beliefs is not revealed by these data. We speculate that this discrepancy develops from their prior experiences with science in high school and earlier. One might suspect that these differences could be due to the maturity or experience of the students (biology majors are generally freshmen in chemistry and sophomores in physics); however, as discussed in the next section, the shifts in beliefs observed in chemistry do not support this logic.

Shifts in beliefs. Figure 2 shows the shifts in CLASS scores over the course of the term (pre-to-post) for the biology majors in the Phys I and Chem I courses. Only categories for which the shifts are statistically different ($p<0.01$) are included. In Chem I, we see the biology majors shifting to be much more novice-like in their beliefs about chemistry – a result consistent with the typical shifts observed in introductory physics courses.[1,2,3] In the Phys I course, however, we observe students beliefs holding steady in most cases. We attribute this to the Phys I instructor's emphasis on

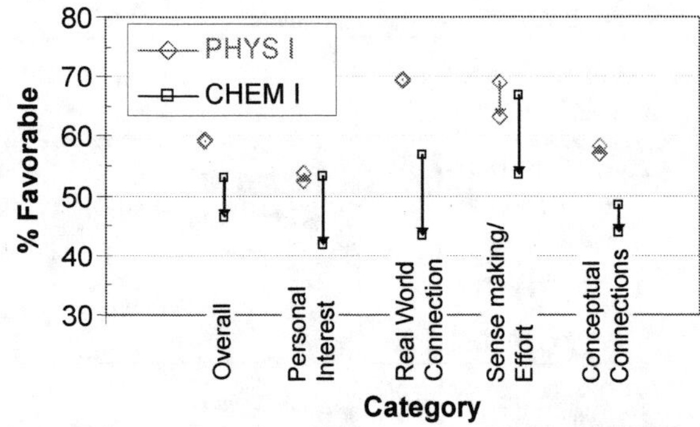

FIGURE 2. Pre-post shift for CLASS-Chemistry and CLASS-Physics scores where there is a statistically significant difference in the shifts ($p<0.01$).

conceptual understanding and reasoning, his infusion of real world examples, and his use of concept tests and peer discussion in class.

Beliefs and Self-rated Interest. In prior work [4], we found a strong correlation between students' beliefs as measured by the CLASS and their self-rated interest as measured by the supplemental questions described above. In Figure 3, we see the strong correlation exists for chemistry as well as physics. In Phys I, where students' beliefs did not regress, 52% of students also stated that their interest increased over the term with only 16% stating that their interest decreased. In Chem I, however, only 32% of students' stated their interest increased, while 31% stated their interest decreased.

In our prior work, we showed that students whose interest increases most often cite the connection between physics and the real world as the reason for their increased interest.[4] Thus, the decline observed in the 'Real World Connection' category for Chem I seems particularly important, and would suggest that efforts to better connect chemistry to the real world may lead to more interest in chemistry among this population of biology majors.

CONCLUSION

We observe significant differences in biology majors' beliefs about physics and their beliefs about chemistry. In all measures, these students have more novice-like beliefs about chemistry, specifically seeing chemistry as *more* about memorizing and *less* about the real world. Since these differences are present at the *start* of these first-term introductory courses, these differences in beliefs were established by some combination of prior experiences. It would be interesting to investigate whether these differences may stem from differences in the way in which physics and chemistry are taught in middle and high school.

As with physics, we see a correlation between students' measured beliefs and their self-rated interest; while these data do not illuminate the cause for this

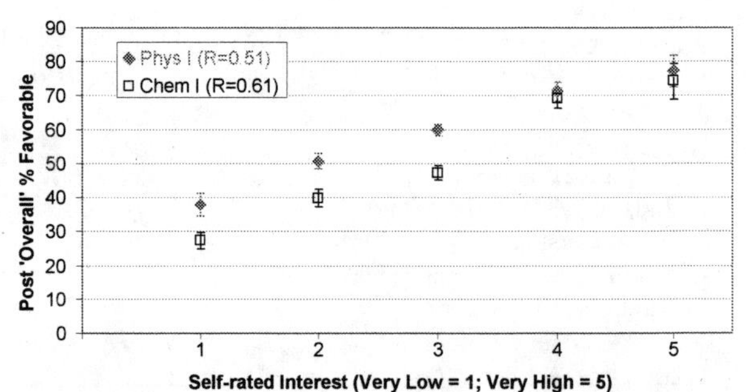

FIGURE 3. Post CLASS 'Overall' % favorable scores versus students' self-rated interest for Phys I (solid) and Chem I (hollow). Error bars show standard error of the mean.

correlation, certainly a plausible explanation is that with more expert-like beliefs, students see physics or chemistry as connected, powerful ideas that are useful for solving a wide variety of relevant problems, and thus more interesting. Establishing the nature of this correlation is planned for future work.

As with many physics courses, we see that introductory chemistry courses, even those that use concept tests and clickers, may lead to significant declines in students beliefs about chemistry and learning chemistry. Physics courses where reforms have succeeded in avoiding this typical decline, may serve as a useful model for reforming chemistry courses. We will test this idea as the Chem I course will be undergoing significant reform efforts over the next few terms, with part of their objective being to reduce this observed decline in students' beliefs.

ACKNOWLEDGMENTS

This work was supported, in part, by CU and NSF. We thank the Phys I and Chem I instructors, Elias Quinn, and the CU Physics Education Research group.

REFERENCES

1. Perkins, K.K. et al. (2005) *Proc. 2004 Physics Education Research Conference*, AIP, New York.
2. Redish, E., Saul, J.M. and Steinberg, R.N. (1998). *American Journal of Physics,* 66, 212-224.
3. Adams, W. K. et al. (2006), Phys. Rev. ST Phys. Educ. Res. 2, 010101. See http://class.colorado.edu
4. Perkins, K.K. et al. (2006) *Proc. 2005 Physics Education Research Conference*, AIP, New York.
5. Redish, E.F. et al. (2003) "Epistemological gains in a large lecture class" AAPT National Meeting Madison, WI; Elby, A. (2001). *PER Suppl. Am. J. Phys.* 69(7), S54-S64.
6. e.g. Bretz, S.L. and N. Grove (2004). "CHEMX: Assessing Cognitive Expectations for Learning Chemistry", 18th Biennial Conference on Chemical Education, Ames, Iowa, http://www.chemx.org/
7. Barbera, J. et al., in preparation.
8. http://www.h-itt.com/

Writing in an Introductory Physics Lab: Correlating English Quality with Physics Content

Dedra Demaree*, Cat Gubernatis†, Jessica Hanzlik*, Scott Franklin**, Lisa Hermsen‡ and Gordon Aubrecht*

*Department of Physics, The Ohio State University, Columbus, Ohio 43210
†Department of English, The Ohio State University, Columbus, Ohio 43210
**Department of Physics, Rochester Institute of Technology, Rochester, New York 14623
‡Department of English, Rochester Institute of Technology, Rochester, New York 14623

Abstract. Members of the Physics and English departments at The Ohio State University and Rochester Institute of Technology are involved in an ongoing study addressing issues related to writing activities in the physics classroom. In summer quarter, 2005, the introductory calculus-based physics lab students wrote essays, some sections with and some without explicit writing instruction. We found a student's essay grade for English correlated strongly with that assigned for physics. In addition, we have studied the location and type of comments made by both physics and English instructors on individual student essays, and the statements students made within their essays. The results from the analysis of our data will be presented.

Keywords: writing to learn, laboratory
PACS: 01.40.Fk

INTRODUCTION

The project discussed in this paper is part of a collaboration between the Physics and English departments at the Ohio State University (OSU) and Rochester Institute of Technology (RIT). We focus here on the first project of our collaboration, which took place summer quarter, 2005, at OSU. We address the questions: is there a correlation between the types of English comments and the types of physics comments made in grading? What type of comments are more frequent? Does harder content engender more comments? Do any of these factors change with instruction and/or practice?

MOTIVATION

At OSU, the College of Engineering surveyed alumni and their employers and found both groups overwhelmingly wished they had been better prepared for writing and communication. The students in this study are predominantly engineering majors, and at this level have been observed to have difficulty with written explanations of physics concepts. In addition to the need for increased writing practice, the need for strong studies to establish the benefits of writing and writing instruction within disciplines has been shown in the literature [1, 2, 3]. Establishing an understanding of how students write in the context of physics, and the relationship between the physics content and the quality of their writing is a step toward approaching these larger issues. Writing activities in physics have been shown qualitatively to improve student writing [4, 5, 6], and the precedent for having an English instructor in the physics classroom has been established [7]. Our method of simultaneously giving a paper to a content and writing expert was first reported by the RIT collaborators [8].

IMPLEMENTATION DETAILS

This project was implemented in OSU's electricity and magnetism segment of introductory calculus-based physics. Two laboratory sections did writing activities during lab. One with 11 students had explicit writing instruction (WI) and the other with 17 students had no writing instruction (NI). The details are in Table 1.

TABLE 1. Division of time on lab activities

Group	Lab	Writing	Extra Instruction
NI	Reg., 1 h	1/2 h, at end of lab	5 min general instruction plus physics help during writing
WI	Reg., 1 h	1/2 h, at end of lab	As above, plus 15 min English instruction

The writing activity consisted of creating a paragraph missing from a pre-written essay. Students were given an explicit prompt including cues for what information was missing. The topics included why a car is safe during lightning, how electrostatic precipitators work, how holiday lights are wired, and how solar particles are trapped in Earth's magnetic field. The missing paragraphs re-

CP883, *2006 Physics Education Research Conference*, edited by L. McCullough, L. Hsu, and P. Heron

quired explanations of some aspect of the phenomena based on content from that day's lab. The students had six labs throughout the quarter, but because slightly different assignments were given the first and last weeks, those essays were omitted from this study.

The weekly writing instruction consisted of lesson plans beginning with higher order concerns and moving to progressively lower order concerns. Higher order concerns are universal issues such as organization, or logical flow, while lower order concerns are sentence or word level issues (e.g., word choice). In one lab, students outlined the information needed to respond to the prompt from the previous essay, then compared their outlines to the essays they wrote. They then discussed what information was missing, and how the order of information could be improved to strengthen their argument. In another lab, students were given a handout with sample sentences from the previous week's essays illustrating problems such as using transitions and equations. After discussion, students corrected the sentences.

Each week, the students' essays were copied, and one copy was given to Cat Gubernatis and graded for the quality of the writing (but not grammar or spelling mistakes). This grade was not biased by the physics content, since she does not have any background knowledge of physics. The other copy was given to Dedra Demaree and graded for physics content. Each grader made comments on the papers as they were graded. The final grade students received for each essay was the average of the English and physics grade. Essay grades were out of 15 points, with 9/15 considered the threshold for passing; most essays obtained between 9 and 15 points.

DATA OBTAINED

The English and physics grading comments were coded by Jessica Hanzlik. Although the graders discussed the importance of students producing a strong argument in their writing, each grader separately came up with their own grading rubrics. Despite this, Jessica found that the English and physics comments were often similar. Most statements could be grouped into five or six subcategories belonging to three main categories. English comments included external and internal language issues (e.g., "need transition," "awkward wording"). Others focused on content issues (e.g., "be more specific," "put this in context"). Both graders used "good" as a common positive comment. Physics comments centered around clarity issues (e.g., "physics not clear"), while others focused on the correctness of the physics.

The location of each physics and English comment was coded for each essay for lab weeks two through five. The basic content of each sentence was also coded. In addition, the students also took the Conceptual Survey of Electricity and Magnetism (CSEM) [9] diagnostic test. The lecture instructor also put a question on the final exam requiring a written explanation of the motion of a charged particle in a magnetic field. This question was graded by Dedra for physics content for the sake of their final exam grade, and later also graded for comparison by Cat for English.

RESULTS

Our results consist of qualitative and quantitative data, based on essay grades, exam grades, surveys, observations, and written comments made by the essay graders. A Mann-Whitney U test showed the CSEM pretest scores for the NI and WI groups were not significantly different ($\rho = 0.846$); therefore they are directly comparable to each other.

The main problems observed in the student writing included clarity, organization, and language. Cat noted that she could not always gain an understanding of the ideas from reading their work. She also noted that students had problems showing the relationships between ideas, equations and diagrams, and physics terminology. It seemed students were often not thinking about these assignments as constructing arguments, but instead thinking of them as just describing facts.

However, student writing became easier to read and was expressed more clearly as the quarter progressed. Students also showed improvement in integrating diagrams and equations in their writing. Only a few students improved greatly, with most obtaining a grade on their final essay within two points of the grade they got on their first essay. We observed that the quality of writing was heavily dependent on students' understanding of the content. Students struggled with the content of the final essay topic: their writing was not as clear, understandable, or well organized.

We consider if explicit writing instruction had an impact on the physics quality in the student writing. The essay grades based solely on physics are graphed in Fig. 1. This shows that although the NI group started with higher physics grades, by the end of the quarter the WI group had higher grades. In weeks one through three, a Mann-Whitney U test showed a 2-tailed significance of no lower than $\rho = 0.4$. However, in week four $\rho = 0.1$, and in week five significance is reached with $\rho = 0.05$. Each week, the physics and English grades correlated well, with a correlation coefficient ranging between 0.51 and 0.79. In addition, 68% of all essays had physics and English grades within 1 point of each other. Due to this, similar results from those shown in Fig. 1 can be shown for the total essay grades (with $\rho = 0.4$ in week 5).

A question on the final exam required explaining particle motion, similar to the week five essay. This specific

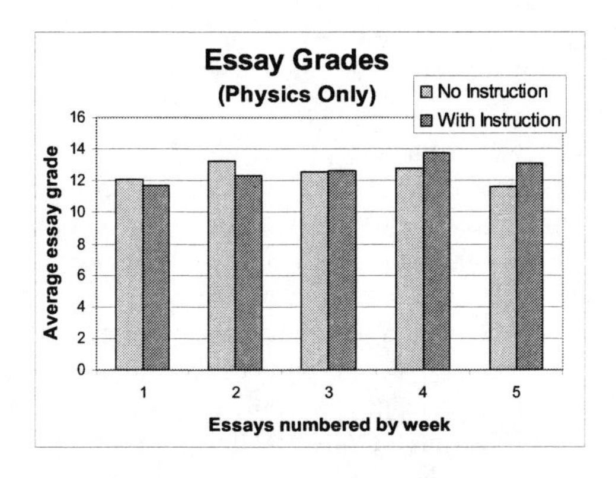

FIGURE 1. Weekly Physics essay grades.

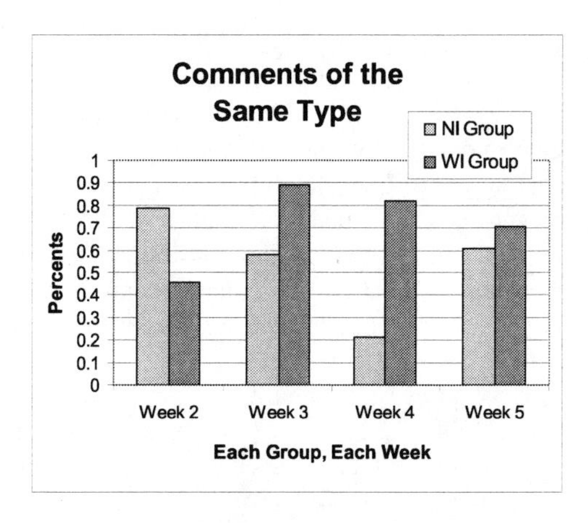

FIGURE 2. Percent of times physics and English comments at same location were of the same type.

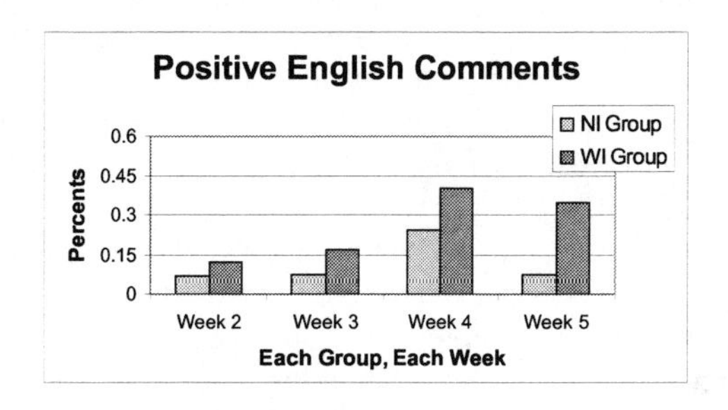

FIGURE 3. Percent of English positive-based comments.

question was covered in detail in lecture. The WI students had much higher physics grades on this problem than the NI students (average for WI = 7.1 out of 15, while the average for NI was 4.9). However, the English grades were nearly identical (10.6 for WI and 10.2 for NI). It should be noted the physics grade was based on points for the inclusion of various details of the problem, while the English grade (which did not count toward their final exam grade) was graded as we had graded in lab.

It seemed the students did not focus on English during the final exam; almost no responses appeared in full sentence form. It is not surprising that students did not transfer the idea of writing from lab to the final exam as the lab and lecture are fairly separate. There was no significant difference between physics grades for the NI and WI students on this problem, but there was a significant difference when the English and physics grades were averaged (the Mann-Whitney U test found $\rho = 0.11$ for the physics grades and $\rho = 0.027$ for the averaged grade). There was no significant difference between the two groups overall on the final exam ($\rho = 0.89$).

First we address the question: is there a correlation between the types of English comments and the types of physics comments made in grading? For every sentence that received either an English or physics comment during grading, 37% of the time there was a comment from both the English grader and the physics grader. For each group, and each week, the percent where both graders commented ranged from 29% to 47%; this agreement was fairly consistent across the two groups and throughout the quarter.

Both graders made comments based on language and clarity, content and correctness, or made positive comments. Grouping comments in that fashion, we found that 63% of the time when both graders commented at the same location, the comments were of the same type. This agreement ranged from 21% to 89% for the two groups each week. It is interesting that except for week 2, the WI group had a higher agreement between the English and physics comments than the NI group. The average agreement in the NI group was 55%, while in the WI group it was 72% (see Fig. 2). It is possible that the writing instruction impacted this agreement, though it is not obvious why it would.

Second we address: what type of comments are more frequent? We looked at the coded comments to see what types of comments occurred for each group throughout the quarter. The percent of positive English comments gradually increased through the quarter, with 9.5% in week 2 and 32% in week 4. This is consistent with both the improvements in student writing, and that the graders wanted to make sure we gave suggestions to correct student work early on, which artificially inflated the increase in positive comments.

Week five is interesting since the WI group received almost three times as high a percentage of positive comments as the NI group. These percents are shown in Fig. 3. Although this essay was difficult for both groups, as is seen by the lower grades in week five and the reduced number of positive comments, the difference between the two groups is striking. This is the week that the difference in essay grades between the two groups achieved statistical significance. The difference in the number of positive comments for each group supports that result.

The percent of language-based English comments decreased steadily throughout the quarter, with a steeper decrease for the WI group, though not as striking as the increase in positive comments. A similar trend is seen in the content-based English comments, with the exception of week five, when students had a lot of difficulty with the content, and the proportion of content-based comments almost doubled. Similar analysis can be done with the physics comments, though no strong trends appear from this data.

Third we address: does harder content engender more comments? One of the authors looked at the sentence content categories that were coded, and rated them as easy, medium, or hard. Easy content would be something like the statement of a physical law when the prompt told students to use that law to show something. One hard sentence content-type was chosen for each essay, and was based mainly on memory of what students struggled with understanding when writing the essays. Considering only the non-positive comments, we calculated the percent of each sentence difficulty type that was commented.

Our prediction was that harder sentences would have a higher percent of negative comments than easier sentences. However, this prediction failed for both groups. In week five's essay, the hardest one, none of the NI students included the content which was coded as hard. However they had a greater percentage of sentences commented on for week five than for the previous weeks. Because they struggled with this content, this suggests that more negative comments were made when students struggled, but not necessarily within any given sentence type within the essay.

Another coding recorded what type of sentence was written in each essay: motivation, observation, speculation, inference, or fact. Speculation was defined as having no basis of support for the statement, so were not conducive to good arguments. Between the WI and NI group the results were fairly similar with observation and inference sentences most common, and the percent of each sentence type fairly steady throughout the quarter. The most notable difference was the amount of speculation: the NI students had considerably more speculative sentences than the WI group. It is also interesting that both groups had more speculative sentences in week five when they struggled with content.

DISCUSSIONS AND CONCLUSIONS

The main implementation problem we encountered was resistance to student writing, and resistance to having a non-physicist aid in instructing their course. Students seemed unwilling to believe that writing was an important skill for their future. If writing were a common part of science pedagogy this resistance would probably be decreased. We also found students seldom took advantage of Cat's presence in the room for help while writing, and they were sometimes resistant to her direct help. Our recommendation is that collaborators from the English Department can be used for designing lesson plans, but the lessons can be taught directly by the physics instructor in the classroom.

We see convincing evidence that writing instruction had a positive impact despite negative student attitudes. The strongest evidence is that the quality of physics by the end of the quarter was significantly better for the WI group than the NI group. In addition, the WI students did better than the NI students on the written final exam problem. It is difficult to conclude if students actually gained more physics knowledge, or if the WI students gained a better ability to explain their knowledge - hence producing better quality physics. The latter is supported by our observations, and is worthwhile even without the former, as improved writing skills are needed.

We also provide evidence for the correlation between English quality and the physics mistakes. This helps establish a connection between the ability to express content knowledge and writing, which provides support for the idea that writing is pedagogically beneficial. The data support the idea that students need explicit instruction in order to take full advantage of writing activities. We suspect that given more practice writing the results from our project would be stronger, and plan to implement writing over a longer time period in the future.

ACKNOWLEDGMENTS

The authors would like to thank Kathy Harper for putting a research question on the final exam.

REFERENCES

1. Ackerman, J., *Written Communication*, **July** (1993).
2. Russell, D. R., *Writing in the Academic Disciplines: A Curricular History*, Southern Illinois University Press, Carbondale and Edwardsville, 2002.
3. Franklin, S. V., and Hermsen, L., *Educational Researcher*, **In Preparation**.
4. Larkin-Hein, T., and Budny, D. D., *ASEE/IEEE Frontiers in Education Converence, Reno, NV*, **October** (2001).
5. Mullin, W. J., *The Physics Teacher*, **May** (1989).
6. John W. Jewett, J., *Journal of College Science Teaching*, **Sept/Oct** (1991).
7. Larry D. Kirkpatrick, A. S. P., *The Physics Teacher*, **March** (1984).
8. Franklin, S. V., and Hermsen, L., *129th AAPT National Meeting: Sacramento, CA* (2004).
9. Maloney, D., O'Kuma, T., Hieggelke, C., and VanHeuvelen, A., *Am. J. Phys.*, **69** (2001).

The Effectiveness of Incorporating Conceptual Writing Assignments into Physics Instruction

Karen Cummings and Michael Murphy
Southern Connecticut State University, New Haven, CT 06515

Abstract. This preliminary study examines the impact of conceptual writing assignments on student understanding of two physics concepts. Writing assignments covered the concepts of Newton's Third Law and the impulse-momentum relationship and were given to students in both high school and college level introductory physics classes. The students in these classes along with students in classes taught in an identical fashion by the same instructors without the addition of writing assignments were tested on their conceptual understanding of the two content areas. The results of this initial study indicate that the efficacy of this approach varied with topic. This study further indicates that students' benefit from the writing assignments was independent of their writing ability.

Keywords: Writing, Conceptual Learning, High School, College

INTRODUCTION

In this paper we report on a small, preliminary study of the effectiveness of conceptual writing assignments in regard to conceptual learning for introductory physics students at the high school and college level. In addition, we probe whether students of greater writing ability benefit more from such assignments than students of lesser writing ability as some researchers would predict.[1]

Although there is a common belief among teachers that writing helps students learn, there is limited evidence that this is true in regard to physics concepts.[2-4] Thus, this study is designed to confirm earlier results and provide a basis for further studies of similar design and refined scope. This study is important because without proof of effectiveness it is difficult to convince physics instructors to incorporate conceptual writing into their classes. These types of assignments take significant time to grade carefully and accurately. The reluctance of instructors to incorporate this type of assignment into their class structure is reflected in the small sample sizes noted in the study discussed here.

POPULATION AND METHODOLOGY

The sample group consisted of high school physics students from two Connecticut public high schools as well as students from Southern Connecticut State University's conceptual physics course.

Two conceptual writing assignments were designed based on accepted models [5-11] for each of the two topics discussed in this paper: Newton's Third Law (N3) and the impulse-momentum relationship. As an example, one of the assignments addressing the relationship between impulse and momentum follows.

After coming home from a long physics class, you decide to dive on your nice fluffy couch and relax. Explain why jumping on the couch is a better, and less painful idea than diving onto the hardwood floor. Be sure to think in terms of momentum, impulse, and force when describing the situation.

All of the writing assignments describe a specific real-world situation and ask students to demonstrate, in a one or two paragraph written response, their understanding of the physics concepts that apply.

Evaluation of students' conceptual understanding was measured using a subset of questions from the *Force and Motion Conceptual Evaluation* by Sokoloff and Thornton[12], and the *Energy and Momentum Conceptual Survey* by Chandraleka Singh[13]. Five questions on the evaluation dealt with Newton's Third Law and three questions covered impulse and momentum. These questions were given to students post-instruction as a measure of conceptual understanding of these two topics. At both the high school and college level, students in classes taught in an identical fashion by the same instructors but without the addition of writing assignments were also tested and serve as a control group. Since the sample size for this study is quite small, we employ t-tests to evaluate the statistical significance of any differences in the mean scores for the groups.

CP883, *2006 Physics Education Research Conference*, edited by L. McCullough, L. Hsu, and P. Heron

Research in other domains indicates that students with greater writing ability might preferentially benefit from assignments such as these.[1] In order to probe this issue, students' writing ability was evaluated using a five point rubric, designed by Robin Lee Harris Freedman in her book Open-ended Questioning.[7] The mean score for each student was calculated from the grades on all of their writing assignments. Students who completed the writing assignments were then divided into two groups based on the quality of their writing. Students who scored an average of '3' or better on a rubric used to determine writing ability were considered to have "high" writing ability while students who averaged less than '3' on the rubric were categorized as having "low" writing ability. Note that these scores do not indicate correctness or incorrectness of the writing in any way. They are purely a measure of writing ability.

RESULTS

Scores for the impulse and momentum part of the study range from zero to three points (there were three J-p questions). Scores for the Newton's third law part of the study range from zero to five points (there were five N3 questions). These results are shown in tables 1-5 below.

Tables 1 and 2. These tables summarize the test scores for one high school and the college group on the impulse-momentum portion of the assessment. P values from t-tests are shown at the bottom of the tables.

Table 1. High School A : Impulse and Momentum

Without Writing Assignments		**With** Writing Assignments	
Score	Frequency	Score	Frequency
0	0	0	0
1	5	1	2
2	9	2	8
3	1	3	5
N =	15	N =	15
Mean	1.73	Mean	2.20
S.D.	0.59	S.D.	0.68
P = 0.03			

Table 2. College: Impulse and Momentum

Without Writing Assignments		**With** Writing Assignments	
Score	Frequency	Score	Frequency
0	26	0	0
1	33	1	3
2	14	2	3
3	2	3	0
N =	75	N =	6
Mean	0.893	Mean	1.50
S.D.	0.798	S.D.	0.548
P = 0.04			

Tables 3-5. These tables summarize the test scores for two high schools and the college group on the Newton's Third Law portion of the assessment. P values from t-tests are shown at the bottom of the tables.

Table 3. High School A : Newton's Third Law

Without Writing Assignments		**With** Writing Assignments	
Score	Frequency	Score	Frequency
0	5	0	3
1	4	1	5
2	3	2	3
3	0	3	1
4	2	4	0
5	1	5	3
N =	15	N =	15
Mean	1.53	Mean	2.07
S.D.	1.64	S.D.	1.79
P = 0.26			

Table 4. High School B: Newton's Third Law

Without Writing Assignments		**With** Writing Assignments	
Score	Frequency	Score	Frequency
0	0	0	1
1	0	1	1
2	1	2	1
3	2	3	0
4	1	4	2
5	1	5	4
N =	5	N =	9
Mean	3.40	Mean	3.44
S.D.	1.14	S.D.	1.94
P = 0.48			

Table 5. College: Newton's Third Law			
Without Writing Assignments		**With** Writing Assignments	
Score	Frequency	Score	Frequency
0	9	0	1
1	16	1	2
2	21	2	2
3	11	3	1
4	10	4	1
5	6	5	1
N =	73	N =	8
Mean	2.21	Mean	2.25
S.D.	1.46	S.D.	1.46
P = 0.47			

The tables shown below articulate the frequency of scores for only those students who were taught using the writing assignments. If students with "high" writing ability benefited more from the writing assignments the mean score should be higher for these students than those with "low" writing ability. However, these tables provide no such indication.

Table 6 High School A: Newton's Third Law			
		Writing Ability	
		High	Low
Newton's Third Law Score	0	2	1
	1	3	2
	2	1	2
	3	1	0
	4	0	0
	5	2	1
N =		9	6
Mean Score		1.83	2
Std. Dev.		1.72	1.94
P = 0.43			

Table 7 High School A: Impulse & Momentum			
		Writing Ability	
		High	Low
Impulse & Momentum Score	1	1	1
	2	5	3
	3	3	2
N =		9	6
Mean Score		2.17	2.22
Std. Dev.		0.753	0.667
P = 0.44			

Table 8 High School B: Newton's Third Law			
		Writing Ability	
		High	Low
Newton's Third Law Score	0	1	0
	1	0	1
	2	1	0
	3	0	0
	4	1	1
	5	1	2
N =		5	4
Mean Score		2.75	4.00
Std. Dev.		2.22	1.73
P = 0.19			

Table 9 College: Newton's Third Law			
		Writing Ability	
		High	Low
Newton's Third Law Score	0	1	0
	1	0	2
	2	1	1
	3	1	0
	4	1	0
	5	0	1
N =		4	4
Mean Score		2.25	2.25
Std. Dev.		1.89	1.71
P = 0.50			

Table 10 College: Impulse & Momentum			
		Writing Ability	
		High	Low
Impulse & Momentum Score	1	3	0
	2	3	0
	3	0	0
N =		6	0
Mean Score		1.50	n/a
Std. Dev.		0.548	n/a

Tables 6-10. These tables summarize the test scores for students of "high" and "low" writing abilities. Tables 6-8 are for students at the high school level. Tables 9 and 10 are for students at the college level. P values from t-tests are shown at the bottom of the tables were applicable.

DISCUSSION AND CONCLUSIONS

Based upon the data shown in tables 1-5, several preliminary conclusions can be drawn. The incorporation of writing assignments related to Newton's Third Law seemed to have no statistically significant effect on students' conceptual understanding. However, the use of writing assignments related to impulse and momentum did impact students' conceptual understanding of this topic. Both of these results were consistent across the three different instructors involved in this study and at both the high school and college level.

These results lead to an unpredicted (preliminary) conclusion. Namely, that writing assignments may be more beneficial in certain content areas than others. This may be an "artificial" result due to writing assignments with a context more closely linked to the context of the associated test questions. However, it could be that some materials is truly more easily learned via writing. Perhaps the closer a student is to correctly understanding an idea the better a conceptual writing assignment works. All of these possibilities offer interesting opportunities for further research.

In addition, tables 6-10 show that scores on the conceptual questions are fairly evenly distributed amongst all students who completed the writing assignments. Again, this result is true for all three instructors and at both instructional levels. Students with high writing ability did not score better than those with lower ability. In fact the mean test score was slightly higher for those of "low" writing ability than those of "high" writing ability in both content areas. However, the very high P-values indicate that the difference in these means are not statistically significant. It will be interesting to repeat this study with a larger sample size to determine if this result becomes significant.

The idea that students of various writing ability benefit equally (or that weaker students benefit more) from conceptual writing assignments is especially interesting because it conflicts with the educational theory known as "multiple intelligences". This theory, originally purported by Howard Gardner, presents the idea that "each individual possesses at least seven different intelligences, each with varied abilities."[7] One of these seven intelligences is 'logical-mathematical,' which is appropriate for much of the traditional instruction of physics; another is 'linguistic.' According to the theory, students who have initial strengths in one of the intelligences will learn better from activities that relate to their individual intellectual strengths. As applied to this study, this means students who excel in linguistic intelligence (those of higher writing ability) should be better able to develop their understanding of physics concepts through the process of writing. Our data do not seem to confirm this hypothesis. This is another result rich in interesting research topics for future studies.

REFERENCES

1. Sue Teele, Rainbows of Intelligence, Exploring How Students Learn, (Corwin Press Inc., Thousand Oaks, California, 2000)
2. Larry D. Kirkpatrick, & Adele S.Pittendrigh, "A writing teacher in the physics classroom," Phys. Teacher, **22**, 159-164 (1984).
3. Mark Vondracek, "Enhancing Student Learning by Tapping into physics They Already Know," Phys. Teacher **41**, 109-112 (2003).
4. Léonard P. Rivard, "A Review of Writing Science to Learn: Implications for Practice and Research" J. Res. Sci. Ed. **41**, 969-983 (1994).
5. Kenneth R. Chuska, Improving Classroom Questions, (Phi Delta Kappa Educational Foundation, Bloomington, Indiana, 1995)
6. Elizabeth Whitelegg and Malcolm Parry, "Real-life Contexts for Learning Physics: Meanings, Issues, and Practice," Phys. Ed. **32**:2, 70 (1999).
7. Robin Lee Harris Freedman, Open-ended Questioning, (Addison-Wesley, New York, 1994)
8. Jay L. Lemke, Talking Science: Language, Learning, and Values, (Ablex Publishing Corporation, Norwood NJ, 1990).
9. Paul Connoly in Writing to Learn Mathematics and Science, edited by Paul Connolly and Teresa Vilardi (Teachers College, Colombia University, New York, 1989) p. 11.
10. John W McBride, Muhammad I Bhatti, Mohamad A Hannan, and Martin Feinberg, "Using an inquiry approach to teach science to secondary school science teachers," Phys. Ed. **39**:5, 435 (2004).
11. William J. Mullin in Writing to Learn Mathematics and Science, edited by Paul Connolly and Teresa Vilardi (Teachers College, Colombia University, New York, 1989) p. 198.
12. R. K. Thronton and D. R. Sokoloff, "Assessing student learning of Newton's Laws: The force and motion conceptual evaluation," Am. J. Phys. **66**:4, 228-351 (1998).
13. Edward F. Redish, Teaching Physics, (John Wiley & Sons, Inc., Hoboken, NJ, 2003).

Characterizing the Epistemological Development of Physics Majors

Elizabeth Gire[*], Edward Price[†] and Barbara Jones[*]

[*]University of California, San Diego
[†]California State University, San Marcos

Abstract. Differences between novice and expert physics students have frequently been reported, yet students' development through intermediate stages has seldom been described. In this study, we characterize undergraduate physics majors' epistemological sophistication at various levels of degree progress. A cross-section of physics majors was surveyed with the Colorado Learning Attitudes about Science Survey. Beginning physics majors are significantly more expert-like than non-physics majors in introductory physics courses; furthermore, this high level of sophistication is constant over the first three years of the physics degree program, with increases at the senior and graduate levels. Based on longitudinal data on a subset of students, we observe negligible average shift in students' responses over periods of up to two years. We discuss implications for how and why physics students' epistemological sophistication develops, including a possible connection between CLASS survey response and self-identification as a physicist.

Keywords: epistemology, physics majors, beliefs, development
PACS: 01.40.–d, 01.40.Fk, 01.40.Ha

INTRODUCTION

The field of science education has a rich research tradition of comparing the skills and ways of thinking of experts to those of novices. Physics education researchers have been especially interested in making these comparisons in the areas of problem solving[1-3] and beliefs about the nature of physics knowledge[4-7]. Studies involving novice-expert comparisons have focused on people at opposite ends of the educational spectrum – students in introductory physics courses and advanced graduate students or faculty. Understanding these differences is important in helping students become more expert-like, but recognizing differences alone may leave out a description of if or how students develop expertise or expert-like views. In this study, we aim to further our understanding of this development process by examining physics majors at intermediate stages of expertise.

We focus on one aspect of students' development – their beliefs about the nature of physics knowledge and how to learn physics. We use the term epistemological sophistication to describe the level of similarity between these beliefs and those commonly held by physicists. These beliefs are thought to be an important factor in what and how students choose to study, and increased sophistication is a hoped-for, but often-unrealized course outcome[7]. Several survey instruments have been designed to assess beliefs about the nature of science and learning[4-8].

STUDY DESIGN

As a measure of students' epistemological sophistication, we administered the Colorado Learning Attitudes about Science Survey (CLASS)[5] during the 2004-2005 and 2005-2006 academic years at the University of California, San Diego to 533 students. Students were surveyed in lower- and upper-division courses for physics majors, several sections of the first course in the introductory physics sequence for engineering students[9] and a first year graduate course (see Table 1). The courses for physics majors are small (n~30) and are uniformly taught in a traditional lecture format. The physics department recommends a four-year program of courses; we surveyed courses from each year that are part of the required core sequence. Thus, students are likely to take these courses in the order recommended by the department. We used a rotating panel study design, surveying courses repeatedly, so that students in different stages of the program were surveyed in successive quarters. We thereby obtained cross-sectional and short-term longitudinal data.

CP883, *2006 Physics Education Research Conference*, edited by L. McCullough, L. Hsu, and P. Heron

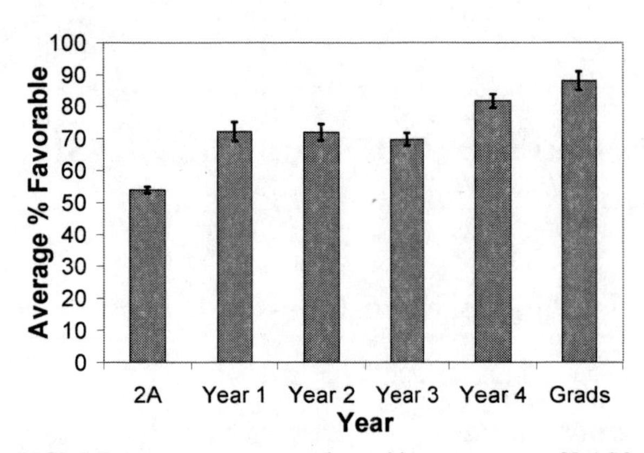

FIGURE 1. Average percent favorable responses to CLASS survey by year in degree program. 2A is an introductory physics course for engineering students, and grads are first year graduate students. Error bars indicate standard error of the mean.

Although many surveys are currently available for measuring students' epistemological stances, we find CLASS the most suitable for a study of physics students at different stages of degree progress. The CLASS items do not address the students' expectations of their performance in any specific course. Instead, the CLASS was designed with the intention to elicit student thinking about a broader context of physics. Although it is likely that students think of their experiences in their current courses while responding to the survey, the wording allows students taking many physics courses to respond more generally about their physics experiences and makes it easier to compare students across courses. In addition, when calibrating the survey, faculty respond to the items in the same frame as the students (not "How would you like your students to answer?"). We find this framing useful in trying to make inferences about how novice students develop more expert-like views.

The surveys were administered in lecture, and were completed by the students in a paper and pencil format. There are, of course, advantages and disadvantages in using this method. In yielding lecture time, instructors give implicit (or in some cases explicit) encouragement to participate in the study and to take the survey seriously. We also find that students who choose to volunteer are very likely to complete the entire survey. However, lecture time is at a premium, so some students are rushed to address all 42 survey items, and some instructors are reluctant to give up any lecture time. Only students who attended lecture were invited to participate in the study, and no course credit was awarded for participation.

The students took the survey during last two weeks of instruction, before the final exams. The only pre-test we administered was in the first course of the degree program (4A) during the winter quarter 2006. We decided to administer the survey only once per course in order to avoid effects of sampling the students too often (in the quarter system, a student could be asked to take the survey six times in a single academic year with both pre- and post-testing). We opted for post-tests over pre-tests because we expect students to be more honest about their current behavior than about what they expect their behavior to be.

Surveys were evaluated by the number of items on which students agreed or strongly agreed with the expert (favorable) response, leading to an overall percent favorable score for each student. Students were grouped by year in the degree program, and ANOVA was used to detect differences between years[10]. With the exception of the introductory course for engineering students, students who had declared majors outside of physics were excluded from the analysis. For the cross-sectional analysis, if a student completed the survey in multiple courses, only the student's first survey was included, so that all included samples are independent. The Games-Howell test was used for post-hoc comparisons between years[11]. The longitudinal component of the study consisted of the students who responded to the survey in more than one quarter. Only students' first and second surveys were included in the analysis. A two-tailed, paired samples t-test was used for making comparisons. Differences at the $p < 0.05$ level were considered to be significant.

TABLE 1. Group Summary

Group (Courses Included)	N	Average % Favorable	Standard Deviation
2A	378	53.9	19.3
Year 1 *(4A, 4B)*	33	72.1	16.9
Year 2 *(4C, 4D, 4E)*	29	71.9	13.6
Year 3 *(100A, 100C & 130A)*	56	69.7	14.7
Year 4 *(130B)*	16	81.7	8.5
Grads *(200A)*	7	88.1	7.5

RESULTS

Figure 1 shows the average number of favorable responses from students in each year of the physics major, the introductory course for engineering students (2A), and the graduate course (200A). Year in the program is based on the department's suggested sequence of courses. Table 1 shows the number of respondents in each course. An analysis of variance indicates statistically significant differences between the average number of favorable responses for these groups, $F(5,513)=26.17$, $p<0.001$. Games-Howell post-hoc testing, summarized in Table 2, indicates that 2A students are statistically different from all years of physics majors and graduate students. Students in Years 1, 2 and 3 are not significantly different from each other, nor are students in Years 1, 2 and 4. However, Years 3 and 4 are statistically different. Additionally, the graduate students are statistically different from all years except Year 4. The graduate students are the most epistemologically sophisticated as measured by CLASS.

TABLE 2. Difference in Average % Favorable, with Post-Hoc Test Results (an * indicates statistical significance at the 0.05 level)

Group i	Group j	Difference in Average % Favorable (i-j)	P
2A	Year 1	-6.56*	<.001
	Year 2	-7.17*	<.001
	Year 3	-5.87*	<.001
	Year 4	-9.84*	<.001
	Grads	-12.30*	<.001
Year 1	Year 2	-0.62	0.998
	Year 3	0.68	0.994
	Year 4	-3.28	0.141
	Grads	-5.74*	0.009
Year 2	Year 3	1.3	0.868
	Year 4	-2.66	0.239
	Grads	-5.13*	0.016
Year 3	Year 4	-3.96*	0.005
	Grads	-6.43*	0.002
Year 4	Grads	-2.46	0.422

Of the 141 physics majors surveyed, 50 responded to the survey more than once, generally within one or two quarters of their initial response. The longitudinal data on this subset of students allows us to directly monitor changes in students' survey responses over time. The average difference between the percent of favorable responses on students' second and first surveys is -0.1%, with a standard deviation of 1.4%. A two-tailed, paired sample t-test shows no statistically significant difference between the percent of favorable responses in students' first and second surveys, $t(50)=0.075$, $p=0.941$. A similar result was found among students in Year 3 only (students transitioning from 100A to 100C). Unfortunately, the data set does not include students transitioning between Year 3 and Year 4.

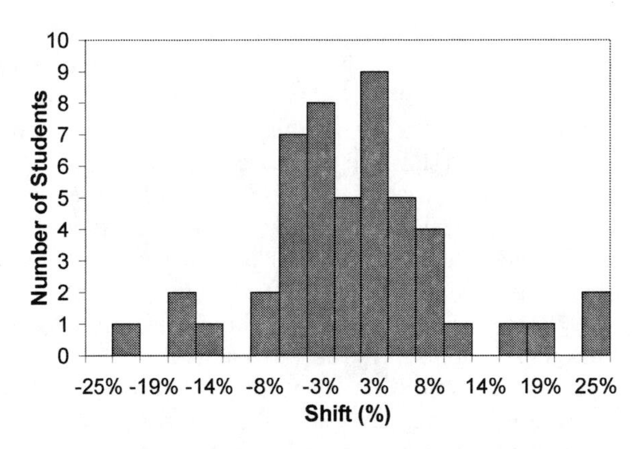

FIGURE 2. Histogram of difference in percent of favorable responses between second and first course in which student responded to survey. The mean difference is -0.1%, with a standard deviation of 1.4%.

DISCUSSION

Based on the data in Figure 1, undergraduate physics majors represent an intermediate stage between novice and expert. All years of physics majors are more expert-like than 2A students, the first three years of majors are less expert-like than graduate students, and students in the fourth year of the major are comparable to graduate students. Interestingly, however, this transition does not seem to be smooth and continuous. Rather, the cross-sectional data indicate a substantial difference between 2A and years 1-3, and no statistical difference between years 1-3. The longitudinal data in Figure 2 indicate that for students in years 1-3, individual students' overall responses to CLASS do not change over time. We thus infer that physics majors begin the degree program with a relatively high degree of physics epistemological sophistication (compared to their non-physics major peers) that does not change during the first three years of the program. As a result, we suggest that the expert-like views assessed by CLASS are, to a significant extent, an inherent, preexisting characteristic of students who choose to be physics majors, rather than a characteristic that is learned or acquired during the degree program. We note there are differences at the fourth year and graduate level.

Although Figure 1 indicates that physics majors have relatively sophisticated ideas about the nature of physics knowledge and how to learn physics when they enter the university, the source of this sophistication is unknown. Does it come from high school physics courses? Perhaps, but many engineering students also study physics in high school. Another possible source is extracurricular investigation of physics. Students who are interested in physics enough to consider it as a major may be more likely to have exposure to popular physics media, which would address many of the nature of physics survey items. Although we do not have data to directly address this hypothesis, it is consistent with the survey data showing that physics majors have more personal interest in physics than engineering students[12].

Although we do not observe an increase in epistemological sophistication in the first three years of study, students in the fourth year are more sophisticated. In light of data suggesting that introductory students can reliably identify expert responses to survey items and distinguish those responses from their own[5], it is possible that this increase in survey score at the senior year results from enculturation and changes in identity. Students in the senior year are typically preparing for their careers after graduation, and many apply for graduate school. During the application process, students identify themselves as potential physicists and seek acceptance into the community of physicists. It is possible that during this process, students' embrace views they recognize as accepted by the physics community.

CONCLUSION

In this study, we find that physics majors come to the university with beliefs about physics that are relatively expert-like. Overall, these beliefs are consistent throughout most of the undergraduate program, with an increase for students in the final year of study. Graduate students have more expert-like beliefs than undergraduates, and physics majors have more expert-like beliefs than engineering students enrolled in introductory physics courses. These results suggest than physics majors' epistemological sophistication, as measured by CLASS, is a preexisting trait rather than something learned at the university.

ACKNOWLEDGMENTS

We would like to thank many instructors who have welcomed us into their classes, and to the students who volunteered to participate. We are grateful to colleagues at the University of Colorado for sharing ideas and preliminary results.

REFERENCES

1. Singh, C. (2002) *American Journal of Physics*, 70(11), pp. 1103-1109.
2. Chi, M., Feltovich, P., Glaser, R. (1981) Cognitive Science, 5, pp. 121-152.
3. Hsu, L., Brewe, E., Foster, T. M. and Harper, K.K. (2004) *American Journal of Physics*, 72(9), pp. 1147-1156.
4. Redish, E., Saul, J.M. and Steinberg, R.N. (1998) *American Journal of Physics*, 66, pp. 212-224.
5. Adams, W. K., Perkins, K. K., Podolefsky, N., Dubson, M., Finkelstein, N. D. and Wieman, C. E. (2006) *Phys. Rev ST: Phys. Educ. Res.* 2(1).
6. Elby, A. (2001a) Epistemological Beliefs Assessment for Physical Science. http://www2.physics.umd.edu/~elby/EBAPS/home.htm
7. Halloun, I. and Hestenes, D. (1998) *Science & Education*, 7(6), pp. 553-577.
8. Lederman, N., Abd-El-Khalick, F., Bell, R., & Schwartz, R. (2002) *Journal of Research in Science Teaching*, 39(6), pp. 497-521.
9. There are few physics majors in the introductory physics course for engineering students; most physics majors enroll in a separate sequence.
10. In our data the groups have unequal variances. ANOVA assumes that groups are normally distributed with equal variances, though it is robust upon departures from these conditions. We confirmed the results of ANOVA with Welch's test, which identifies differences between groups and does not require equal variances. In all cases, Welch's test agreed with ANOVA.
11. Games-Howell was chosen because it does not require equal sample sizes or variances.
12. Perkins, K. K., Gratny, M. M., Adams, W. K., Finkelstein, N. D., and Wieman, C. E. (2006) *PERC Proceedings 2005*, 818, pp. 137-140.

Modeling Aspects Of Nature Of Science To Preservice Elementary Teachers

Paul Ashcraft

School of Science, Penn State Erie, The Behrend College, Erie, PA 16563

Abstract. Nature of science was modeled using guided inquiry activities in the university classroom with elementary education majors. A physical science content course initially used an Aristotelian model where students discussed the relationship between distance from a constant radiation source and the amount of radiation received based on accepted "truths" or principles and concluded that there was an inverse relationship. The class became Galilean in nature, using the scientific method to test that hypothesis. Examining data, the class rejected their hypothesis and concluded that there is an inverse square relationship. Assignments, given before and after the hypothesis testing, show the student's misconceptions and their acceptance of scientifically acceptable conceptions. Answers on exam questions further support this conceptual change. Students spent less class time on the inverse square relationship later when examining electrostatic force, magnetic force, gravity, and planetary solar radiation because the students related this particular experience to other physical relationships.

Keywords: Nature of Science, scientific method, physics education research
PACS: 01.40.Fk, 01.40.gb, 01.40.jc

INTRODUCTION

As an instructor of a freshmen level physical science class, my students usually fell into two categories, elementary education majors and students who were taking the last science course of their lives. The scientific method and the nature of science ([1], NOS) is part of my class' curricula because pre-service teachers will be teaching both topics in their future elementary classrooms, since these topics are on the national standards and many state standards [2]. These topics are included in the standards to promote scientific literacy and to help non-scientists understand science's role in society [3]. Often times these topics were quickly covered, making sure certain points were emphasized that could be asked objectively on a future exam. A student could memorize the definition of *scientific fact*, *hypothesis*, *theory*, and *scientific law*, memorize the text's five steps of the scientific method [4] and easily answer the exam questions on those topics.

Novice and preservice teachers tend to exhibit behaviors in the classroom that can only be modeled on their past experiences as students [5, 6, 7]. My students, modeling my class in their own future classrooms, would have their students learn vocabulary words and answer questions on tests written at their students' reading level. Yet, teachers are expected to have a scientifically accepted view of NOS and teach science using an inquiry approach [1,8]. Many teachers do not have a mature view of NOS and some preservice teachers only receive explicit instruction on NOS in their science teaching methods classes [9]. However, university science faculty can model future teacher's teaching methods by explicitly teaching NOS and by using constructivist teaching methods involving students using prior knowledge acquired through personal experiences or a shared social experience [10] as a base for a better understanding of scientifically accepted concepts. This case study will examine a class that models both a passive Aristotelian classroom and guided inquiry in order to understand a physical concept.

THE CASE STUDY

The following case study describes a teaching method where the students participated in a passive, Aristotelian classroom, considering a concept that has been experienced by everyone. After discussion, a conclusion was reached using accepted "truths" or principles. Later the same concept is investigated, using constructivist methods, the scientific method and the collection of data. The data supports conclusions that are very different than expected, yet applicable to many other physical concepts. Student comprehension improved, evidenced by comparing

CP883, *2006 Physics Education Research Conference*, edited by L. McCullough, L. Hsu, and P. Heron

reasons for choosing multiple-choice answers of similar questions given before and after the constructivist intervention.

Context

This study took place in a general education physical science course at a medium sized, rural, four-year Midwestern university. Forty of the 45 students enrolled in two classes participated in this study. One class of 17 was almost entirely sophomore, female, elementary education students; the other class was about one half education majors and juniors and about one quarter male and seniors.

During the first week of the semester, the class discussed the definition of a *scientific fact*, how these facts support *hypotheses*, how *theories* are a large synthesis of supported hypotheses, and finally how a *scientific law* is a theory that has not yet been refuted but may someday be proven wrong. Included in the discussion was the fact that better measuring or observing procedures can change scientific facts; for example, the Hubbell telescope radically transformed the knowledge base in astronomy. Another classroom example of the non-static nature of science was the Newtonian concept of gravitation replacing Aristotelian concepts.

Aristotelian Class

Prior to the class discussion of the inverse square law, the students answered the following problem outside of class and submitted their responses electronically:

> Choose an answer and please tell me why you chose that answer (in about 30 words). What was your thinking to come up with that answer? Why did you discount other answers?
>
> You are standing in a room that is lit by a single light bulb. You stand five feet from that light bulb. You move so you are now ten feet from that light bulb. How has the brightness of the light hitting your face changed when you moved from five feet away to ten feet away?
>
> a) you will receive four times as much light
> b) you will receive two times as much light
> c) you will receive the same amount of light
> d) you will receive one half as much light
> e) you will receive one quarter as much light *
> * Denotes correct answer

Student responses to this qualitative assignment are listed in Table 1. The participants overwhelmingly answered "D" while 8 percent correctly answered "E" and another 7 percent answered "C" because they read the question differently. ("The amount of light doesn't change; the light doesn't care where you are standing.") One student responded "C or D," leading to 2.5 responses to "C" and 32.5 responses to "D."

During the first inverse square law discussion, the students paired up, read their answers to their partner from the hard copy they brought to class, and discussed the problem. The two students came to a consensus, and those answers were tallied. Since the overwhelming majority chose "D", the pair's discussion confirmed that answer, and there were no dissenters in the classroom discussion. The instructor presented reasons for all answers (from the student's emails), some of which were conflicting. The class concluded that there is an inverse relationship between distance from the source and the amount of light you receive; if you double the distance from the light, one receives half the light as before.

Although there were no dissenters during the class discussion, after the class ended, one student told me that he still thought the answer was "E," that the light received had an inverse square relationship to distance. I asked him why he didn't say anything in class; he told me that everyone was so confident that he didn't want to say anything. Thus we should not assume that just because students appear to agree with the general consensus in these types of classroom exercises, that they actually do.

Data Collection Through Modeling and Experiment

In the next class period, following the above discussion, a biologist described the methodology of his current research and gave an example of the scientific method at work. He emphasized teamwork in science, how creativity plays a large part in research, and how assumptions and hypotheses are constantly being tested. His most important research involved an experiment that produced totally different results than expected; one that he had his student replicate over and over (until the professor replicated it) because data went against current theory. He emphasized that while his research group followed certain guidelines, that there was not one and only one scientific method to add to scientific knowledge.

During the third class period we revisited the radiation and distance relationship and considered the question using the scientific method:

> We clearly stated the question: How does the amount of light received change as you

change the distance from a constant light source?

We proposed a hypothesis: The amount of light is inversely related to distance.

We considered what consequences would happen if this hypothesis was true: If you double your distance from the source, you will receive one half the light as before; if you halve the distance, you will receive twice the light. Using this logic, if you tripled your distance, you would receive one-third the light and if you quadrupled the distance, you'd receive one-quarter the light.

Students then participated in a thought experiment that considered a globe that could be lit. The students considered the following scenario: the light bulb inside the globe was a point source located at the sphere's center that emitted light equally in all directions, so the light bulb emitted radiation evenly upon the surface of the globe. In order to calculate the amount of radiation received for every square centimeter on that globe, the class calculated the surface area of that sphere and divided the total radiation emitted by that surface area.

Next, spheres of different sizes were examined. As the radius doubled, the radiation per unit area was one-fourth the original. When the radius tripled, the radiation was 1/9th what it was before; if the distance quadrupled, the radiation reduced to 1/16th of the original value.

The class rejected their hypothesis that the radiation was inversely related to distance from the source because the data supported a hypothesis that the radiation had an inverse square relationship to distance from the source.

In the next week, the students answered an assignment question that was literally the same as before with the exception that numerical values were given to the amount of radiation received at the two locations. The student responses were markedly different (second line, Table 1). Thirty-eight subjects responded with 79 percent now replying that one received one-quarter of the radiation when one is twice the original distance, while 13 percent still believed that you'd only receive one-half the radiation.

Later the class tested this relationship measuring gamma rays (a type of electromagnetic radiation) in three-minute intervals when the radioactive sample was different distances from the Geiger tube. The students assumed one measurement was a "true" measurement and calculated theoretical counts per minute expected at both shorter and longer distances. Differences between empirical and expected data were between -10 and 30 percent, which lead to discussions of random error, sampling error, and assumptions in measuring.

One week later two similar, but open-ended, questions (one qualitative, one quantitative) were asked of the students on their exam. The instructor translated the student responses to equivalent multiple-choice question answers. These results, listed on the third and fourth lines of Table 1, show that 88 percent responded correctly on the qualitative question and 65 percent correctly calculated the correct quantitative response using the inverse square relationship. Only 3 and 8 percent, respectively, responded that there is an inverse relationship between distance and radiation received, the common initial misconception of the class. Responses listed as "other" include two students listing answers that did not appear to be logical and six students calculating the quantitative question using a direct relationship (or using a constant ratio between distance and radiation received).

DISCUSSION

The students, by modeling aspects of NOS by working in a scientific manner, experienced the scientific method. Guided inquiry activities produced data that was then examined for trends. Once a trend was recognized, explanations for this new concept were conceptualized and data justified their new conceptualization. After explicit instruction on NOS, with explicit reflection and discussion about the processes the students engaged in, these students should then be more likely to model this teaching method in their own future classrooms. While one could argue that this experience could be recalled or even reviewed during the students' method class instruction on NOS, explicit reflection and discussion must also happen in the classroom context in order for gains to be made

TABLE 1. Number of Student Reponses to Questions Dealing with the Inverse Square Relationship between the Distance from Radiation Source and Amount of Radiation Received. Note: MC denotes multiple-choice question. OE denotes open-ended question. * denotes correct answer.

	A	B	C	D	E*	Other
Qualitative assignment (MC)	0	0	2.5	32.5	3	0
Quantitative assignment (MC)	0	2	1	5	30	0
Qualitative exam question (OE)	0	2	0	1	35	2
Quantitative exam question (OE)	2	1	0	3	26	8

in understanding NOS [11, 12].

Was the time spent searching for the answer worth the amount learned from it? The same instructor usually uses 100 minutes of normal lecture time while the new method takes slightly more time, estimated at 125 minutes. Over the semester, the class spent less time examining the (inverse square) relationship between electrostatic force and distance, between magnetic force and distance, between gravitational force and distance, and between the amount of solar radiation received and planetary distance from the Sun. Thus, although it appears that an extensive amount of time was spent on this topic, time spent on other similar topics later was minimized.

A more disturbing aspect was the time spent modeling Aristotelian behavior, along with the class recognizing and then solidifying their misunderstanding of a physical relationship. As teachers, we often feel the need to teach, to correct, and to steer a student onto the correct path of thinking. As professors, we often feel the need to profess our expertise in our chosen field. Yet, if we allow the students to express themselves and their misconceptions (and even let them write them down in their notes), the students themselves will recognize their misconceptions and change their conception of their world, resulting in longer-lasting learning [13].

In summary, while this method is more time consuming and student-centered, students performed better on exam questions about inverse square relationships after modeling scientific methods than previous classes did by listening or watching a demonstration of science. This model can be used for other content and/or other courses since the act of asking students what they believe and why they believe it, can apply to any field. While this method asks the students their preconceptions (and misconceptions) as the starting point of a learning cycle, NOS and the scientific method are modeled through guided inquiry. If the student has scientifically valid preconceptions, the modeling of the scientific method makes those correct conceptions more concrete. If there are misconceptions, one can use the scientific method to provide evidence against the misconceptions and towards another conception, and lead students to a more scientifically acceptable conception. This particular inquiry led most students to a different-than-expected answer, causing them to examine the problem closer to explain their data. This understanding of the problem and its solution is valuable later in the semester when the students encountered analogous situations.

This type of instruction has great potential for helping students develop an understanding of the nature and practices of science, clearly they have engaged in practices that are similar to what scientists do. The next step would be to include activities that help these pre-service teachers explicitly reflect on their own learning and to compare the evolution of their own ideas to the evolution of the ideas within the scientific community. Other curricula [14] has "Learning about our own Learning" activities embedded within the curriculum, as well as activities that guide students thorough explorations of the nature of science, the historical development of ideas in science, and how students' own learning process compares the learning process of scientists.

REFERENCES

1. N.G. Lederman, F. Abd-El-Khalick, R.L. Bell, and R.S. Schwartz, *J. Res. Sci. Teach.* **39**, 497-521 (2002).
2. National Research Council [NRC], *National Science Education Standards*, National Academy Press, Washington, DC 1996.
3. J. D. Miller, *Pub. Und. Sci.* **13**, 273-294 (2004).
4. P.G. Hewitt, J. Suchocki, and L. A. Hewitt, *Conceptual Physical Science* (3rd ed), Pearson Ed. Inc., San Francisco, CA, 2004.
5. D. Lortie, *Schoolteacher: A sociological study*, U. of Chicago Press, 1975.
6. V. Richardson, "The role of attitudes and beliefs in learning to teach" in *Handbook of Research on Teacher Education* edited by J. Sikula, Prentice Hall, Upper Saddle River, NJ (1996).
7. C.S. Weinstein, *J. Teach. Ed.* **40(2)**, 53-60 (1989).
8. NRC, *Inquiry and the National Science Education Standards*, National Academy Press, Washington, DC, 2000.
9. F. Abd-El-Khalick and V.L. Akerson, *Sci. Ed.*, **88**, 785-810 (2004)
10. E. von Glasserfeld, *Synthese*, **80**, 121-140 (1989).
11. V. L. Akerson, F. Abd-El-Khalick, and N.G. Lederman, J. *Res. Sci. Teach.* **37**, 295-317 (2000).
12. R. Khishfe and F. Abd-El-Khalick, *J. Res. Sci. Teach.* **39**, 551-578 (2002).
13. P. Hewson, M. Beeth, and N.R. Thorley, "Teaching conceptual change" in *International Handbook of Science Education* edited by B.J. Frasier and K.G. Tobin, Kluwer, London, 1998.
14. F. Goldberg, S. Robinson, and V. Otero, Physics for Elementary Teachers, 2004

Beyond Concepts: Transfer From Inquiry-Based Physics To Elementary Classrooms

Danielle B. Harlow and Valerie K. Otero

249 UCB, School of Education, University of Colorado, Boulder, CO 80309

Abstract. Physics education researchers have created specialized physics courses to meet the needs of elementary teachers. While there is evidence that such courses help teachers develop physics content knowledge, little is known about what teachers transfer from such courses into their teaching practices. In this study, we examine how one elementary teacher changed her questioning strategies after learning physics in a course for elementary teachers.

Keywords: Physics Education Research, School Science, Elementary School Science, Teacher Training.
PACS: 01.40.-d, 01.40.E-, 01.40.eg, 01.40.Fk, 01.40.G-, 01.40.gb, 01.40.J-, 01.40.jh

INTRODUCTION

It is increasingly evident that PER does and will continue to play an important role in the education of teachers at all levels [1]. One indication of this is that PER researchers have developed multiple inquiry-based curricula with elementary teachers in mind [2,-4]. These curricula use guided inquiry activities to facilitate teachers' conceptual development of physics ideas and model desired teaching strategies. Furthermore, they align with recommendations of how to best prepare K-12 teachers which suggest that teachers should have the opportunity to learn science in the ways they will be expected to teach [5].

Evidence supports that inquiry-based curricula result in greater learning gains on tests of conceptual understanding [6] and that such courses lead teachers to be more open to the prospect of teaching by inquiry [7]. However, little has been done to understand what teachers actually transfer from such courses into their teaching practices. Understanding the usefulness of inquiry physics courses such as those developed by the PER community requires investigating what teachers actually *do* when teaching science to elementary students.

We investigate how the Physics for Elementary Teachers (PET)[1] curriculum impacts elementary classrooms. Participants in a professional development course adopted from PET were video taped teaching science and interviewed both before and after taking the course. We examined the data of one teacher to identify aspects of the physics course that she may have transferred. Here, we describe how the course impacted one aspect of her teaching practices: the types of questions she asked her students.

Questioning as a Teaching Practice

Harlen [8] categorized teachers' questions as either productive or unproductive with respect to science learning. Productive questions are those which lead to scientific activities such as observing, describing and explaining observations while unproductive questions lead students to looking for answers in books or from their teacher. In elementary classrooms, students begin to acquire ways of doing science and thinking about science. When teachers ask questions which prompt students to *do* something or to reflect on their own personal experiences, they promote a view of science as a way of knowing and as something that connects to their students' own experiences rather than as a collection of facts that are found in books. Unfortunately, the vast majority of teacher questions lead students to recall or look up science ideas in books or other secondary sources. In the analysis that follows, we present evidence that after taking a physics class, one teacher changed her questioning style to one which promoted science as a way of knowing rather than as a collection of facts.

[1] Physics for Elementary Teachers has recently been renamed. This curriculum is now "Physics and Everyday Thinking."

CP883, *2006 Physics Education Research Conference*, edited by L. McCullough, L. Hsu, and P. Heron

DATA AND ANALYSIS

The participants in this study were practicing K-5 teachers who enrolled in *Magnetism and Electricity for Elementary Teachers* (MEET), a 15-hour professional development course adapted from the *Physics for Elementary Teachers* (PET) curriculum. Five participants were videotaped for the entire course and were also videotaped teaching science in their elementary classroom both before and after taking MEET. In this paper, we focus on Ms. Doty whom we chose as the initial case study teacher because she was an excellent and enthusiastic science teacher who expressed apprehension at the prospect of teaching magnetism and electricity topics.

At the time of the study, Ms. Doty had been teaching elementary school for 12 years. During the initial interview, she claimed that she struggled with teaching science because of her lack of science knowledge, but that, despite her low levels of confidence, she enjoyed teaching science. Like many teachers in her district, she relied heavily on the district created science kits which contain materials and instructions for science activities and relevant science literature for students to read. Ms. Doty typically completed four kits a year (each kit took 2-3 weeks to complete) and was comfortable teaching with the kits she regularly used. This year, however, Ms. Doty moved to a new grade level (3rd/4th combination) and was expected to teach topics in magnetism and electricity for the first time. She claimed not to know anything about either topic and did not know where to begin figuring what to do on her own.

Data about Ms. Doty's teaching and learning were collected in three stages. First, she taught two representative science classes prior to beginning MEET. In these lessons, she used a science kit entitled "The Web of Life," in which students studied how living organisms depend upon one another. In the second stage of data collection, Ms. Doty was observed as she learned about magnetism and electricity in MEET. Finally, after taking MEET, she taught two lessons based on the topics covered in MEET. In her previous 12 years of teaching, Ms. Doty had taught neither magnetism nor electricity. She chose to teach a unit on magnetism.

All pre and post observations were analyzed to determine questioning patterns before and after MEET. First, all *teacher questions* were identified in the transcripts of the pre and post observations of her science teaching. A statement was determined to be a *teacher question* if it appeared that the student(s) interpreted her statement as one that required a verbal response. This was indicated by the student(s) actually providing a subsequent verbal response. All *teacher questions* were further examined to identify the purpose of the question. Questions were first categorized as either questions about science (SCI) or questions asked for the purpose of classroom management (such as making sure that students understand directions or checking to see how far along they are). The questions about science were further differentiated by the source that the students drew on to answer the teacher's question: Secondary Sources (Sci-SS), Science Experiences (Sci-SE) or their Own Ideas (Sci-OI) as shown in Table 1. We used only the three types of *science questions* in the analysis.

TABLE 1. Coding Scheme – Science Questions

Code	Description	Examples
Sci-SE	**Science Experiences**: Students either do something or reflect on prior experiences to answer SE questions.	• Does that magnet work in water? • Have you ever left bread out for a long time? What happened?
Sci-OI	**Own Ideas:** Students use their own thoughts and creativity to answer OI questions.	• What do you think is happening inside the magnet?
Sci-SS	**Secondary Sources**: Students draw upon secondary sources such as books or the teacher to answer SS questions.	• Does anybody remember what the word hypothesis means? • How many species of turtles are there?

Questions were categorized by taking the context and the students' response into consideration. For example, the students in this classroom study had previously studied anoles (a lizard) and had observed a pair of living anoles in their classroom. The teacher question, "What do anoles eat?" would be a SE question if the students were expected to draw on their observations of their classroom anoles. However, this same question asked in a context in which the students are expected to look up the information on a fact sheet would be considered a SS question.

FINDINGS

The results of coding the five observed lessons are presented in Figure 1. Pre-1 was an experiment about decomposition and Pre-2 was the cumulative activity of a life science unit. Post 1 was an experiment about magnetism and Post-2 and Post-3 were the cumulative lesson in magnetism, which spanned two class periods.

In a post interview, Ms. Doty stated that one thing she had changed about the way she taught after taking MEET is that she let the students do more of the talking. This is supported by the increase in the number of science questions she asked her students.

Figure 1 shows that Ms. Doty asked an average of 112 science questions during the post observations (Post-1, Post-2, and Post-3), more than twice as many questions as she asked during the pre observations (Pre-1 and Pre-2) which averaged 42 science questions per class meeting. This observation alone is important because it implies that Ms. Doty collected more information from her students and allowed them to talk more during science lessons after taking MEET.

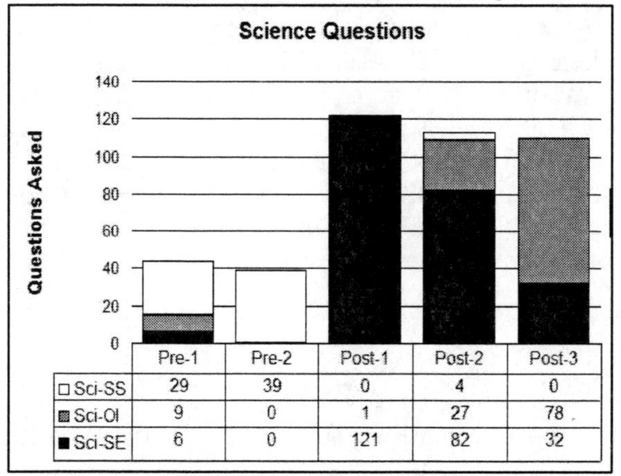

	Pre-1	Pre-2	Post-1	Post-2	Post-3
□ Sci-SS	29	39	0	4	0
▩ Sci-OI	9	0	1	27	78
■ Sci-SE	6	0	121	82	32

FIGURE 1. The frequency and types of science questions asked by the teacher during each observed science lesson.

Figure 1 also provides evidence that Ms. Doty asked different types of science questions during the post observations than during the two pre observations. Ms. Doty's two pre lessons were dominated by questions that asked students to draw upon secondary sources. In contrast, during the post observations she asked students to draw on their science experiences and their own creativity almost exclusively. Following, we describe this difference in more detail.

Pre-Observation Questioning Style

In her initial interview, Ms. Doty claimed that she liked for her students to be able to figure things out on their own. However, she also expressed an understanding of science as vocabulary and as a collection of facts that are obtained by authority. When talking about why she liked to teach with district-created science kits, Ms. Doty stated, "[The kits] give you the vocabulary. And I think you have to have that entire knowledge." Furthermore, when Ms. Doty described a science lesson she created, it was heavily literature based: students used encyclopedias to gather information about the planets.

Her two lessons prior to MEET reflect this conflict between wanting to let students figure things out and needing for students to know science vocabulary and facts. During Pre-1, Ms. Doty's students buried dead crickets in wet and dry soil to compare which crickets decomposed first. Despite the experimental nature of the activity, 66% of the science questions Ms. Doty asked her students represented science as a collection of facts and terms. She began the lesson by stating that the main topic of the day was decomposition and asked if anybody knew what the word meant. After her students suggested answers, Ms. Doty concluded, stating, "We've been working on using our dictionaries and I wanted you to see that teachers use dictionaries as well." She then read the dictionary definition. A similar pattern of discussion occurred throughout this lesson.

Furthermore, although the students spent two weeks observing live organisms in their classroom, the cumulative activity (Pre-2) for the unit emphasized science as a collection of facts. The students used books and fact sheets to create posters which summarized interesting facts about lizards, crocodiles, snakes and turtles. Ms. Doty moved between groups asking about what they had learned. At one group she asked, "What type of food do snakes eat?" The students offered many answers including "wild boars," "eggs," "anything with meat in them," and "dead snakes." After the students answered her question, Ms. Doty responded, "Sara, do you want to read – it's right here, guys {points to fact sheet} -meat eating strategies- what do they eat? Lizards, birds, what's that one right there – that should be in your garden what's that i-word right there?" to which Sara responded "insects." Ms. Doty wanted her students to read the answer from the fact sheet and her question successfully motivated this activity.

Post-Observation Questioning Style

After the post observations, Ms. Doty claimed that, during the magnetism lessons (Post 1-3), she tried, for the first time, to teach without telling her students everything because, "I just think you learn so much better when you do figure it out for yourself. That really made sense to me. [Before MEET] I was against that idea so that's kind of funny. It was like 'oh no they have to- I have to give them all the information in the end. They have to know.' Well they're not going to remember if I do give it to them."

The data collected during post observations support the claim that Ms. Doty was letting her students figure things out because the majority of her questions were about her students' ideas and experiences rather than about ideas they learned from secondary sources. During lesson Post-1, the students in Ms. Doty's class conducted experiments at stations and Ms. Doty circulated around the room asking students what they were discovering and asking them questions to prompt further investigation. Ms. Doty asked 122 science questions during this lesson, more than any other observed lesson and 121 (99%) of these were coded as Sci-SE, indicating that Ms. Doty was encouraging her students to use their own science experiences to answer questions. She appeared to value their contributions, saying things like, "I'm curious about something you said there" and asking them to elaborate on their own ideas.

The second lesson was the final activity of the magnetism unit and spanned two class periods, (Post-2 and Post-3). The students magnetized a nail and observed that the magnetized nail attracted paperclips. They then drew, described, and refined models of what they thought made the magnetized nail different than an ordinary nail. Ms. Doty asked students to share their thinking and suggested ways they might test the alternative models presented.

Of the 223 questions that Ms. Doty asked of her students over these two class periods, only four (2%) were coded as Sci-SS, and 219 (98%) were coded as either Sci-SE or Sci-OI. Unlike the final activity of the Web of Life unit, this lesson asked students to draw upon and develop their own ideas to answer the vast majority of their teacher's questions.

DISCUSSION

The evidence presented here points to a substantial change in Ms. Doty's questioning strategy. Initially her science questions asked her students to draw largely on secondary sources. After MEET, Ms. Doty asked questions which prompted students to *do* science or reflect on their experiences and own ideas to answer her questions. The primary goal of MEET was for teachers to develop a conceptual understanding of magnetism and of current electricity. The *teaching* of magnetism was not explicitly addressed in MEET. Yet there is reason to believe that Ms. Doty transferred aspects of the course pedagogy into her teaching of elementary school science.

In MEET, Ms. Doty and her classmates collected and interpreted evidence through laboratory and computer activities and developed a simplified domain model of magnetism. Throughout the course, ideas were expected to be supported with evidence collected during experiments rather than by outside sources. As a learner, Ms. Doty initially found it frustrating to not be given the correct answer; however, her experience in the course led her to attempt this same method in her own teaching. In her final interview she stated that without having had the experience of developing a model of magnetism on her own she would not have risked asking her students to do the same. Furthermore she stated that she learned in MEET that teaching science "is about letting the kids do the work, letting the kids do the experiments, letting the kids figure things out and you just figuring out ways – how are you going to guide this particular class." This idea is reflected in her change in questioning style.

CONCLUSIONS

Understanding what teachers actually do in the context of teaching science in their elementary classroom is vital to understanding what teachers take away from our courses and how they use it in their teaching. The teacher described here transferred an important aspect of the course pedagogy: asking students to draw upon science experiences to make sense of the world rather than looking to other sources of authority.

ACKNOWLEDGMENTS

This research is supported by the National Science Foundation Grant #0096856.

REFERENCES

1. See the collection of invited papers in the 2005 PERC proceedings which addressed the 2005 theme, "Connecting Physics Education Research to Teacher Education at all Levels: K-20.
2. L. McDermott, *Physics by Inquiry*, New York: John Wiley & Sons, Inc., 1996.
3. AAPT, *Powerful Ideas in Physical Science*, College Park, 1995.
4. F. Goldberg, S, Robinson, and V. Otero, *Physics for Elementary Teachers (PET)*, It's About Time, 2005.
5. NRC, *National Science Education Standards*, Washington DC: National Academies Press, 2006.
6. NRC, *Improving Undergraduate Instruction in Science, Mathematics, and Technology,* Washington DC: NAP, 2005.
7. G. Aubrecht, *Grounding Inquiry-based teaching and learning methods in physics experiences,* PERC proceedings 2004
8. W. Harlen, *Primary Science: Taking the Plunge*, Portsmouth: Heinemann, 2005.

Learning and Dynamic Transfer Using the 'Constructing Physics Understanding' (CPU) Curriculum: A Case Study*

Charles B. Mamolo and N. Sanjay Rebello

Department of Physics, 116 Cardwell Hall, Kansas State University, Manhattan, KS 66506-2601

Abstract. This research investigated the ways in which students interact with Constructing Physics Understanding (CPU) curriculum in an instructional unit focusing on waves and sound. The research was conducted at University of San Carlos, Philippines with six students. We draw on the constructivist philosophy and employ a phenomenographic approach in our analysis of student conversation during the activity. The dynamic transfer model developed previously was the analytical framework used to map out students' development of ideas as they worked through the instructional materials in this CPU unit.

Keywords: Physics education research, curricula, teaching methods, strategies, theory of testing and evaluation
PACS: 01.40.Fk, 01.40.Gm, 01.40.Jp

INTRODUCTION

The Constructing Physics Understanding (CPU) curriculum [1] is a popular curriculum used in furthering science education. The *Wave and Sound* unit (Cycles I and III) was pilot tested at the University of San Carlos, Philippines. This paper reports on six senior undergraduate teacher education students' learning while interacting with Cycle I. The six students were divided into two groups. The CPU pedagogy cycle is shown in Figure 1.

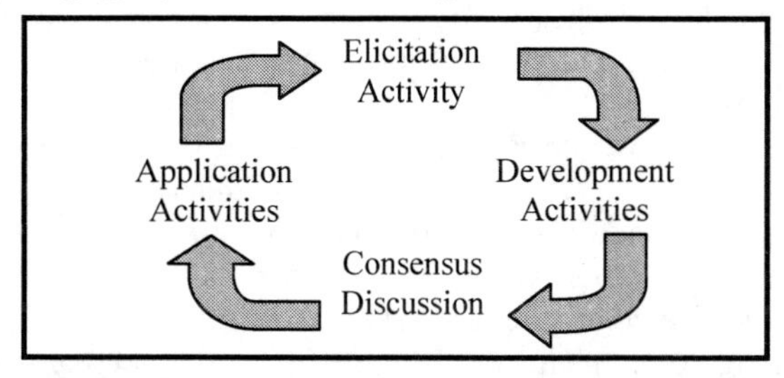

FIGURE 1. Outline of the CPU Pedagogy Cycle

Each cycle begins with an *elicitation activity* that builds upon prior experience to formulate an initial explanation for a phenomenon. Following the elicitation activity, each group of students tests and modifies their initial ideas by working through a series of several *development activities.* In Cycle I the students worked on four activities. In this paper we will report on one of these activities –*Wave and Sound I: How do waves behave on a spring?* At the end of the development activities, students contribute to a *consensus discussion activity.* During this consensus discussion the instructor engaged the students in Socratic dialogue and gave short lectures when necessary. During the *application activities* students applied the class consensus ideas in a wide variety of situations. The pedagogical cycle described above was videotaped. In addition, all students' activity sheets, individual journals and written group consensus ideas were used to triangulate the information presented in this paper.

RESEARCH QUESTION

The research question that we seek to address within the relatively narrow context of a single CPU Unit – *Waves & Sound* is the following:

> *What are the processes through which students generate and share ideas as they attempt to reach a consensus in a group while completing the activities in the unit?*

We recognize the relatively narrow scope of the study in that it based on results of small group of students at a single university working on a small segment of the curriculum. Nevertheless, we believe it provides some interesting and useful insights of the potential issues that students could face as they work through with this type of curricular material.

* This work is supported by NSF grant REC-0133621

CP883, *2006 Physics Education Research Conference*, edited by L. McCullough, L. Hsu, and P. Heron

METHODOLOGY

The CPU pedagogy, as briefly described earlier, affords opportunities for students to interact with the materials and with their peers. Consistent with the notion of Vygotsky's Zone of Proximal Development (ZPD), the CPU unit builds from one activity to the next to form scaffolding. [2] A phenomenographic approach was used to examine the variation in experiences of a student's interaction with the CPU materials and with the other members of their group [3, 4] in order to answer the research questions listed above.

We divided the study into two phases. In Phase I, we analyzed the first cycle of the *Wave and Sound* unit. We compared students' class consensus ideas in Cycle I with the CPU target ideas. To gain a better understanding of these ideas we analyzed the transcripts that led to the development of these consensus ideas. Our analysis focused on the process of attaining the consensus ideas rather than the ideas themselves. In Phase II, to gain insights into students' intellectual development, we conducted a detailed investigation of a group's interaction with an activity in Cycle I – *Spring Activity*. In both phases we used the 'dynamic transfer model' [5] in the analysis of the transcripts. The model will be elucidated in later examples.

RESULTS AND DISCUSSION

Phase I

The unit of analysis in Phase I was the class consensus stage of Cycle I. As noted in Figure 12, the Class Consensus phase occurs after the completion of the Development Activities. Ideally (not shown in Figure 1) the *Class Consensus* (CC-I) would occur only after each group had arrived at *Group Consensus* on wave properties emerging from the development activities that comprised the *Spring Activity*, *Tuning Fork Activity*, *Ripple Tank Activity* and *Simulator Activity*. However, the *Ripple Tank Activity* did not go as planned, so we decided to proceed with CC-I to ensure that the *Group Consensus* was ultimately productive. Thus, during CC-I they redid the *Ripple Tank Activity* and completed investigations on the *Simulator Activity* that enabled them to come to a group consensus regarding wave properties. We note here that the two groups worked independently except for the ripple tank activity since there was only one ripple tank available at that time.

We have classified students' disagreements into two types: **activity-disparity** and **group-disparity**. An **activity-disparity** is a disagreement of results between two activities for which both groups have the arrived at the same conclusion. For example the *Simulator Activity* supported the idea that "wave speed is directly proportional to wavelength" while the *Ripple Tank Activity* supported the idea that "wave speed is inversely proportional to wavelength." In another example the *Simulator Activity* provided evidence that "frequency does not affect wave speed" but the *Ripple Tank* and *Spring Activities* provided evidence that "frequency affects wave speed."

The second type of disagreement -- **group-disparity**, is between groups that draw different conclusions from the same activity. For example in the *Spring Activity*, Group 1 found that as "amplitude increases the wave speed increases" while Group 2 found that "wave speed is amplitude invariant." Another example is the idea that "frequency affects wave speed" as supported by the *Spring Activity* for Group 1 while "frequency does not affect the wave speed" as was found in the same activity for Group 2. Both groups which performed the same activity reached different results.

One of the objectives of the class consensus activity was for the students to arrive at an agreement on wave properties and resolve any disparities. We have identified major themes that helped resolve the disparities. These themes were not discussed by the students chronologically but were chosen throughout the class consensus transcript using the dynamic transfer analytical framework. Figure 2 shows an analysis of two students' knowledge associations. The unit of analysis was the group because students within a group were extending each others' ideas. In a sense, the knowledge was "owned" by the group and identifying individual knowledge structures would be futile.

As shown in Figure 2 the external inputs were the disparities, velocity-wavelength-frequency equation, linear equation and the question that asked students to find out which of the wave properties were constant. The shaded bubbles are the *source tools* activated from long term memory. The students did not mention these concepts explicitly but they were implied from their statements. The first association constructed by the students was between the *source tool* of 'equation manipulation' and the *target tool* '$v = \lambda f$' resulting in an output '$v / \lambda = f$'. This reformulation became the new *target tool* and was next associated with a *source tool* -- 'dimensional analysis.' Association II gave an output which emphasized the unit '1/cm'. This output was in turn associated with the question, "Which is constant?" and gave the output that 1/cm is constant because it is not measurable and therefore not real and hence constant. The conversation is quoted below.

Student 1: Student 2 was telling us about the formula (writes the formula). So if we transfer λ so that's $v/\lambda = f$, and if we separate these two, we will have: v (1/cm) = f. And 1/λ, as you said (referring to student 2) is not possible because we cannot measure distance of 1/cm or...(pause) that's what Student 2 is telling me. So, because this (1/cm) is not possible... (Asks student 1)?

Student 2:... this (1/λ) should be constant. This is not measurable...isn't this not real?

Student 1: Yeah...this is not real, ...

The shaded area in Figure 2 represents the model that students first associate as *'not measurable'* with *'not changing,'* i.e. *'constant'* Thus, a *'not measurable'* quantity must be *constant*. Hence, wavelength (λ) is constant. Further association with this new knowledge made them conclude that v (velocity) and f (frequency) are varying. This example shows that students possess prior knowledge and are capable of constructing their knowledge and the CPU pedagogy gives them an opportunity to do so.

We employed the same kind of analysis throughout the class discussion transcripts that helped us identify the disparities as well as the themes that resolved these disparities.

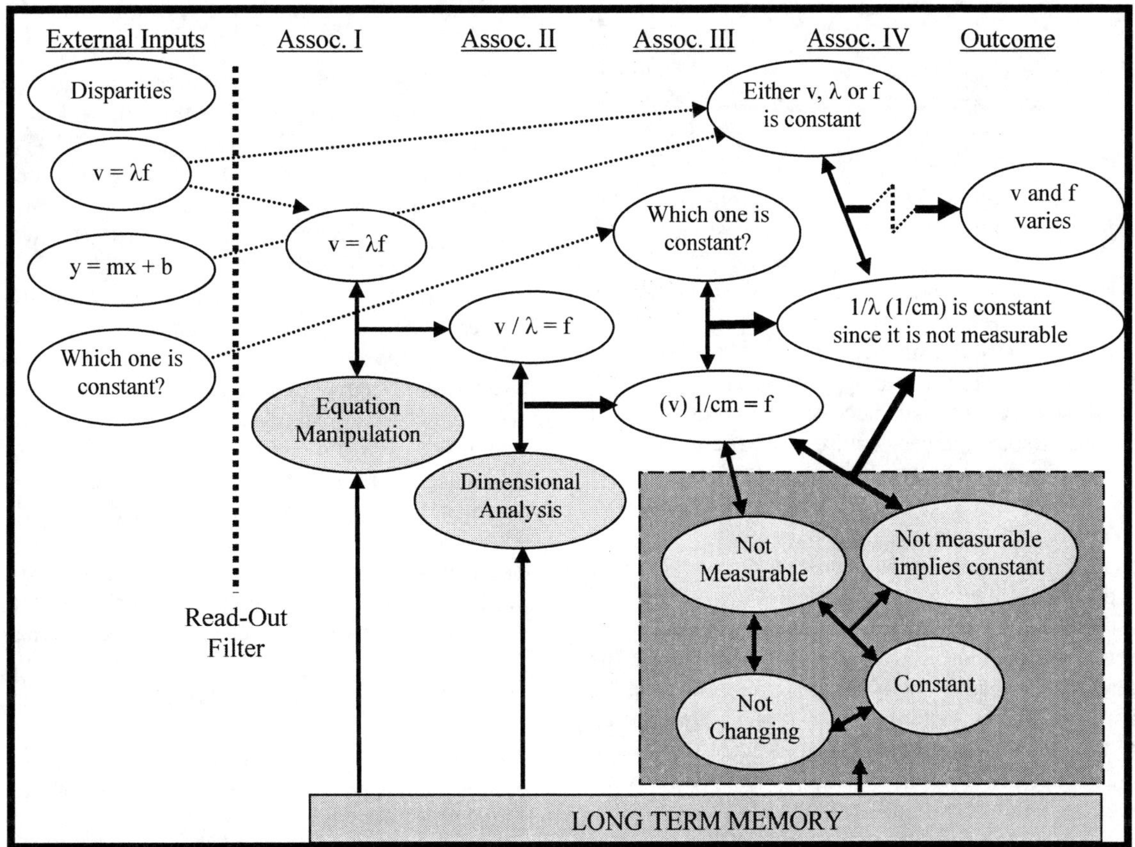

FIGURE 2. An example of two students read out from external inputs into target tools and associations from memory.

Phase II

In Phase II our goal was to study the process which produces the disparities that we noticed in Phase I. Specifically, we decided to study the *Spring Activity* and focus on a single group. The *Spring Activity* created the **group-disparity** mentioned in Phase I although the students did not redo the activity. However, they cited the *Spring Activity* as evidence for the wave properties during another class consensus.

We will discuss the themes emerging from our analysis of the *Spring Activity*. These themes could either be productive or unproductive in facilitating student conceptual learning. One important theme is that **students extend others' ideas that fit their knowledge structures**. For instance this occurred when a student extended another student's illustration to fit her concept that an increase in amplitude increases the wavelength. Two other emergent themes were **students' predictions being controlled by prior knowledge or activities** and the **filtration of inputs into target concepts** that would limit students' perception of an activity. Figure 3 illustrates these themes.

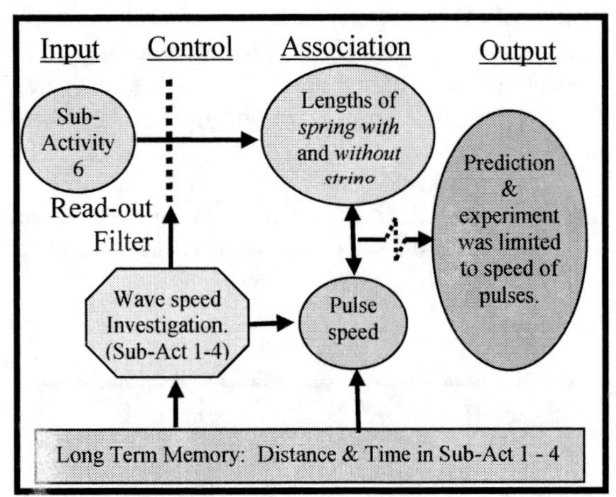

FIGURE 3. Influence of Sub-activities 1 through 4 on the interpretation of Sub-activity 6.

The external input was sub-activity 6 in the *Spring Activity*. A sub-activity is a section of the larger unit or activity that has a specific focus. For instance, in sub-activity 6, the students were asked to predict the effect on wave properties when having a spring with a string and a spring without a string as a pulse went through it. Prior to this sub-activity, in sub-activities 1 through 4, they had made successful investigations of the relationship between pulse speed and various factors influencing it. For the purpose of our analysis it is not necessary to describe in detail each sub-activity, rather we emphasize that in each of the sub-activities 1-4, the focus was on measuring the pulse speed of either a longitudinal or transversal disturbance. Students' recollections of these pulse speed investigations in sub-activities 1-4 became the controlling factor mediating the knowledge construction in sub-activity 6, which had nothing to do with the pulse speed investigations. Rather, the goal in sub-activity 6 was to investigate difference between waves traveling in two different media – a spring and a string. Although these goals are explicitly mentioned in the activity, students' recollections of the pulse speed investigation prompted them to focus on the length of the spring and string in sub-activity 6 and associate these lengths with the pulse speed. As a result of these associations, students' later investigations were limited to wave speed while they did not notice other relevant factors affecting wave properties such as change in the medium. This is an example of students being unable to accomplish the target idea of an activity because their experience with prior activities prompted them to focus only on or "filter out" certain aspects of the new activity – aspects that were not relevant to the goals of the new activity. This type of filtering can lead to activity disparity, if occurring uniformly across groups. If it occurs unevenly between groups it can lead to group disparity.

LIMITATIONS OF STUDY

As mentioned earlier, this case study is rather limited in scope. However, we do believe that it is valuable in that it uncovers some interesting phenomena that can influence the ways in which students work collaboratively using an activity-based curriculum based on constructivist pedagogy.

CONCLUSIONS

In addressing the research question raised at the beginning of the study we find that students' working together through the CPU unit can potentially lead to creation of disparities of ideas generated between activities and groups. These disparities often occur primarily due to the lack of adequate working equipment (e.g. the ripple tank) and guidance in using the equipment. Such disparities could potentially pose barriers to achieving the target ideas.

The activities do appear to foster students' intellectual development as students work collaboratively to extend each others' ideas leading to both productive and unproductive outcomes. We also find that perceptions of questions and instructions may be filtered out based on prior knowledge or previous activities, and can affect the outcome of the activity. These observations point to the fact that instructors need to facilitate student intellectual development through interactive strategies such as Socratic dialog.

REFERENCES

1. Goldberg, F., *Constructing Physics Understanding in a Computer-Supported Learning Environment*, in *The Changing Role of Physics Departments in Modern Universities: Proceedings of ICUPE*, E.F. Redish, Ridgen, J. S., Editor. 1997, AIP Publishing.
2. Vygotsky, L.S., *Mind in Society: The Development of Higher Psychological Processes*. 1978, Cambridge: Harvard University Press.
3. Trigwell, K., *Phenomenography: variation and discernment*, in *Improving student learning: Proceedings 1999 7th International Symposium*. 2000: Oxford, UK. p. 75-85.
4. Marton, F., *Phenomenography- a research approach to investigating different understanding of reality*. Journal of Thought, 1986. **21**: p. 29-39.
5. Rebello, N.S., et al., *Dynamic Transfer: A Perspective from Physics Education Research*, in *Transfer of Learning from a Modern Multidisciplinary Perspective*, J.P. Mestre, Editor. 2005, Information Age Publishing Inc.: Greenwich, CT.

Studying Transfer Of Scientific Reasoning Abilities

Eugenia Etkina, Anna Karelina, and Maria Ruibal Villasenor

Graduate School of Education, Rutgers University, New Brunswick, NJ 08904

Abstract. Students taking introductory physics courses not only need to learn the fundamental concepts and to solve simple problems but also need to learn to approach more complex problems and to reason like scientists. Hypothetico-deductive reasoning is considered one of the most important types of reasoning employed by scientists. *If-then* logic allows students to test hypotheses and reject those that are not supported by testing experiments. Can we teach students to reason hypothetico-deductively and to apply this reasoning to problems outside of physics? This study investigates the development and transfer from physics to real life of hypothetico-deductive reasoning abilities by students enrolled in an introductory physics course at a large state university The abilities include formulating hypotheses and making predictions concerning the outcomes of testing experiments. (The work was supported by NSF grant REC 0529065.)

Keywords: Transfer, hypothetico-deductive reasoning.
PACS: 01.40.Fk; 01.40.gb; 01.50.Qb.

INTRODUCTION

Introductory physics courses are expected to promote scientific literacy, critical thinking skills and help students develop abilities necessary for the 21st century workplace. In their future work our students will need to be able to pose their own questions, to design experiments to test hypotheses or to solve a problem, to collect and analyze real data, and to communicate the details of the experimental procedure [1, 2]. The goal of the study described in this paper is to investigate whether students who learn physics in a specially designed environment [3] that has one of its primary goals the development of scientific abilities [4] can transfer these abilities to other content areas besides physics. In this study we focus on the ability to engage in hypothetico-deductive reasoning, and in particular, on the abilities to formulate a testable hypothesis and to make a prediction of the outcome of the experiment based on the hypothesis.

THEORETICAL BACKGROUND

Hypothetico-deductive reasoning Hypothetico-deductive reasoning has been recognized as one of the reasoning approaches widely used in the practice of science [5]. This reasoning involves devising a hypothesis to explain some observational facts, designing an experiment whose outcome can be predicted using the hypothesis, using *if, then* logic to predict the outcome of the testing experiment based on the hypothesis, carrying out the experiment, comparing the outcome with the prediction, and rejecting the hypothesis if there is a mismatch of the outcome and the prediction.

Examples of hypothetico-deductive reasoning can be found in abundance in the history of physics. One of them is by Rutherford: "On consideration, I realized that this scattering backwards (*observational fact*) must be the result of a single collision... It was then that I had an idea of an atom with a minute massive center carrying charge (*hypothesis*). I worked mathematically what laws the scattering should obey, and I found that the number of particles should be proportional to the thickness of scattering foil, the square of the nuclear charge, and inversely proportional to the fourth power of velocity (*prediction*) These deductions were later verified by Geiger and Marsden in a series of beautiful experiments (*testing*) ("Development of theory of atomic structure", cited in Holton and Brush [6]).

One of the important aspects of hypothetico-deductive reasoning is the understanding of the difference between a hypothesis and a prediction. A hypothesis is some general explanatory statement and a prediction is a description of an outcome of a particular experiment that should occur if the hypothesis is true.

It turns out that the development of hypothetico-deductive reasoning is not only important for students' future work but for their present learning. Results reported by Coletta and Philips [7] suggest that

CP883, *2006 Physics Education Research Conference*, edited by L. McCullough, L. Hsu, and P. Heron

students who have a higher level of this reasoning have higher learning gains.

Transfer: Physics as an experimental science yields itself to the development of hypothetico-deductive reasoning naturally. However, as most of our students in the future are going to reason outside of content of physics, it is useful to know whether they can transfer this reasoning ability to a different content area. Transfer refers to the ability to apply knowledge, skills and representations to new contexts and problems [2, 8, 9 10]. Research shows that achieving transfer is difficult [11]. There are several theoretical models of transfer, and in the proposed study we will address the Direct Application model which considers "the ability to directly apply one's previous learning to a new setting or problem" [8]. Transfer can be near or far. When a situation in which a person needs to apply knowledge or skills is close to the situation in which she learned the knowledge or skills, the transfer is near. If the content or context is different, or the new task is given much later than the training task(s), the transfer is far [9].

To facilitate transfer, instructors may focus students' attention on pattern recognition among cases and induction of general schemas from a diversity of problems [12]. Another strategy is to engage students in meta-cognitive reflection on implemented strategies [13, 14].

ISLE LEARNING SYSTEM

The Investigative Science Learning Environment [3] models some of the processes that scientists use to construct knowledge. Students start each conceptual unit by analyzing patterns in experimental data and construct possible explanations or mathematical relationships. Then students test their constructed ideas by using them to predict the outcomes of new experiments, and possibly revising their ideas if the outcomes do not match the predictions. Hypothetico-deductive reasoning plays a central role in this process. In addition to practicing it in large room meetings (our word for lectures) where the instructor leads them through the steps of the reasoning, students engage in it in laboratories, where they often need to test hypotheses constructed in large room meetings or test hypotheses that are based on students' alternative ideas. Students work in groups to design their own experiments. Write-ups for *ISLE* labs do not contain instructions on how to perform the experiments but instead guide students through various aspects of a typical experimental process and reasoning and after the experiments is done ask students to reflect on the steps they took. Students' use self-assessment rubrics that scaffold their work on experiments and help them write lab reports [15]. There are several specific rubrics related to the ability to design experiments to test hypotheses and to make testable predictions. Two of them relevant to the study are shown in Table 1. The rubrics can also be used to score student work to find whether they acquire the desired abilities. Rubrics have been validated and trained raters achieve a high degree of consistency using them [4].

The *ISLE* approach combined with open-ended tasks supplemented with reflection and rubrics has many elements that have shown to promote transfer. Thus we hypothesize that students, who learn physics through *ISLE* and use scientific ability rubrics to self-assess their work, should not only acquire scientific abilities but also be able to transfer them. This hypothesis is based on the assumptions that students acquire the understanding of the abilities, can recognize patterns in laboratory tasks with the help of rubrics, abstract the abilities from the tasks, and map these patterns into new situations. To test whether students transfer the abilities described above we conducted the following study.

DESCRIPTION OF THE STUDY

The study was conducted in a large enrollment (190 students) introductory physics course for science majors. There were two 55-min lectures, one 80-min recitation and a 3-hour lab each week. Prior to the experiment, students conducted four labs and had a practical exam. In two of the labs and in the practical exam they had to test hypotheses using hypothetico-deductive reasoning. In large room meetings the instructor emphasized the importance of this reasoning and reflected on every instance when this reasoning was used.

For example, in week 2 of the first semester, students had to design an experiment to test a proposed hypothesis [in this case the hypothesis was incorrect]. The write-up for the experiment is shown below.

Design an experiment to test the following hypothesis: An object **always** moves in the direction of the unbalanced force exerted on it by other objects.

a) State what hypothesis you will test in your experiment.
b) Brainstorm the task. Make a list of possible experiments. Decide what experiments are best.
c) Draw a labeled sketch of your chosen experiment.
d) Write a brief description of your procedure.
e) Construct a free body diagram of the object.
f) List assumptions you make. How could they affect the outcome?
g) Make a *prediction* about the outcome of the experiment *based on the hypothesis you are testing.*

h) Perform the experiment. Record the outcome.
i) Make a judgment about the hypothesis based on and the experimental outcome.
j) Discuss in your group the difference between a hypothesis and a prediction.

During the first 55-min exam of the semester (week 5), one of the 12 questions that students had to answer was as follows: Write a paragraph describing when in your future work you might need to test an idea by using it to predict an outcome of a new experiment. First choose an idea; then use it to make a prediction of an outcome of a possible experiment, and then describe a possible outcome that will rule it out.

According to the classification of transfer by Barnet and Ceci [9], the transfer we were examining was far in terms of knowledge domain, physical context (exam hall instead of a physics lab), functional context (writing an answer to a question versus designing and performing an experiment), social context (individual versus group) and modality (exam versus a lab). To have some control groups we later posed the same questions for the beginning students in a PhD program (n=9) and advance graduate students in the Graduate School of Education (n=12) and undergraduate students in a third year biology course (n=8) (who had a physics course prior to this). None of these subjects were taking physics at that time. We had a group of 12 physics students but had to discount this sample because we were not sure whether the data were reliable.

Three graders scored student responses using the rubrics for two abilities: an ability to identify the idea to be tested and an ability to make a reasonable prediction based on the idea (see Table 1). The discrepancies in the scores were discussed and the scoring continued until we agreed on all of the scores.

A representative example of students' responses from the experimental group is shown below. We also provide annotations in Italics to help the reader see how we scored the responses.

Student A

When studying the populations of the Chesapeake Bay Blue Crab, I've noticed the total population over a span of 5 years has decreased (*observed fact*). I formulate an idea that the population is decreasing due to overcrabbing by local fishermen (*idea, or hypothesis*). I predict that by placing regulations on the fishermen to reduce the crab catch, the population of the Blue Crab will increase dramatically over next 3 years (*prediction*). Once the regulations have been in place, we've noticed that the population of the Blue Crab has drastically decreased even further (*outcome of testing experiment*). Due to this outcome, we've concluded that the decrease in Blue Crab populations is NOT due to overcrabbing by local fishermen (*judgment*).

FINDINGS

Students' responses in the experimental group showed that many of them could describe an idea to be tested but fewer could make a prediction of an outcome of the experiment based on the idea. The most common difficulty we found was that students used the word "testing" as a substitute for "trying". Responses in control groups varied a great deal. Graduate students in both subgroups could formulate an idea to test but had a difficulty coming up with predictions of the outcomes of testing experiments. Most of the undergraduate students could not identify an idea and could not make predictions. Results of the scoring are shown in Figure 1.

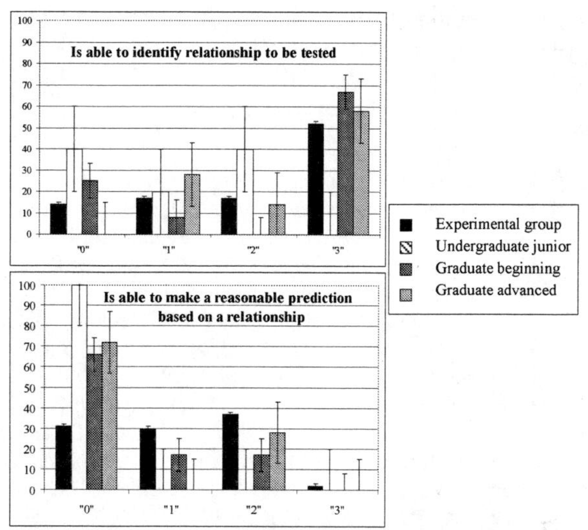

FIGURE 1. Percentage of students in a group who received rubrics scores of 0, 1, 2, and 3. The scores were based on two rubrics: an ability to identify an idea to be tested and an ability to make a prediction based on the idea under test.

DISCUSSION

We found that students in the experimental group could successfully identify an idea to be tested and a relatively high percentage of them could make a prediction based on the idea under test. The latter turned out to be much more difficult for traditionally taught undergraduate students and for traditionally taught graduate students. Could this result be due to an unfamiliar vernacular or due to the nature of the question itself? One of the subjects in a group of physics students at the end of his physics course (we did not include this group in the study due to a low reliability of data collection) wrote: "Why are you asking me this question? I have never seen a question like this before". Another said: "Did we discuss questions like this in class? I do not remember anything like this". These responses are quite alarming. If we agree that hypothetico-deductive reasoning is important for the practice of science and that most of our students in introductory courses will be related to some aspect of science in the future (even only as consumers and citizens), then we should place more emphasis on helping students develop this reasoning and transfer it to areas out of physics. Examples of tasks that develop this reasoning in an introductory physics courses can be found in [16]. We speculate that the reasons for the success were repetitive attention to these aspects of learning in large room meetings and labs, meta-cognitive processes encouraged by the use of self-assessment rubrics and followed up in lectures, and the fact that students had to struggle in labs to design experiments. We plan to have a control study to find if these results are due to the maturation of the subjects or if we observed some real transfer.

REFERENCES

1. National Science Foundation, "*Shaping the Future: New Expectations for Undergraduate Education in Science, Mathematics, Engineering, and Technology,*" NSF, Directorate for Education and Human Resources Review of Undergraduate Education, May 15, 1996.
2. J. D. Bransford, A.L. Brown, and R.R. Cocking, "*How People Learn: Brain, Mind, Experience, and School,*" (Washington, DC: National Academy Press, 1994).
3. E. Etkina, and A. Van Heuvelen, "Investigative Science Learning Environment: Using the Processes of Science and Cognitive Strategies to Learn physics", *Proceedings of Physics Education Research Conference*, (Rochester, NY: PERC publishing, 2001, pp.17-21).
4. E. Etkina, A. Van Heuvelen, S. White-Brahmia, D. T. Brookes, M. Gentile, S. Murthy, D. Rosengrant, and A. Warren, "Scientific Abilities and Their Assessment", Phys. Rev. ST Phys. Educ. Res. **2**, 020103 (2006).
5. A. Lawson, "What does Galileo's discovery of Jupiter's Moons tell us about the process of scientific discover", Science & Education, **11**, 1 (2002).
6. G. Holton, & S. Brush, Physics, the Human Adventure. (New Brunswick, NJ: Rutgers University Press, 2001).
7. V. P. Coletta and J. A. Phillips, "Interpreting FCI Scores: Normalized Gain, Preinstruction Scores, and Scientific Reasoning Ability", AJP, **73**, 1172 (2005).
8. J. D. Bransford, and D. T. Schwartz, "Rethinking Transfer: A Simple Proposal With Multiple Implications," in *Review of Research in Education* edited by A. Iran-Nejad and P.D. Pearson, (Washington DC: AERA, **24**, 61-100, 1999).
9. S. M. Barnett and S. J. Ceci, "When and Where Do We Apply What We learn?: A taxonomy for far transfer," Psychological Bulletin, **128**, 612 (2002).
10. J. P. Mestre, "Probing Adults' Conceptual Understanding and Transfer of Learning Via Problem Posing," Journal of Applied Developmental Psychology, **23**, 9 (2002).
11. M. L. Gick and K. J. Holyoak, "Schema Induction and Analogical Transfer," Cognitive Psychology, **15**, (1983).
12. D. Gentner, J. Loewenstein, and L. Thompson, "Learning and Transfer: A General Role for Analogical Encoding," Journal of Educational Psychology, **95**, 393, (2003)
13. G. Salomon, & D. N. Perkins, "Rocky Road to Transfer: Rethinking Mechanisms of a Neglected Phenomenon," Educational Psychologist, **24**, 113, (1989).
14. R. Catrambone, and K. L. Holyoak, "Overcoming Contextual Limitations on Problem Solving Transfer," Journal of Experimental Psychology: Learning, Memory, and Cognition, **15**, 145, (1989).
15. Rubrics can be found at http://paer.rutgers.edu/scientificabilities

TABLE 1. Rubric descriptors for two scientific abilities.

Scientific Ability	0 - Missing	1 - Inadequate	2 - Needs some improvement	3 - Adequate
Is able to identify the idea (or the hypothesis) to be tested.	No mention is made of an idea.	An attempt is made to identify the idea but is described in a confusing manner.	The idea to be tested is described but there are minor omissions or vague details.	The idea is clearly stated.
Is able to make a prediction based on the idea (or the hypothesis).	No attempt to make a prediction is made.	A prediction is made but it doesn't follow from idea being tested.	A prediction follows from the idea but does not contain assumptions.	A prediction is made that follows from the idea and incorporates the assumptions.

Resource Plasticity: Detailing a Common Chain of Reasoning with Damped Harmonic Motion

Eleanor C Sayre[1], Michael C Wittmann[1,3], and John E Donovan II[2,3]

[1]*Department of Physics and Astronomy, University of Maine, 5709 Bennett Hall, Orono, ME*
[2] *Department of Mathematics and Statistics, University of Maine*
[3]*College of Education and Human Development, University of Maine*

Abstract. As part of ongoing research into cognitive processes and student thought, we have investigated the interplay between mathematics and physics resources in intermediate mechanics students. We present evidence from a reformed sophomore-level mechanics class which contains both tutorial and lecture components. In the context of writing Newton's Second Law for damped harmonic motion, students discuss the signs of the spring and damping forces. Using a grounded theory approach, we identify a common chain of reasoning in which a request for reasoning is followed by elaborative sense-making and checks for consistency, finishing with an optional appeal for group consensus. Our analysis provides evidence for a description of student thinking in terms of Plasticity, an extension of Resource Theory.

Keywords: cognitive modeling, damped harmonic motion, advanced students
PACS: 01.30.lb, 01.40.Fk, 01.40.Ha

INTRODUCTION

Intermediate Mechanics is a particularly rich course in which to study the interaction between students' ideas about mathematics and their ideas about physics: students often enter with a good intuitive grasp of the physics, but have not yet applied sophisticated mathematics. At the University of Maine, intermediate mechanics is a one-semester course with two lectures and one tutorial[1] each week. About half of the students are concurrently enrolled in Differential Equations; the other half have already taken it.

We follow a three-pronged approach to studying students' ideas about mathematics and physics: data collection of students using mathematics and physics, which drives cognitive model building, which informs curriculum development.

Our primary data comes from three sources. Video data of short interviews track small groups of students through the semester. Video data and field notes of homework help sessions (HHS) examine groups of students as they work through their homework with the help of a TA. It is important to note that the role of the TA is to facilitate students' understanding of the topic, and not merely explore their existing ideas, in contrast to the role of the researcher in the interviews. We supplement our video and field note data with exams and homeworks.

We use a grounded theory approach to model student ideas about specific topics (such as coordinate systems, air resistance, and Lagrangians). These models drive more general theoretical work expanding Resource Theory[2-5] and connecting it to other theories in Mathematics Education Research (such as the Theory of Reification[6] and the RBC model for abstraction[7]). This theoretical work informs curriculum development.

Following the general outline for the Project, this paper will motivate and explain a general theoretical development with a specific example taken from damped harmonic motion.

THEORY DEVELOPMENT

Resource Theory is a constructivist schema theory which bridges neuro-cognitive models of the brain and results from education research to describe the phenomenology of problem solving.[4] Resources are small, reusable pieces[3] of thought that make up concepts and arguments.[2] Resources have internal linked structure[8] made up of other resources. To be considered a resource, an idea must have sufficient duration and stability to be reused. Resources can be epistemological, metacognitive, or content-oriented.[9]

CP883, *2006 Physics Education Research Conference*, edited by L. McCullough, L. Hsu, and P. Heron

The physical context and cognitive state of the user determine which resources are available to be *activated* The activation of resources occurs when their invocation, express or implicit, is used to support or form an argument. Once activated, they form a network with graph-like structure.[8] If their internal structure is explorable (but currently not explored) by the user, they may be called concepts. If their internal structure is no longer explorable, they may be called primitives. A large body of literature has identified both concepts and some kinds of primitives.[4]

"Plasticity" is a continuum which extends Resource Theory to describe the development of resources and connect their use to observables. The two directions in the continuum are *more solid* and *more plastic*. A solid resource can be considered a durable concept: its connections to other resources are plentiful, and its internal structure is unlikely to change under typical use. Existing literature details solid resources and their properties well.[2,4,5,8,9] Plastic resources, in contrast, are less durable in time or less stable in structure. While a solid resource may remain unchanged for years, a plastic one may only last "until the exam." Consider the path that you take to drive home from work. To you, this path is quite solid, and you find yourself getting home without much thought. In contrast, your colleague who visits you for the first time must explicitly look for landmarks, check the route against your directions, and figure out how to get to your house. For your colleague, getting to your house is plastic.

The more plastic a resource is, the less likely the user is able to apply it to new situations. More explanation is needed to justify and explain its use. The more solid a resource is, the more likely the user is to refer to the resource in diverse contexts without explaining its internal structure. The plasticity of a resource is independent of its veracity.

We use the RBC model[7] to inform and improve the plasticity continuum, noting that it was originally intended to describe the reification and abstraction of concepts. The RBC model proposes three epistemic actions through which abstraction occurs and which may be inferred from behavior. These three actions - *recognizing*, *building-with*, and *constructing* - are dynamically nested. Recognizing, the simplest action of the three, occurs when a student realizes that a "familiar mathematical notion, process, or idea ... is inherent in a given mathematical situation."[10] These recognized cognitive objects are the resources in Resource Theory. Recognition is thus synonymous with activation. Content resources such as these need not be restricted to mathematics; physics is another appropriate subject area. The specifics of which resources are recognized gives insight into students' thought structure. Ease of recognition is therefore a marker of solidity.

Once a familiar idea has been recognized, a student may build-with that idea to solve a local goal, such as solving a problem or justifying a statement. Several resources may need to be recognized and built-with at once. Under Resource Theory, activated resources form a web or graph[8] that may be built on the fly. Thus, recognizing is akin to activation, and building-with is akin to building graphs on the fly. Because building-with and recognizing are two separate actions, the RBC model can describe behavior when students mention an idea, but don't appear to know what to do with it.

In contrast to building-with, constructing has purpose and duration beyond solving a local goal. Constructing creates a less-local, more abstract entity. As a construction becomes more durable, it becomes more consolidated and is no longer necessarily built on the fly. It becomes a resource in its own right, and therefore can be recognized or built-with in later local goals. The new-formed resource may be quite plastic, but as further constructions are added to it and as it compiles further, it can become more solid. Thus constructing is a mechanism for increasing the solidity of specific resources. Extremely solid resources – *rigid* resources -- have been so tightly compiled that their internal structure is not readily accessible to the user.

The process of abstracting and consolidating resources can be of long duration and difficult to show in a brief paper. Instead, we focus on an example of recognizing and building-with which shows different levels of plasticity in one encounter.

EXAMPLE

In this example, taken from a Homework Help Session, a student ("Bill") works through the sign of a velocity-dependent drag force in damped harmonic motion (Figure 1). Immediately prior, Bill asks to go over why the drag and spring forces are negative in the statement of Newton's Second Law:

$$\sum \vec{F} = \sum k\vec{x}\ \sum c\vec{v} \tag{1}$$

On his first attempt, he is unable to explain why the forces are negative, and he confuses himself while trying to figure it out. He starts over, explaining that

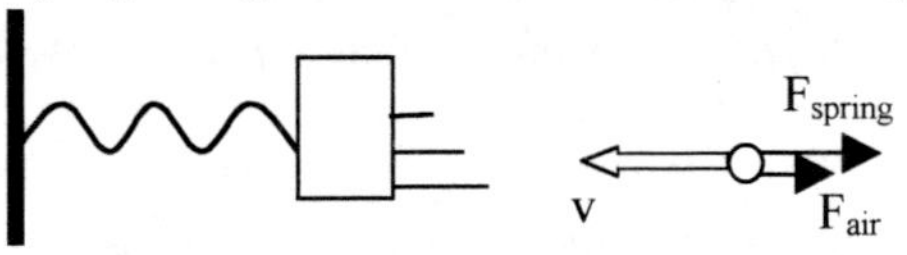

FIGURE 1. An object undergoing damped harmonic motion. The object is to the left of equilibrium, and moves leftwards.

the drag force opposes velocity and the spring force opposes the displacement, but he is unable to connect his description to the algebraic signs of the forces. While correct, his explanation confuses him again. To refocus discussion, the TA asks about –cv first, holding –kx for later. On his third attempt, Bill's explanation is longer, and he considers leftwards movement and rightwards movement separately. He references an implied coordinate system in which the positive direction is to the right. Bill's explanation takes 37 seconds:

1 TA: So if we think about the cv term, why is
2 it minus cv?
3 Bill: cv... (writes –cv) cv... um... before...
4 (draws equilibrium line and base) this velocity
5 is going to be this way (draws arrow pointing
6 right) and the air resistance is going to be that
7 way (draws arrow pointing left). Alright, so
8 the velocity is positive and the force is
9 negative and when the velocity becomes
10 negative (draws arrow pointing left) the force
11 is positive (draws arrow pointing right) so it
12 changes the sign around. Of this (points at
13 –cv). In here. Alright? Is that right? If this
14 is the....

Bill's third attempt is correct, complete, and does not confuse him. After figuring out the sign of the air resistance force, discussion moves to reversing the implied coordinate system and figuring out the sign of the spring force. When finding the sign of the spring force, Bill employs a similar argument, but he uses it much more readily and facilely. For a minute analysis of the argument and its implications about the plasticity of two resources, *force sign* and *coordinate systems*, we examine Bill's successful reasoning about the air resistance force.

Common Chain

In this chain of reasoning, a request for reasoning is followed by elaborative sense-making and checks for consistency. It finishes with an optional appeal for group consensus.

Bill uses this chain five times in a productive six-minute episode, explicitly asking for agreement four times, and explicitly receiving it three times. Other students in other clips also use the chain to formulate and explain their reasoning.

In this example of the chain, we learn about the plasticity of two resources, *force sign* and *coordinate systems*.

Request for reasoning

The chain starts when one participant (here, the TA) asks another (here, Bill) about a previous statement, focusing the discussion (ll 1-2). The chain does not necessarily start with a refocusing question; it can also start with a request for a definition or a statement that an existing definition is incomplete.

Sense-making: elaboration

Bill responds by elaborating on his previous statement, making sense of the physics as he goes (ll 3-13). His hesitancy at the start (line 3) (which is more apparent in the video than the transcript), together with his earlier difficulties with these questions, signify that he is making sense of the situation as he goes, rather than repeating a pat explanation. Bill is performing a build-with action. His sense-making indicates that his ideas about the signs of these forces are not solid – they are neither predetermined nor readily available. Furthermore, the detail of his description indicates that his *force sign* resource is plastic.

Sense-making: consistency check

Bill nests a consistency check within his elaboration (ll 7-12). By so doing, he explicitly tests if these newer ideas about the sign of the air resistance force are consistent with differently articulated previous work involving coordinate systems and directionality. Because Bill is facile enough with these ideas to test their consistency, they are sufficiently compiled to be resources, not mere fluid ideas. However, because he needs to test explicitly, *force sign* is plastic to him.

Justification through activity

Bill further justifies his response through reference to an activity: choosing and using a coordinate system (ll 8-12). His tacit use of a coordinate system as justification implies that Bill's *coordinate systems* resource is more solid than his *force sign* one. The implied nature of his coordinate system is further evidence of its solidity.

Social norm: agreement

Bill finishes his chain with an explicit social call for agreement (ll 13-14), signaling that he sees his reasoning as sufficient. Because his call is explicit, he does not see his reasoning as inherently self-obvious, further implying that his *force sign* resource is plastic.

As previously noted, the chain does not always conclude with an explicit verbal call for agreement. Sometimes, the other participants voice their assent without the call. At other times, the chain simply ends without verbal agreement. From the video data we have, which show a top-down view of the work surface and the participants hands, but not their faces, we cannot tell if non-verbal consensus is reached. Had we frontal views, we could capture more gestures and perhaps evidence of non-verbal consensus.

Discussion

The scope of this mechanics problem – the sign of the drag force in damped harmonic motion – is extremely small. However, it is often difficult for sophomore level physics majors and therefore strongly emphasized in the curriculum and homework. This interaction starts because Bill recognizes both his own confusion and the problem's importance to the course. Perhaps in other settings, different contexts would lead to different resources' activation, and this chain may not be employed. For example, had Bill activated his *knowledge-from-authority* epistemological resources instead of his *knowledge-as-invented-stuff* resources,[9] he might have responded to the TA's request for reasoning very differently. He might have responded that the damping term is negative because the professor said so. Such a response precludes our observation of the extensive sense-making used in the chain.

The scope of this interaction – 37 seconds within a six-minute episode – is also quite small. In that short time frame, we see evidence of some resources' plasticity, but we cannot expect to see their level of plasticity change. In the longer episode, Bill reuses the chain to reason about the sign of the spring force, -—kx, and to reverse his tacit coordinate system, making leftwards the positive direction. Each time, his reasoning is shorter and less verbally detailed, indicating solidification. Semester-long tracking of some resources may uncover more evidence of plasticity changing.

We have chosen to use plasticity and the RBC model to describe this interaction. Another description can be made using epistemic games, in which the chain of reasoning is identified as a specific game. However, the RBC model combined with plasiticity provides a richness of description about the state of the resources involved as well as convienient language for their development during an interaction, a class period, or a semester-long course.

CONCLUSIONS

In nearly-novel sitations[11] such as this one, students can reason successfully with the tools available to them given enough opportunity. These resources are used with varying levels of assuredness and detail, indicating different levels of plasticity.

As teachers, we are interested in helping students solidify appropriate resources. As researchers, we are interested in detailing the interactions between resources with differing plasticity to give us better insight into the working of student minds. In both roles, connecting student behaviors to differing levels of plasticity allows us to hone our observational skills and build better models of our students and their learning, which may lead to better curricular design.

REFERENCES

1. B.S. Ambrose, *Am.J.Phys.* **72**, 453 (2004).
2. D. Hammer, *Am.J.Phys* **67** (Physics Education Research Supplement), S45 (2000).
3. A. A. diSessa, in *Constructivism in the computer age.*, edited by G. Forman, P. B. Pufall and et al. (Lawrence Erlbaum Associates, Inc, Hillsdale, NJ, USA, 1988), p. 49.
4. A. A. diSessa, *Cog.Instr.* **10**, 105 (1993).
5. M.Sabella and E.F.Redish, unpublished.
6. A Sfard, *Educational Studies in Mathematics* **22**, 1 (1991)
7. P. Tsamir and T. Dreyfus, *Journal of Mathematical Behavior* **113**, 1 (2002).
8. M. C. Wittmann, *Physical Review Special Topics - Physics Education Research* **2,** 020105 (2006).
9. D. Hammer and A. Elby, *Journal of the Learning Sciences* **12**, 53 (2003).
10. T. Dreyfus and P. Tsamir, *Journal of Mathematical Behavior* **115**, 3 (2004).
11. E. C. Sayre, M. C. Wittmann, and J. R. Thompson, in *Physics Education Research Conference Proceedings 2003*, edited by K. C. Cummings, S. Franklin and J. Marx (Springer New York, LLC, Secaucus, NJ, 2003).

Depictive Gestures as Evidence for Dynamic Mental Imagery in Four Types of Student Reasoning

A. Lynn Stephens and John J. Clement

School of Education and Scientific Reasoning Research Institute
University of Massachusetts, Amherst, MA 01003-9308

Abstract. We discuss evidence for the use of runnable imagery (imagistic simulation) in four types of student reasoning. In an in-depth case study of a high school physics class, we identified multiple instances of students running mental models, using analogies, using extreme cases, and using Gedanken experiments. Previous case studies of expert scientists have indicated that these processes can be central during scientific model construction; here we discuss their spontaneous use by students. We also discuss their association with spontaneous, depictive gestures, which we interpret as an indicator of the use of dynamic and kinesthetic imagery. Of the numerous instances of these forms of reasoning observed in the class, most were associated with depictive gestures and over half with gestures that depicted motion or force. This evidence suggests that runnable, dynamic mental imagery can be very important in student reasoning.

Keywords: Models; modeling; analogies; imagery; pedagogy; imagistic simulation.
PACS: 01.40.ek; 01.40.Fk; 01.40.Ha

INTRODUCTION

Previous case studies of expert scientists have indicated that running mental models, using extreme cases, using analogies, and using Gedanken experiments can be central reasoning processes in scientific model construction [1-3]. It has been argued that the ability to generate and evaluate mental models is a crucial aspect of science [4, 5], and that students need to be helped to assimilate prior experience into accepted models [6]. Research continues to indicate the importance of mental modeling in experts and students [7, 8]. We believe there is much more that can be learned about the variety of reasoning processes students employ in the classroom to evaluate mental models.

METHODOLOGY

In the present paper, we report on a transcript of classroom activity where inquiry-based methods of teaching and learning were employed. As part of a larger project, we have already coded a number of transcripts of classroom activity for imagery indicators [9, 10]. We coded additional transcripts for the presence of depictive hand motions, only. These hand motions appear to depict an image in the air, and are taken as one indication that mental imagery is being used [11]. Here, we analyze one particular class for the presence of the four expert reasoning processes and for their co-occurrence with depictive hand motions, which will be the main form of imagery indicator considered in this paper. We are still using the coding process as a way to refine definitions of categories for these processes, so in all cases, coding for processes was done jointly by the two authors and disputes were used as a mechanism for refining and clarifying the coding criteria. The four thinking processes—running an explanatory model, using extreme cases, using analogies, and using evaluative Gedanken experiments—as defined here, are not intended as mutually exclusive categories. In some circumstances, more than one can apply to the same case.

We cannot present the full case study here, but merely attempt to illustrate a methodology for identifying types of student reasoning and examining whether they depend on mental simulations run via the use of dynamic imagery (*imagistic simulations*). We will define each process, describe an example of each, and discuss our coding for gesture in each example. We will also review the results of coding the whole transcript, which, in this case, showed a strong correlation between the four thought processes and depictive gestures.

In the examples below, the gestures are coded as Shape Indicating [G-S], Movement Indicating [G-M], or Force Indicating [G-F], according to the images they appeared to depict. We refer to the Movement

CP883, *2006 Physics Education Research Conference*, edited by L. McCullough, L. Hsu, and P. Heron

Indicating and Force Indicating gestures collectively as *Action* gestures. Students' use of movement or force terms in conjunction with their gestures was considered added evidence that could strengthen our choice of coding, and we note such terms in boldface. (See the list of imagery indicators in [11].)

EXAMPLES OF FOUR EXPERT REASONING PROCESSES

This case study is of a lengthy discussion that occurred in a college preparatory physics class that was using an innovative curriculum [12]. The class was in a middle-class suburban high school in the northeastern United States. The discussion, which was videotaped, lasted about 45 minutes.

The teacher wanted students to consider whether a table exerts an upward force on objects resting on its surface. A common conception prior to instruction is that inanimate objects cannot exert upward forces against gravity. The target model for the lesson was one in which objects exert normal forces that are equal and opposite to the weight of objects resting on them. The teacher began by introducing an analogy. He placed a book on his desk and called students' attention to it, then drew two figures on the chalkboard. One was a simple line drawing of a book on a table (the target), and another of a hand pressing downward on a spring (the base). He hoped that all of the students would believe that the spring pushed up on the hand and that he could use this as an anchoring case for the lesson. It became clear that, although many of the students did believe that the spring would exert a force on the hand, a large number did not believe that the table was exerting an upward force on the book. The teacher intended to introduce a number of *bridging analogies* [13], designed to bridge the distance between the spring/hand case and the table/book case. However, these students preempted him, producing their own bridging cases and reasoning about them.

A minute or so into the discussion, S4 agreed that a spring could exert a force back against a hand, but argued that the table is not equivalent:

S4: . . . and if you kind of release the spring, it would force your hand up, where the table's not going to push the book up.

T: I see . . . there's no way we could use the table to uh, send things flying off into outer space now, the way a spring might. Hmm.

The four examples that follow occasionally refer back to this exchange, although they all occurred somewhat later in the discussion.

Running an Explanatory Model

An explanatory model is defined here as a mental model of a system that projects some initially hidden feature into the system and that offers an explanation for why it behaves the way it does.

S13: If you put something heavy on the table and it collapsed, that is because the table is not exerting enough force. If you put something on the table that was like, really [G-F] heavy, and the table [G-M] **collapsed**, I would argue that that was because the table is not exerting enough force on whatever is on top of it.

Saying "if you put something heavy on a table it will collapse" can be interpreted as running a mental model of a table, although this model is not yet explanatory. The model predicts that the table will collapse under certain conditions, those conditions being when it is subjected to a weight great enough. However, to go on to reason that the *cause* of this behavior is that the table does not exert enough force to hold up the weight is to project the initially hidden feature of forces into the table, producing an *explanatory model* of this system.

The gestures suggest the presence of mental imagery that has both visual and kinesthetic components. Even though the student did not specify what was being put on the table, he appeared to be reacting to the weight of something "really heavy." He also appeared to depict the motion of the table as it collapsed.

Extreme Cases

We say that an extreme case has been run when a subject, in thinking about a target situation A, shifts, without being prompted, to consider situation E (the extreme case) where some variable from situation A has been maximized or minimized.

Early in the discussion, when arguing that the table *does* exert a force, S15 had proposed that the table warps slightly and pushes back against the object. Now, over half an hour later, S5 returned to that point to argue that the table does not have enough power to "exceed" the weight of an object to move it in the other direction, "and as soon as (*the weight*) gets too great then the table collapses." S15 then recast this statement as an extreme case of warping in order to argue *for* the normal force:

S15: (*S5's*) idea is compatible with the warped table theory. The idea is that the [G-S] elephant sitting on the table is too much [G-S] for the material that the table is made out of, and it [G-F] **punctures** the thing; it [G-S] warps it too much.

The shape depicted by the final gesture was a deep upward curve, much deeper than a table could normally form without breaking. By pushing the warped table to an extreme, the student had transformed the first case into the second. Numerous gestures give indication of the presence of visual imagery, and the force gesture accompanied by the force term *punctures* suggests the presence of kinesthetic imagery.

Analogies

For our purposes, analogies occur when a subject, in thinking about a situation A (the target), shifts, without being prompted, to consider a situation B (the base) which differs in some significant way from A, and intends to apply findings from B to A.

About a half hour into the discussion, a student made an analogy between two cancellations: the cancellation of *velocities* between a powerboat and a current (which would cause the boat to remain still with respect to the shore) and the cancellation of *forces* between gravity and the normal force (which would cause the book to remain at rest on the table). The student reasoned about what would happen to the boat if the current were taken away, and by analogy, what would happen to the book if the force of the table were taken away. (Although the analogy would have been stronger had the student said—or meant—*forces* from the current and the boat's engine, taking his words at face value still indicates the use of an analogy; a vector representation of the cancellation of velocities and the cancellation of forces is visually equivalent.) In the student's words:

> S14: The book is pushing down, say with, um, it's a little far fetched, but (*pause*) with the velocity of the engine, it's **pushing** [G-F] down. And the table's pushing up with the velocity of the current. If you take the current away, then the engine (*unintelligible*), if you take the [G-F] **force** of the table away, then the book would just fall [G-M] down.

The use of the force term *pushing* reinforces the impression that the student was indicating a force with his gesture even though he had just referred to the "velocity of the engine." The student's gestures created a stronger parallel between the two situations than did his words, as he reasoned about a scenario that was proving confusing to many of his classmates.

Evaluative Gedanken Experiments

An evaluative Gedanken experiment is defined here as the act of considering an untested, observable system designed to help evaluate a scientific concept, model, or theory—and attempting to predict aspects of its behavior. In these experiments, an element of a theory is tested as it is applied to the untested system. These experiments, as discussed in [2], can be quite complex, come in many varieties, and can incorporate the other three types of reasoning.

In this example, which occurred immediately after the analogy example above, S15 used the same analogy. Rather than taking the current (normal force) away, however, he imagined taking the engine (force of gravity) away, and predicted what would happen to the boat (book):

> S15: But by the same analogy, then, if gravity disappeared, right, the **force** of the [G-F, *sudden thrust downward*] engine on the book, even the book would just [G-M, *flings arms upward and outward*] **fly** off into space.

If the engine disappeared, the current would move the boat, and by analogy, if gravity disappeared, the normal force would send the book off into space. (The table would suddenly unwarp.) The case of gravity disappearing is an untested system and the student attempted to predict an aspect of its behavior—what would happen to a book on a table in such a situation. The case appears to have been constructed to evaluate an aspect of the theory of normal forces.

For this example, we have included descriptions of the gestures, as we have not, for reasons of space, done elsewhere. The descriptions are intended to convey to the reader the energetic quality of these gestures, which we take to be indications of the student's use of imagery that had kinesthetic, as well as visual, components.

FINDINGS

In the 45 minutes of transcript, we coded (with each category including the one below it):

- 22 instances of the four expert reasoning processes, where
- 17 of the 22 instances were paired with depictive gestures;
- 12 of these 17 instances involved depictive gestures that were action gestures (indicating force or motion), and
- 5 of these 12 instances involved, specifically, force-indicating gestures.

In addition to the 17 instances where expert reasoning processes were paired with depictive gestures, 7 other utterances co-occurred with depictive gestures, leading to a total of 24 utterances co-occurring with depictive gestures, as shown in Table 1. Many of these utterances were accompanied by multiple gestures; we coded 53 individual depictive gestures made by the students.

TABLE 1. Utterances Co-Occurring with Depictive Gestures

Expert reasoning processes paired with depictive gestures	17
Other utterances paired with depictive gestures	7
Total utterances paired with depictive gestures	24

An example of the coding is given in Table 2. The transcript line numbers hint at the density of these reasoning processes in this classroom discussion. Note the occurrence of action gestures with most, if not all, of these incidents. (The partially obscured gesture was not included in the totals.)

In summary, most instances of these reasoning processes were associated with depictive gestures and over half with gestures that depicted motion or force. This case study provides evidence indicating that runnable mental imagery (imagistic simulation) can be closely associated with, and may be centrally important in, four types of student reasoning with mental models. This supports the importance of fostering the development of effective, runnable mental models as a goal in science pedagogy and underlines the possible value of gestures as a window onto student imagery. Our long-range hypothesis is that runnable mental models engender a form of sense making in science that contrasts sharply with a knowledge of memorized facts or rules.

TABLE 2. Coding Summary for One Section of Transcript

	Line 98	*Line 105*	*Line 112*	*Line 114*	*Line 122*	*Line 125*
Reasoning Processes	Gedanken Exp. Extreme Case	Running Model	Analogy	Gedanken Exp. Running Model	Running Model	Extreme Case
Gestures	[G-S] 2 [G-M] 4	[G-M] 1? (ptly obscured)	[G-M] 1 [G-F] 2	[G-M] 1 [G-F] 1	[G-M] 1 [G-F] 1	[G-M] 4

ACKNOWLEDGMENTS

This material is based upon work supported by the National Science Foundation under Grant REC-0231808, John J. Clement, PI. Any opinions, findings, and conclusions or recommendations expressed in this paper are those of the author(s) and do not necessarily reflect the views of the National Science Foundation.

REFERENCES

1. J. J. Clement, Imagistic Simulation in Scientific Model Construction, in *Proceedings of the Twenty-Fifth Annual Conference of the Cognitive Science Society, 25*, edited by R. Alterman and D. Kirsh, Mahwah, NJ: Erlbaum, 2003, pp. 258-263.
2. J. J. Clement, "Thought Experiments and Imagery in Expert Protocols" in *Model-Based Reasoning in Science and Engineering*, edited by L. Magnani, London: King's College Publications, 2006.
3. D. L. Craig, N. Nersessian, and R. Catrambone, "Perceptual Simulation in Analogical Problem Solving" in *Model-Based Reasoning: Science, Technology, Values*, edited by L. Magnani and N. Nersessian, New York, NY: Kluwer Academic Publishers, 2002, pp. 167-189.
4. L. Darden, *Theory Change in Science: Strategies from Mendelian Genetics*, New York: Oxford, 1991.
5. R. N. Giere, *Explaining Science: A Cognitive Approach*, Chicago: Chicago University Press, 1988.
6. R. Driver, "The Fallacy of Induction in Science Teaching" in *The Pupil as Scientist?* edited by Milton Keynes, Open University Press, 1983.
7. D. Gentner, "Psychology of Mental Models" in *International Encyclopedia of the Social and Behavioral Sciences,* edited by N. J. Smelser and P. B. Bates, Amsterdam: Elsevier Science, pp. 9683-9687, 2002.
8. J. J. Clement and M. S. Steinberg, "Step-Wise Evolution of Mental Models of Electric Circuits: A 'Learning-Aloud' Case Study," *The Journal of the Learning Sciences, 11*(4), 389-452 (2002).
9. J. J. Clement, A. Zietsman, J. Monaghan, "Imagery in Science Learning in Students and Experts" in *Visualization in Science Education*, edited by J. Gilbert, New York: Springer, 2005, pp. 169-184.
10. A. L. Stephens and J. J. Clement, "Designing Classroom Thought Experiments: What We Can Learn From Imagery Indicators and Expert Protocols," *Proceedings of the NARST 2006 Annual Meeting*, San Francisco, CA, 2006, #211184.
11. J. J. Clement, "Use of Physical Intuition and Imagistic Simulation in Expert Problem Solving" in *Implicit and Explicit Knowledge: An Educational Approach*, edited by D. Tirosh, Ablex Publishing Corp, 1994.
12. C. Camp, J. J. Clement, D. Brown, K. Gonzalez, J. Kuduke, J. Minstrell, K. Schultz, M. Steinberg, V. Veneman, and A. Zietsman, *Preconceptions in Mechanics: Lessons Dealing With Conceptual Difficulties*, Dubuque, Iowa: Kendall Hunt, 1994.
13. J. J. Clement, "Using Bridging Analogies and Anchoring Intuitions to Deal with Students' Preconceptions in Physics," *Journal of Research in Science Teaching, 30* (10), 1241-1257 (1993).

When And How Do Students Engage In Sense-Making In A Physics Lab?

Anna Karelina, and Eugenia Etkina

Graduate School of Education, Rutgers University, New Brunswick, NJ 08904

Abstract. The Rutgers PAER group developed and implemented *ISLE* labs in which students design their own experiments being guided by self-assessment rubrics. Studies reported in 2004 and 2005 PERC proceedings showed that students in these labs acquire such scientific abilities as an ability to design an experiment, to analyze data, and to communicate. These studies concentrated mostly on analyzing students' writings evaluated by specially designed scientific abilities rubrics. The new question is whether the ISLE labs make students not only write like scientists but also engage in discussions and act like scientists: plan an experiment, validate assumptions, evaluate results, and revise the experiment if necessary. Another important question is whether these activities require a lot of cognitive and metacognitive efforts or are carried out superficially. To answer these questions we monitored students' activity during labs. (The work was supported by the NSF grants DUE 0241078 and REC 0529065.)

Keywords: Design labs, sense-making.
PACS: 01.40.Fk; 01.40.gb; 01.50.Qb.

INTRODUCTION

Investigative Science Learning Environment (*ISLE*) is a physics learning system which focuses on helping students acquire abilities used in the practice of science [1]. These abilities include an ability to design an experiment to investigate a phenomenon or to test a hypothesis, to collect and analyze data, to evaluate the effects of assumptions and uncertainties, and many others [2]. Students construct and test physics concepts by following a scientific investigation cycle in large room meetings and design their own experiments in labs [2]. Do they develop scientific abilities in such an environment? Previous studies that answered this question positively, used, as sources of data, students' writing in labs and on exams assessed by specially designed scientific abilities rubrics [3]. Do *ISLE* labs make students not only write like scientists but also engage in discussions and act like scientists? To answer this question, we monitored students' activity during labs.

Earlier, educational researchers monitored students' activity in a classroom and in laboratories to study metacognition [4-8]. Lippmann et al. studied the differences in student metacognition and sense-making in the labs where students designed their own experiments and where the experiments were prescribed by a write up. They found that focusing on sense-making episodes in labs is a productive and a reliable way to code student activities.

DESCRIPTION OF THE STUDY

The research was conducted in the labs that were integrated in two introductory physics courses for science majors. The population in the two courses is roughly similar but they are offered on two campuses of Rutgers University. Both courses have a 3-hour lab as a part of the course credits. During the time of the study the experimental course followed the *ISLE* curriculum [1] and the control course had traditional labs supplemented by reflective questions at the end. Below we describe the differences between the labs.

Control Course: Cookbook+Explanations Lab. In these labs students perform experiments by following well-written, clear and concise guidelines (Appendix A shows an excerpt from one of the write-ups) which instruct them on what and how to measure and how to record the data. The adjusted equipment and elaborate writes-up eliminate all possible difficulties such as parasite effects, wrong assumption effects and large uncertainty. In some labs students have to devise their own mathematical method to analyze data. After each part of the lab, students have to answer conceptual and reflective questions. TAs provide immediate help to the students when they have a question.

CP883, *2006 Physics Education Research Conference*, edited by L. McCullough, L. Hsu, and P. Heron

Experimental course: design lab. *ISLE* laboratories are described in detail in [2, 3]. In these labs students design their own experiments. Write-ups do not contain instructions on how to perform the experiments; instead they guide students through various aspects of a typical experimental process (Appendix B shows an excerpt from an *ISLE* lab). At the end of each experiment students answer reflective questions that focus on different aspects of the procedure that they invented. In addition students use scientific abilities rubrics for guidance and self-assessment [2, 3]. TAs serve as facilitators.

Sample. We observed the behavior of 14 groups of students (one group per lab). Nine of the groups were in *ISLE* labs and 5 groups were in cook-book+explanation labs. Each observation lasted for an entire lab.

CODING AND LIMITATIONS

An observer sat with a student lab group timing and recording all student activities and conversations. After the lab was over, the field notes were rewritten and a complete transcript of each lab session was constructed. The analysis of the first transcripts revealed patterns in student activities that lead to devising codes for 4 categories of activities for the experimental course and 5 for the control course. Subsequent observations were analyzed using the coding scheme to note any behaviors that did not fit into the codes (we assumed that the presence of the observer did not affect student behavior). The last observations were made after the coding scheme was devised and we could not find any behaviors that did not fit the coding categories.

General codes. Our coding scheme turned out to be very similar to the one described in [4, 5]. Lippmann had making-sense, logistic and off-task codes. We observed similar types of activities. The only difference was that we subdivided the logistic code into two sub codes (procedure and writing).

Making sense – students' discussions about physics concepts, experimental design, the data, and the write-up questions.

Writing – students' descriptions of the experiment, data recording, calculations, and explanations.

Procedure – students gathering equipment, mounting set-up, and taking data.

Off-task – any activity that did not relate to the laboratory task.

In the control course we had to use one more code: **TA help** to note considerable time students spent listening to a TA explaining and answering questions.

b) Sense-making codes

We focused on the sense-making because it represents verbalization of the students' cognitive processes. The content of sense-making discussions was classified further according to the activities matching the descriptions of different scientific abilities (sense-making subcodes).

D – *Design*: Discussing experimental design and set-up, planning the experiment, etc.

M – *Model*: Choosing the mathematical model and the parameters to be measured.

A – *Assumptions*: Discussing assumptions in the mathematical model and their effects.

U – *Uncertainties*: Discussing sources and calculating values of experimental uncertainties.

Min – *Minimizing:* Discussing how to minimize uncertainties and the effects of the assumptions.

R – *Revising*: Discussing reasons for the discrepancy and the ways to improve the experimental design to get the discrepancy less than the uncertainty.

Examples of sense-making discussions related to the effects of assumptions on the experimental results.

ISLE Lab: Effect of assumptions

S1: I think we can ignore the friction.

S2: But we cannot ignore it. We account for it.

S1: No, it is too small.

TA: How can you check this?

S1: Let's measure the friction.

S2: How?

S1: Do you remember that lab where we measured it? We can tilt the track and measure the angle when the car starts sliding. *They tilt and observe that the car slides at an extremely small angle which they cannot measure.*

S2&S1: So, we can ignore the friction!

Cookbook+explanation lab: Effect of assumptions

S1: What temperature should we plug into the equation?

S2: 0°C

S1: How can you be sure that it is zero degrees?

S2: It should be. It is always 0°C.

S3: No. Ice can have much lower temperature.

S1&S2: Lets ask TA if we should take zero degrees.

TA: Yes, you can assume that it is 0°C.

Observer note: The lab manual: "Add some ice chips at 0°C to the tap water in the calorimeter". Thus, if students paid more attention to the write-up, they would not have this discussion.

Notice here that the beginnings of the discussions are very similar: students are exchanging their unfounded opinions and their discussion brings them nowhere. However the TAs' intrusions are very different. In the *ISLE* lab the TA triggers the next level of discussion by suggesting that students check their ideas. In the second episode, the TA answers the question. This makes any further sense-making unnecessary.

FINDINGS OF THE STUDY

Duration: Design lab versus a cookbook+explanation lab. Each lab in the control course had more experimental tasks, but it took students about half the time to complete each lab compared to *ISLE* labs (average of 80 min/lab versus 160 min). In *ISLE* labs students spent a great deal of time planning, discussing, and writing a detailed lab report. The experiments themselves took more time because students needed to improve them as they progressed (Fig.1). In *ISLE* labs, the students' interactions with TA were minimal (TAs did not provide explanations). Discussions similar to those shown above were considered as sense-making for both courses.

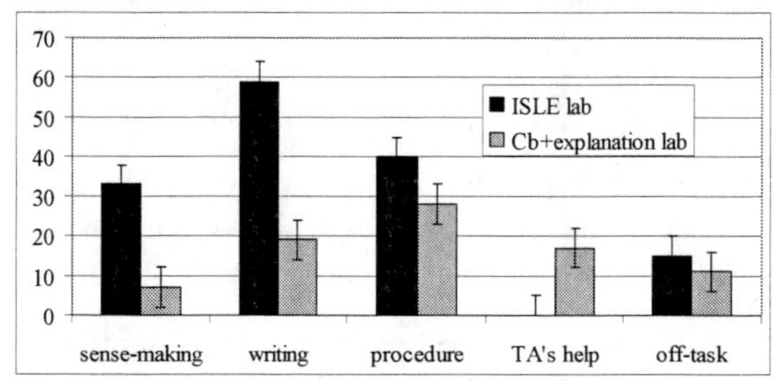

FIGURE 1. The time spent on different activities (in minutes). The data are averaged over the 14 groups sample.

Sense-making and scientific abilities. Figure 1 shows that students engaged in sense-making for about 33 minutes in *ISLE* labs - for 20% of the lab duration. In cookbook+explanation labs sense-making lasted for about 5-8 minutes, i.e. 9% of the actual lab time. The conceptual questions took on average 3 min. The time lines with the smallest increment of 1 minute supply detailed information on student activities. The typical time lines are shown on Fig. 2.

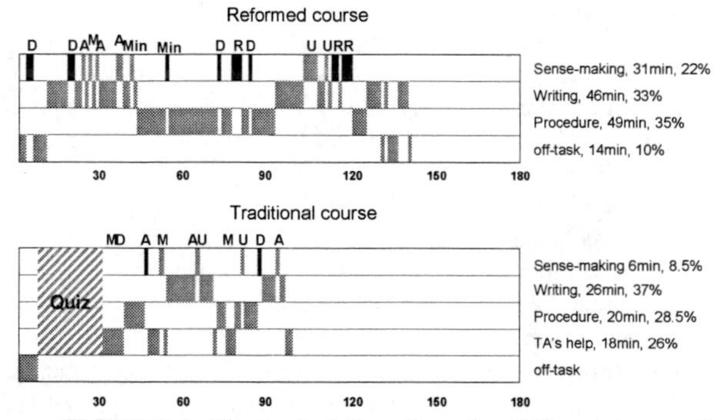

FIGURE 2. The typical time lines for different types of labs. Black color notes episodes of sense making that were not prompted by a write-up or a TA. The letters on the top of the timelines show the sense-making subcodes for each discussion episode.

The detailed analysis of the sense making episodes reveals differences between the courses. In the control course students engaged in sense making for a very short time. Few of their statements could be coded as related to scientific abilities. The TA explained the experiment design and the mathematical model. Although students had to answer questions about assumptions and uncertainties, they engaged in this activity superficially spending less than a minute on the discussions. That happened probably because students considered these questions unrelated to the experimental procedure. For *ISLE* students write-ups' questions about assumptions and uncertainties were crucial because they had to make a decision how to conduct the experiment and whether they needed to repeat it.

Another important difference was how often students switched to sense-making mode without prompting by TAs or questions in the manual. Such episodes of self-triggered sense-making are shown on the time lines by black color (Fig. 2). As experimental design required independent decisions. It is not surprising that these episodes happened more often in *ISLE* labs. The self-triggered sense-making often happened during periods of planning, executing, and revising an experiment.

Outcome of sense-making. A detailed analysis of the time lines reveals that sense-making episodes caused different student behavior in different labs. If we set aside the episodes of sense-making alternating with writing when students answered the manual questions and discussed their writings, we can see that in *ISLE* labs, sense-making discussions led to procedural changes, i.e. attempts to improve and revise the experiment or carrying out the next steps (Fig. 3a). In cookbook labs, in about 70% of such episodes the students' discussions led to asking a TA who provided an immediate answer (Fig. 3b). Thus *ISLE* labs make students pose questions and answer them, whereas in cookbook labs students seldom pose rare questions and do not tend to search for answers.

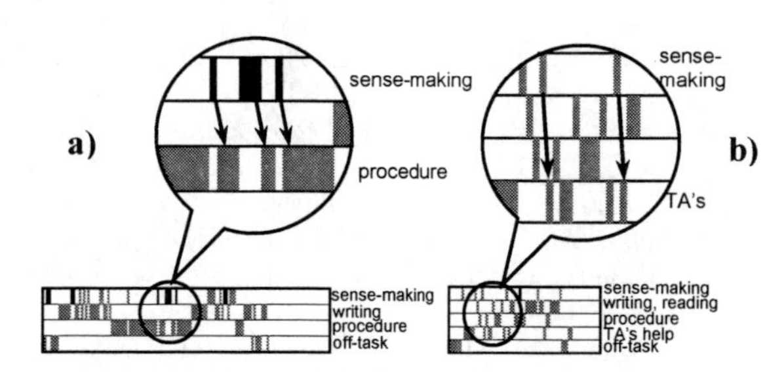

FIGURE 3. Sense-making outcomes: a) in *ISLE* labs sense-making causes a discussion which leads to a decision and to an execution; b) in cookbook labs sense-making causes TA's help mode with a TA answering and explaining.

DISCUSSION

Those who have tried to implement labs where students design their own experiments know how difficult it is. Students get frustrated with the design process, especially at the beginning. They want clear directions and clean experiments. Is this struggle worth the effort? Our findings show a dramatic difference between the behaviors and discussions of students in the labs where they had to design their own experiments and where the design was provided to them. *ISLE* students act very much like scientists in labs. They design an experiment spending time on careful planning and paying attention to details. They consider assumptions inherent in a mathematical model and try to devise a procedure to minimize the effect of assumptions. They consider experimental uncertainty while comparing results of two independent experiments and make a decision whether they have to improve and repeat the experiment. They spent a great deal of time writing lab reports to communicate the details of the experiment. We have observed that all these activities took significant amount of time and discussion.

The traditional labs, even supplemented with conceptual and reflection questions, did not engage students in developing these scientific abilities. They did not spend time choosing strategy, validating results and improving design. The discussions about additional conceptual questions are very brief so they do not make large difference.

Maybe the design labs are worth the effort.

REFERENCES

1. E. Etkina and A. Van Heuvelen, *Proceedings of the 2001 Physics Education Research Conference*. Rochester, NY, PERC publishing, 17-21 (2001).
2. E. Etkina, A. Van Heuvelen, S. White-Brahmia, D. T. Brookes, M. Gentile, S. Murthy, D. Rosengrant, and A. Warren, Phys. Rev. ST Phys. Educ. Res. **2**, 020103 (2006).
3. S. Murthy and E. Etkina, 2004 Physics Education Research conference, edited by J. Marx, P. Heron, and S. Franklin, *AIP Conference Proceedings*, **790**, 133-136 (2005).
4. Rebecca Lippmann and the Physics Education Research Group, "Analyzing students' use of metacognition during laboratory activities," AREA Meeting, New Orleans, April 2002.
5. R. Lippmann Kung, A. Danielson, and C. Linder," Metacognition in the student laboratory: is increase metacognition necessarily better?", EARLI symposium August 2005.
6. V. Otero, "The process of learning static electricity and the role of the computer stimulator," Ph.D. Thesis, San Diego University, 2001.
7. A. F. Artzt and E. Armour-Thomas, *Cognition and Instruction*, **9**, 137-175 (1992).
8. P. Adey and M. Shayer, *Cognition and instruction*, **11**, 1-29 (1993).

Appendix A: Cookbook+Explanation Lab.

Thermal inertia: The heat capacity of aluminum

1. Use the following table to record your data (omitted here).
2. Fill the calorimeter about ¾ full with tap water.
3. Record the mass of the calorimeter and water ($m_{water+Cal}$).
4. Record the initial temperature of the water (T_{water}).
5. Place a hot piece of aluminum, which you can obtain from the boiling water bath, in the calorimeter and cover it immediately. Notice that you need to record the temperature of the aluminum piece for later calculations.
6. Shake the water making sure you move the piece of aluminum so that the water/aluminum system may come to thermal equilibrium.
7. After **3 minutes** record the final temperature (T_{equ}).
8. Obtain the mass of the calorimeter, water and aluminum ($m_{Al+Water+Cal}$).
9. Devise a method to determine the specific heat of aluminum, and compare it with the specific heat of aluminum in the text book. Find percentage difference.

Questions

1. What assumptions did you have to make to derive the formula for the heat capacity of aluminum?
2. Are there any ways to reduce the error it this minilab?
3. Suppose you put a block of iron and block of Styrofoam in the freezer and allow them to stay for a little while. If you gripped both blocks which one would feel warmer? Explain why.

Appendix B: Design (ISLE) lab.

Specific heat capacity of unknown object

Design two independent experiments to determine the specific heat capacity of the given object. The material the object is not known.

Equipment: Water, ice, beaker, hot plate, Styrofoam container with a lid, weighing balance, and thermometer.

First, recall why it is important to design *two* experiments to determine a quantity. Play with the equipment to find how you can use it to achieve the goal of the experiment. Come up with as many designs as possible. Choose the best two designs. Indicate the criteria that you used for the decision. For each method, write the following in your lab-report:

a) A verbal description and a labeled sketch of the design.
b) The mathematical procedure you will use.
c) All assumptions you have made in your procedure.
d) Sources of experimental uncertainty. How would you minimize uncertainties?
e) Perform the experiment. Try to minimize experimental uncertainties. Record your measurements in an appropriate format.
f) Calculate the specific heat capacity.
g) After you have done both experiments, compare the two outcomes. Discuss if they are close to each other within your experimental uncertainty. If not, explain what might have gone wrong. If your results are not close, perform the experiment again taking steps to improve your design.

Reframing Analogy: framing as a mechanism of analogy use

Noah S. Podolefsky and Noah D. Finkelstein

Department of Physics, University of Colorado at Boulder

Abstract. In a series of large-scale (N>100) studies of analogy use in college physics, we have explored how, when, and why analogies affect student reasoning. In the first of these studies, we demonstrated that analogies affect student reasoning when taught in a large enrollment physics course [1]. In the present follow-up study, we demonstrate that teaching EM waves concepts implicitly via analogy leads to greater conceptual change compared to teaching explicitly without analogies. Students were divided into two groups, one taught using analogies (string and sound waves) and the other taught without analogies (EM waves only). On a targeted concept question given before and after instruction, students who were taught with analogies outperformed those taught without analogies demonstrating that analogies can affect student reasoning in productive ways, even when taught implicitly. We propose framing as a mechanism to begin to explain why analogies can be productive when used implicitly.

Keywords: analogy, framing, electromagnetism, waves
PACS: 01.40.Fk, 01.40.Gb, 01.40.Ha

INTRODUCTION

Analogies are commonly used by both physics teachers and by practicing physicists. The purpose of an analogy is to ground an unfamiliar concept in terms of a more familiar one. An often cited example is the planetary model of the atom. While the original utility of this analogy was for physicists to explain the results of experiment [2], it is also used to teach an introductory, albeit incomplete, model of the atom to students. An analogy can be defined as a mapping from a base domain (e.g., the solar system) to a target domain (e.g., the atom), akin to a mathematical isomorphism. [3] In the planetary model of the atom, the sun maps to the nucleus, planets to electrons, the gravitational force to the Coulomb force, etc.

The ubiquity of analogies in physics textbooks [4,5] as well as the findings of some researchers (e.g. [6]) suggest teaching with analogies can sometimes be a productive instructional strategy. It has been demonstrated that analogies amount to more than colorful language [7]. In fact, they can generate inferences, shaping learners' conceptions of a target domain. In the example above, the solar system analogy generates the inference that electrons in an atom are tiny spheres that revolve around the nucleus. Some researchers, however, have found that teaching with analogies is only sometimes productive. [8,9] Thus, analogies do generate inferences, but only under certain conditions. These researchers hypothesize that teaching with analogies can fail to change students' conceptions for a number of reasons, including: 1) students may have an inadequate understanding of the base domain to map; 2) the base domain or the relationship between base and target may be too abstract; 3) students may not know how to make productive mappings [8,9].

These hypotheses have led to a common view that effective teaching with analogies should include explicit instruction as to which mappings are productive for understanding the target domain, and which mappings may be harmful [5]. In the present study, we took a different, but more common, approach and taught analogies implicitly. That is, we explicitly taught students the base domains to learn about a target, but did not explicitly instruct students on which mappings to make.

To date studies of analogy in physics have only demonstrated success when taught to small numbers of students (N<20) with a focus on explicit instruction (e.g., [8,9]). We build on this previous research on analogy in a series of studies, including large-scale (N>100) studies and student interviews, to ask whether and how analogies might be productively taught in the canonical large enrollment physics courses. These studies focus on student learning of electromagnetic (EM) waves, a topic with which students have difficulty [10].

Two analogies commonly used to teach EM waves are a wave on a string and sound waves. In our first large-scale study, we taught students about EM waves

CP883, *2006 Physics Education Research Conference*, edited by L. McCullough, L. Hsu, and P. Heron

using either a wave on a string or a sound wave as an analogy [1]. We tested the effectiveness of the analogies using concept questions given after instruction. We found that the analogies did affect student reasoning. For instance, on a multiple choice concept question similar to Figure 1, we found that students' answers were associated with which analogy they were taught (string or sound). The distracters students chose reflected characteristics of the particular analogy they were taught. For instance, students who were taught a string analogy tended to choose a distracter that characterized an EM wave as two dimensional, which is characteristic of an oscillating string. Thus, different analogies generated correspondingly different inferences about EM waves. In fact, EM waves have characteristics of both waves on a string and sound waves.

In interviews, we found that students had difficulty answering the same EM wave concept question when taught only about EM waves. Suggesting string and sound wave analogies verbally to students did not substantially help these students in answering the concept question. However, we found that providing iconographic string and sound wave representations as analogies to EM wave representations did help students make significant progress towards answering the concept question correctly.

Based on these prior findings, we hypothesize that the optimal approach to teaching EM waves would be to use both string and sound analogies. In the present study, we taught students about EM waves using both string and sound analogies, or without analogies. We compared the effectiveness of these two teaching methods with a concept question targeted specifically to the concepts in the tutorial, and with follow-up interviews. Here, we focus on results from the concept question. We found that students who were taught EM wave concepts implicitly via analogy performed better than students who were taught the same EM wave concepts explicitly but without analogies. Thus, we confirmed our earlier findings that analogies generate inferences when taught in a large scale physics course. Further, we demonstrated that the findings of small scale studies (i.e. that analogies generate inferences) are reproducible in a large-scale study.

EXPERIMENTAL METHODS

The experiment involved N=156 students enrolled in a second semester, calculus-based physics course at a large university. The course consisted of three lectures and a single 50-minute recitation each week. The lectures made extensive use of clickers and concept questions, but were otherwise traditional (as described in [11]). Recitations were led by two or three teaching assistants, where students used the *Tutorials in Introductory Physics* [12].

For the present experiment, students were divided into two groups, denoted the analogy (N=72) and no-analogy (N=74) groups. In both groups, students completed a modified tutorial on EM waves that borrowed heavily from the *Tutorials*. The tutorial consisted of three sections. Section 1 covered basic wave concepts such as wavelength, frequency, and amplitude as well as traveling vs. standing waves. Section 2 covered plane wave concepts, focusing on the idea that they are three dimensional (3D) waves. Section 3 covered EM wave representations and forces on charges due to the electric and magnetic fields of an EM wave. For students in the analogy group, section 1 focused on a wave on a string, and section 2 grounded the study of waves in sound waves. The only section of the analogy tutorial that discussed EM waves was section 3. The no-analogy tutorial was very similar in length and content to the analogy tutorial, and used sections nearly isomorphic to the analogy tutorial to teach these concepts, but always in the context of EM waves. Notably, only approximately 1/3 of the analogy tutorial dealt explicitly with EM waves, while the entire no-analogy tutorial was dedicated to EM waves. Teaching assistants were told not to use analogies when teaching in the no-analogy sections, but were encouraged to use analogies when teaching in the analogy sections.

To compare the effectiveness of the analogy and no-analogy tutorials, a challenging concept question on EM waves was given in lecture on the days immediately prior to and after recitation. Students were asked the concept question shown in Figure 1. Since students from both groups attended the same lecture, students were told not to discuss the question with their in-class peers until after the entire class had finished responding. Individual student responses were collected electronically and only students who attended the recitation and responded to the questions both before and after and were included in the study.

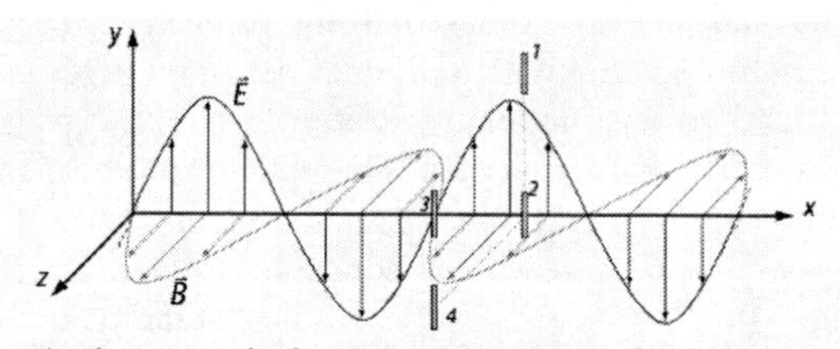

An electromagnetic *plane* wave propagates to the right in the figure above. Four antennas are labeled 1-4. The antennas are oriented vertically. Antennas 1,2, and 3 lie in the x-y plane. Antennas 1,2, and 4 have the same x-coordinate, but antenna 4 is located further out in the z-direction.

Which choice below is the best ranking of the *time-averaged signals* received by each antenna? *(Hint: the time-averaged signal is the signal averaged over several cycles of the wave.)*

A) 1=2=3>4 B) 3>2>1=4 C) 1=2=4>3 D) 1=2=3=4 E) 3>1=2=4

FIGURE 1. Multiple choice EM wave concept question.

RESULTS

The results for the EM waves concept question are shown in Figure 2. The vertical axis shows the gain, or difference in responses between the concept question given before (pre) and after (post) recitation for the analogy and no-analogy groups. The two answers with positive gain were the correct answer (1=2=3=4) and main distracter (1=2=4>3). In both groups, more than 86% of students chose one of these two answers on the post-test. We found that the gain on answering with 1=2=4>3 was 24% in the analogy group and 38% in the no-analogy group. The gain on answering with 1=2=3=4 was 21% in the analogy group and 7% in the no-analogy group. Thus, the analogy group had a greater gain on the correct answer, and the no-analogy group had a greater gain on the main distracter ($p<0.1$, 1-tailed z-test).

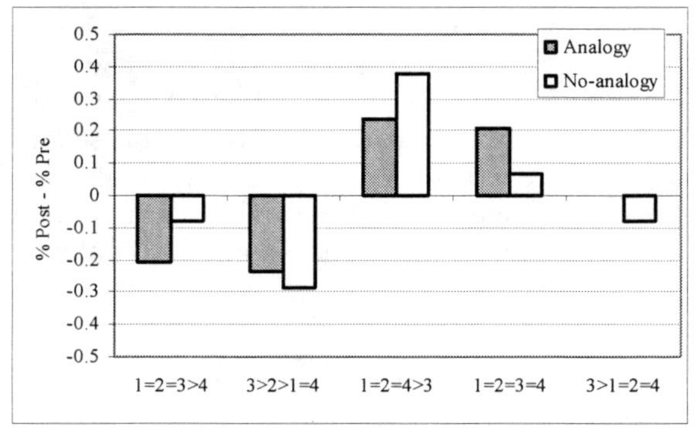

FIGURE 2. Results on the EM wave concept question. Vertical axis is the gain (% Post - % Pre) for the analogy and no-analogy groups. Multiple choice answers are shown below the horizontal axis. Gain on 3>1=2=4 for the analogy group was zero.

DISCUSSION

Interpretation of Results: Based on the above results, we point out two key findings: 1) Students changed their conceptions of EM waves whether taught with or without analogies, and 2) The students in the analogy group made greater progress towards the correct answer compared to students in the no-analogy group. These findings demonstrate that analogies, taught implicitly, successfully generated inferences about EM waves.

The concept question above requires students to understand at least two characteristics of EM waves. First, EM waves are plane waves. Thus the time averaged signal is equal at antennas 1, 2, and 4. Second, this particular EM wave is a traveling wave, so the time-averaged signal at antenna 3 is the same as that at 1, 2, and 4. In the no-analogy tutorial, these ideas were all addressed explicitly for EM waves. Thus, students in the no-analogy group were taught explicitly that EM waves can be traveling waves and that they are plane waves. In the analogy tutorial, these ideas were taught for analogous situations, but not EM waves. Students learned that waves on a string can be traveling waves, and that sound waves are (locally) plane waves. In order to answer the EM waves concept question correctly, students in the analogy groups had to apply their knowledge of string and sound waves to EM waves.

We find more specific details in our results as well. Note that both the correct answer and the main distracter include the characteristic 1=2=4. We will refer to this characteristic as the plane-wave feature. The additional feature in the correct answer is the traveling-wave feature (i.e. 3 is equal to the others). What we have found is that most students chose an answer that contained the plane-wave feature – in both groups, the gain on the correct answer and main distracter combined was 45%, and more than 86% of students chose one of these answers on the post-test. Thus, this feature maps strongly from sound waves to EM waves since, in the analogy group, it was never taught explicitly for EM waves. The traveling-wave feature appears to be more difficult to teach, either by analogy or directly. There may be several reasons for this difficulty, one of which is the appealing idea for students that antenna 3 lies at a node. Teaching with analogies seems to have marginally better success at overcoming this difficulty.

What we found is that the analogies affected student reasoning about EM waves when taught implicitly. It is possible that teaching the analogies explicitly may have resulted in a more pronounced effect. However, implicit teaching of analogies is a common practice, and our findings suggest that these practices can affect student reasoning.

A Mechanism of Analogy: The fact that we did not instruct students explicitly on how to use the analogies leads us to ask how students knew to reason about EM waves (i.e., make mappings) as they did. We believe that existing theories of analogy (e.g. [8]), while essential for creating productive analogies, do not explain the way people commonly use analogies implicitly. To begin to explain this, we posit a complementary view of analogy. According to this view, analogies work because the base domain provides a frame, or conceptual structure, which can be applied to the target domain. A frame exists as a compiled set of ideas that can be cued, or activated, together [13,14]. In the analogy tutorials, we used several iconographic representations to teach string and sound waves. In each case, students were taught that a sine wave could be used to represent each type of wave. By using the sine wave in different contexts, students may have compiled more and more meaning into that particular wave representation [15]. Thus, the sine wave frame consists of generic features of waves, such as traveling and 3D. When the same sine wave representation was then used to reference the EM wave, the existing sine wave frame was cued, and could be applied to the unfamiliar target domain of EM waves. The result is that content specific to EM waves, such as the electric field, takes on the structure of the frame, producing a model of electric field propagating as 3D waves. One reason this approach may work, in spite of being taught implicitly, is that creating and cueing frames is largely an unconscious, but extremely powerful, cognitive process. Such unconscious processes may require less cognitive work from students than conscious processes [14].

While we found analogies led students to greater progress towards the correct answer on the concept question, these ideas remain difficult for students to master, and the effect of using analogies is not so pronounced. We offer two hypotheses as to why this might be. One is time on task. Although we found that a single 50-minute recitation was sufficient to change some student conceptions of EM waves, it might not have been sufficient to promote learning of more difficult concepts. Further, students in the analogy group spent less time learning EM waves specifically compared to the no-analogy group. Given more time with EM waves, students prepared with analogies may have performed substantially better.

A second hypothesis we will propose recognizes the distinction between everyday and classroom knowledge [16]. In student interviews conducted as part of this study, we have observed that students are not always able to apply everyday knowledge of string or sound waves, but are able to apply classroom knowledge (i.e. drawn from particular wave representations) to learn about EM waves. For instance, we observe that when presented with a sine wave representation of sound, some students state that the sound can only be heard in the plane of the sine wave. The usefulness of analogies rests on the assumption that students will ground their understanding of unfamiliar ideas in terms of the familiar. The tutorials used in this study were designed to draw on students' familiar, everyday knowledge, and couple this to abstract classroom knowledge. However, in the context of a physics course, students may not use the resources that analogies are meant to draw on. That is, even though students have had everyday experiences with waves on a string, they may not bring this knowledge to bear in a classroom setting. Both the analogies and formal representations may be treated as equally abstract by students. This hypothesis suggests future studies on whether and how students use everyday knowledge when reasoning by analogy in a classroom setting.

CONCLUSION

In a large scale study of student use of analogy, we have demonstrated that analogies generate inferences when taught. We found that teaching EM waves implicitly via analogy led to better student performance than teaching explicitly without analogies. Student reasoning about EM waves was affected when taught with both methods, but we found students who were taught with analogies made better progress towards the correct answer on a pre-post concept question. We have begun to explore how teaching of analogy can work, and posit that the underlying mechanism may be the creation and cuing of cognitive frames and use of representations.

ACKNOWLEDGMENTS

This work has been supported by the National Science Foundation (DUE-CCLI 0410744 and REC CAREER# 0448176), the AAPT/AIP/APS (Colorado PhysTEC program), and the University of Colorado. We wish to extend sincere thanks to Profs. Jamie Nagle and Shijie Zhong, and the PER at Colorado Group for their essential and significant contributions to this work.

REFERENCES

1. N.S. Podolefsky and N.D. Finkelstein, *Phys. Rev. ST Phys. Educ. Res.* to appear.
2. E. Rutherford, *Philos. Mag.*, 6, 21 (1911)
3. D. Gentner, *Cognitive Science*, 7:155-170 (1983)
4. M.K. Iding, *Instructional Science.* 25:233-253 (1997)
5. S.M. Glynn, in S. Glynn, R. Yeany and B. Beritton (eds.) *The Psychology of Learning Science*, Erlbaum, 1991
6. S.M. Glynn and T. Takahashi, *Journal of Research in Science Teaching*, 35:1129-1149 (1998)
7. L.J. Atkins, Ph.D. Thesis, Univ. of Maryland, 2004.
8. D. Gentner and D.R. Gentner, in *Mental Models*, Gentner and Stevens (eds.) Lawrence Erlbaum, 1983
9. D. E. Brown and J. Clement, *Instr. Sci.* 18, 237 (1989)
10. B.S. Ambrose, P.R.L. Heron, S. Vokos, L.C. McDermott, *Am. J. Phys.* 67(10), 891 (1999)
11. N.D. Finkelstein and S.J. Pollock, *Phys. Rev. ST Phys. Educ. Res.* 1, 010101 (2005)
12. L.C. McDermott, and P.S. Schaffer, *Tutorials in Introductory Physics*, Prentice Hall, 2001
13. D. Hammer, A. Elby, R.E. Scherr, E.F. Redish, in J. Mestre (ed.) *Transfer of Learning*, Information Age Publishing, 2005
14. G. Fauconnier and M. Turner, *The Way We Think*, Basic Books, 2003
15. W.M. Roth and G.M. Bowen, *Learning and Instruction* 9:235-255 (1999)
16. J.S. Brown, A. Collins, and P. Duguid, *Education Researcher*, 18(1): 32-42 (1989)

What Factors Really Influence Shifts in Students' Attitudes and Expectations in an Introductory Physics Course?

Jeffrey Marx† and Karen Cummings‡

†McDaniel College and ‡Southern Connecticut State University

Abstract. To gauge the impact of instruction on students' general expectations about physics and their attitudes about problem solving, we administered two different, but related, survey instruments to students in the first semester of introductory, calculus-based physics at McDaniel College. The surveys we used were the Maryland Physics Expectation Survey (MPEX) and the Attitudes about Problem Solving Survey (APSS). We found that the McDaniel College students' overall responses were more "expert-like" post-instruction: on the MPEX, the students' *Overall* agree/disagree score started at 59/18 and ended at 63/17, and on the APSS, the students' agreement-score went from 63 to 79. (All scores are out of 100%.) All of the students to whom we administered the MPEX and a significant sub-group to whom we administered the APSS realized these improvements without experiencing any explicit instructional intervention in this course aimed toward improving attitudes and expectations. These results contrast much of the previously reported findings in this area.

INTRODUCTION

Students' attitudes and expectations about their own learning, a course's content, and the structure of scientific knowledge strongly influence how and what they will learn in a science class. Since the late 1990s, some physics education researchers have focused on characterizing students' attitudes and expectations, as well as developing curricular strategies and materials intended to help students realize more sophisticated attitudes and expectations. The article "Student Expectations in Introductory Physics" by Redish, Saul, and Steinberg helped catalyze this movement by showing that entire groups of students frequently averaged *lower* on the Maryland Physics Expectation Survey (MPEX) after one semester of introductory physics instruction, even in cases when the course in question had a strong research-based foundation.[1]

One oft-cited example of a course successfully designed to positively impact students' attitudes was reported by Elby in 2001.[2] In his Virginia high school physics class, Elby purposely exposed his students to materials that would help raise their awareness of their own thinking and the inherent organization of the library of physics knowledge. Redish, too, reported gains on the MPEX after incorporating specific materials and approaches into an algebra-based course.[3] We also note that in Redish et al., one group of students' *Overall* MPEX scores resisted an erosion of attitudes. These students, under the tutelage of Pricilla Laws at Dickenson College, were exposed to *Workshop Physics*, which is a coherent and refined research-based curriculum.[4]

In 2002, we developed and administered (along with Stephanie Lockwood) a survey designed to probe students' attitudes about problem solving–Attitudes about Problem Solving Survey (APSS).[5] Some of the items of this 17-item survey were adopted from the MPEX, while others we included were unique to this instrument. The only set of pre/post-instructional data we gathered for reference 5 was for a group of students from Rensselaer Polytechnic Institute (RPI), and we found that those students experienced a negative shift in attitudes about solving physics problems. In light of the broader data published about the MPEX, this was not surprising. What was interesting and relevant for this paper was that McDaniel College students' post-instructional score was at a level that we felt was quite high (73% expert-like).

CP883, *2006 Physics Education Research Conference*, edited by L. McCullough, L. Hsu, and P. Heron

POPULATION, COURSE STRUCTURES, AND TIMELINE

This study involved students enrolled in *General Physics I* at McDaniel College in the Fall semesters of 2001 through 2005. This course is part of a calculus-based sequence and the only introductory physics course offered at McDaniel. (Details about McDaniel College can be found in reference 6.)

We collected demographic and background information in four semesters: 2001 (gender, only), 2002, 2004, and 2005. Nearly all students in this course are science majors. Over half are biology or biochemistry majors, many of whom are declared to be on the pre-medical-school track. Most of the students are sophomores, but all class years were well represented. The gender ratio was nearly one-to-one (M/F:49/51). Not surprisingly, the majority of students entered this course with significant previous exposure to formal science instruction. The average student for whom we have data (N = 98) self-reported over seven semesters of science in high school and over four semesters of science in college.

There were two phases to our study. In Phase I (Fall semesters 2001 and 2002) we collected matched MPEX data from one section of *General Physics I* at McDaniel College, which was taught by one of us, JM. We were interested to determine how these students' expectations changed over the duration of one semester of instruction, which was not specifically designed to influence attitudes. At that time, JM taught a "research-inspired" course focused on augmenting students' conceptual understanding of the material. Class time included Interactive Lecture Demonstrations and frequent in-class exercises designed to help students understand the basic physical concepts by applying them to simple exercises. The students' class grade depended on their scores on the homework problems (only three representative problems out of a particular set were graded); laboratory work; and in-class examinations that stressed, to varying degrees, mathematical solutions, as well as pictorial and verbal expressions of physical concepts. The research-based textbook was *Physics for Scientists and Engineers*, by Serway and Beichner.[7] As we mentioned previously, in the Fall of 2002 we also administered the APSS to McDaniel students, post-instruction only.

Phase II of our study began in the Fall of 2004. In this phase we included the other section of *General Physics I* in our study. That section was taught by Dr. Apollo Mian [AM], a professor interested in incorporating research-based curricular materials into his class, but who is not a physics education researcher. Both course sections were very similar. JM and AM used the same research-based textbook (Cummings, et al., *Understanding Physics* [8]); covered essentially the same material at the same pace; used similar, and in many cases, exactly the same, research-inspired curricular materials in lecture, as describe above for Phase I; assigned the same homework problems and posted the same solutions to our respective websites; and shared a common set of laboratory sections. The basis for students' grades in both sections was similar to Phase I. However, there were differences between the two sections that are relevant for this paper.

Bolstered by the high post-instructional average from 2002, and with hopes of realizing even more sophisticated post-instructional attitudes, JM augmented his class instruction by devoting time to modeling, and helping students model, effective problem solving. This included having students learn to draw an effective sketch of the problem (later in the course referred to as a "non-mathematical representation" [9]); record knowns and unknowns; identify assumptions; explain the major physical principles, which would serve as the starting point for the solution; and list all of the relevant "starting" equations (which came from a short equation sheet that JM provided at the beginning of the term).[10] To help reinforce this tactic for solving problems, JM graded three problems from each homework set for the correct mathematical solution, but also for how well students applied this approach, which accounted for one-third of a student's homework grade. Also, formal exams had at least one problem in which points (10% of the exam grade) could be earned only by applying "good problem-solving techniques." AM did no such modeling or reinforcement, and grades he awarded on homeworks and exams were solely based on a student's mathematical solutions. We refer to JM's sections as Problem-Solving (*PS*) intensive (N = 33), and AM's sections as Non-Problem-Solving (*N-PS*) intensive (N = 28).

ANALYSIS

Figure 1a below displays the MPEX data from Phase I arranged on an agree/disagree plot.[1] This highlights how a class' responses to the various clusters on the survey, as well as their *Overall* score, changed from pre-instruction to post-instruction. Figure 1b places the McDaniel College *Overall* score within the larger context of some of the published pre/post-instructional data for the MPEX.

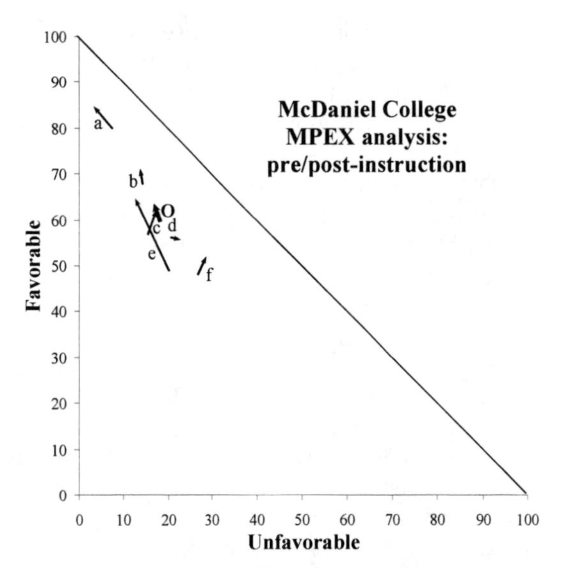

Figure 1a: McDaniel College students' expectations as measured by the MPEX. The arrows indicate the shift in score from pre-instruction (tail of arrow) to post-instruction (head of arrow). Individual thin arrows represent the various MPEX clusters: *a - Reality, b - Effort, c - Math, d - Concepts, e - Coherence, f - Independence*. The thicker arrow, *O*, represents the *Overall* score.

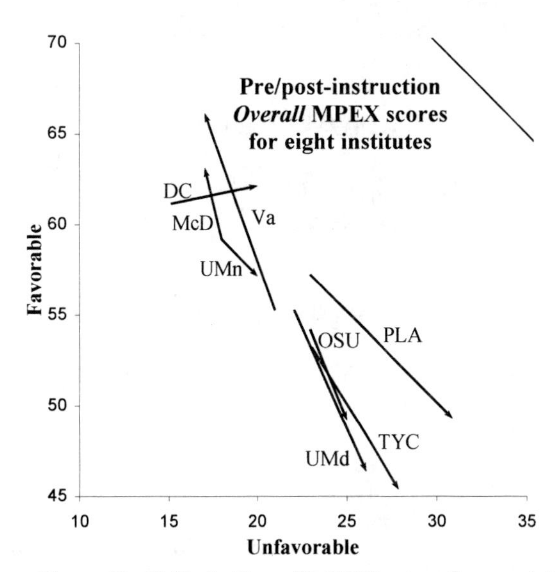

Figure 1b: Shifts in *Overall* MPEX scores for a variety of institutes, including McDaniel (McD). All of the data except McD and Va are adapted from reference 1. The Va data are adapted from reference 2. Notice that the scale of the agree/disagree plot has been increased 4× for legibility.

Figure 2 displays the matched pre/post-instructional scores on the APSS for two sections of McDaniel students during Phase II of our study (labeled *PS* and *N-PS*). Also included are the pre/post-instructional averages from RPI, and the post-instructional averages for McDaniel College and Southern Connecticut State University (SCSU) for the Fall semester of 2002.

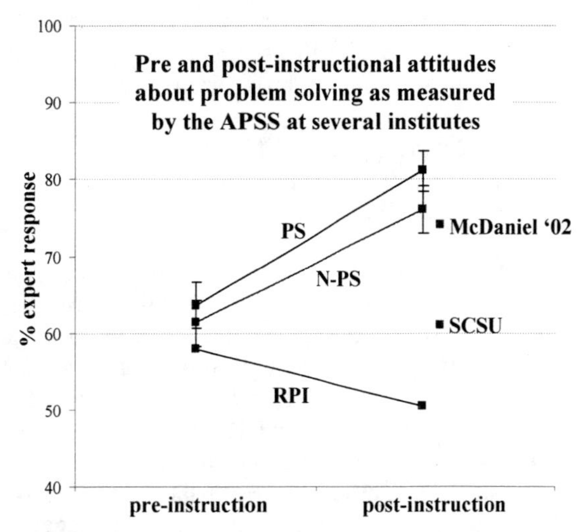

Figure 2: Pre/post-instructional data for the APSS. Error bars are the standard error. The points for which we have only post-instructional data are shifted slightly to the right for clarity purposes, only. The RPI data, and the post-instructional-only data for McDaniel and SCSU are adapted from reference 5.

DISCUSSION AND CONCLUSIONS

Our analysis revealed several interesting features. On the MPEX, the McDaniel College students avoided the typical negative shift in expectations from pre-instruction to post-instruction one might associate with a class not designed to address such issues. Furthermore, unlike the Rensselaer students who showed a negative change in attitudes about problem solving, the McDaniel students experienced a significantly positive shift, and much of this was realized without any instruction designed to motivate such change. It is worth noting that the datum point in Figure 2 representing the McDaniel College students' post-instructional score in 2002 was after instruction by JM, who, at the time, used an instructional strategy more like the *N-PS* style than the *PS* style. (The SCSU students lie at the middle of the post-instructional results; however, we hesitate to draw any conclusions about their pre-instructional attitudes, since they are a less traditional group of students than either the RPI or McDaniel cohorts.)

So, returning to the title of this paper we wonder: "What factors really influence shifts in students' attitudes and expectations in an introductory physics course?" Unfortunately, at this point we lack sufficient data to identify a mechanism or complex of mechanisms responsible for the overall positive shifts in attitudes for the McDaniel students we have reported here. Lacking such insight, we find it difficult to confidently draw any strong conclusions; but, in a moment we will propose two hypotheses stemming from our analysis. For now, we can say that our results stand apart from much of the previously reported, peer-reviewed work. So, one can not presume *a priori* that students' attitudes will necessarily decay in a course that fails to explicitly address attitudes and expectations.

Our first hypothesis is that sophisticated pre-instructional attitudes, as measured by the MPEX, may inoculate students against the deleterious effects of a non-attitudinally-oriented course. Ignoring the exceptional Virginia MPEX data, which was the result of a carefully planned intervention, the three institutes with the highest pre-instructional scores–Dickenson (DC), McDaniel, and the University of Minnesota (UMn)–all saw the smallest negative shifts, and in some cases, positive shifts. Anecdotally, other unreviewed data we have seen supports our proposition. Additionally, it seems inoculation may be more complete for some areas than others. For example, the McDaniel students' *Concepts* score did not present a particularly favorable shift. Regarding attitudes about problem solving, high pre-instructional attitudes may only weakly influence gains in that area. Unfortunately, our APSS data does not strongly compliment our hypothesis. The pre-instructional data for RPI seems barely distinguishable from the McDaniel data. Lacking MPEX data for RPI prevents us from broadening or rejecting our hypothesis.

Our second hypothesis is that the two groups of students who most resisted MPEX-measured degradation–McDaniel and Dickenson–were self-selecting for and/or acculturated in similar academic environments. Both institutes are small, private, nationally competitive, liberal arts colleges in bordering states with modestly-sized physics departments. Furthermore, both colleges have at least one full-time faculty member who is a physics education researcher devoted to teaching introductory physics. We posit this combination of traits may attract a particular kind of student and then implicitly reinforce the importance of achieving attitudinal gains. Reference 1 also includes information about an un-named, *public* liberal arts college (PLA). Unfortunately, because we know so little about this population and the sample was so small (N = 12), we hesitate to reject our hypothesis based on that group. Because of a shortage of data, it is difficult to determine if our APSS results compliment our second hypothesis.

Before concluding, we shall discuss observations regarding the McDaniel APSS results. Although a considerable fraction of the attitudinal gains seems to have occurred "naturally," the *PS* instructional approach produced a measurable effect on this population. We hope these results will help motivate introductory physics instructors interested in augmenting their students' attitudes about problem solving to consider implementing similar methodologies. We feel our approach would have a similarly positive influence on students who would normally achieve no, or small, positive attitudinal gains with respect to solving physics problems.

To conclude, our McDaniel APSS pre/post-instructional data further demonstrates that courses openly designed to influence students' expectations and attitudes are worth developing and can have demonstrable results. However, we have also uncovered that instructional environments that fail to explicitly address students' attitudes and expectation do not necessarily doom students to the negative shifts that we and others have previously reported in the literature. Finally, more research is clearly required to support or reject either of our hypotheses regarding factors that influence shifts in attitudes.

ACKNOWLEDGMENTS

We thank Apollo Mian for taking the time to administer the APSS in his class.

REFERENCES

1. Redish, E., Saul, J., and Steinberg, R., Am. J. Phys. 66, 212 - 224 (1998)
2. Elby, A. Am. J. Phys., PER Sup. 69, S54 - S64 (2001)
3. Redish, E., "Teaching Physics with the Physics Suite" John Wiley and Sons (2003)
4. Laws, P., "Workshop Physics," Wiley (1999)
5. Cummings, K., Lockwood, S., Marx, J., AIP Con. Proc. 720, 133 - 136 (2004)
6. Marx, J., Mian, S., and Pagonis, V., 2004 Phys. Ed. Confer. Proc., AIP Confer. Proc., 790, 125 - 128 (2005)
7. Serway, R., and Beichner, R., "Physics for Scientists and Engineers: Fifth Edition," Saunders (2000)
8. Cummings, K., et al., "Understanding Physics," Wiley (2004)
9. De Leone, C., and Gire, E., AIP Con. Proc. 818, 45 – 48, (2006)
10. This approach began at the U. of Oregon with Sokoloff and Johnston, and was later adopted and refined by Kolitch. See Johnston, M., Am. J. Phys., PER Sup. 69, S2 – S11 (2001) for a related discussion.

Reformed Physics Instruction Through The Eyes Of Students

Maria Ruibal Villasenor and Eugenia Etkina

Graduate School of Education, Rutgers University, New Brunswick, NJ 08904

Abstract. This paper reports on a qualitative study of students' responses towards innovations in an introductory physics course: their attitudes toward the change; their perceptions of the learning methods and the subject; and the relationships among these variables. We found that students' ideas about learning affected their reposes to the reforms.

Keywords: curriculum reform, student responses, qualitative research.
PACS: 00.01.40.FK 00.01.30.Cc 00.01.30.lb

INTRODUCTION

For the past five years the Rutgers PAER group has been modifying "Physics for the Sciences" - an algebra-based course for science majors at Rutgers University. Presently the course follows the *Investigative Science Learning Environment* [*ISLE,* 1] that facilitates authentic science learning.

Sadly, anecdotal student reactions indicated that the transition from traditional courses to this innovative learning environment was difficult. We decided to conduct a study to answer the following questions: How do students respond to new teaching methods? Do they perceive differences between "traditional" science classes and constructivist approaches? If they see the differences, do they appreciate the changes? What factors affect their responses to new teaching?

Answers to these questions will help facilitate the adoption of new teaching and learning practices.

PHYSICS FOR THE SCIENCES

"Physics for the Sciences" (193/194) follows the *ISLE* system where students construct their own knowledge and acquire scientific abilities by emulating the research practices of physicists. They start every conceptual unit by observing physical phenomena; collecting data, analyzing them and inventing multiple explanations. Then, using hypothetico-deductive reasoning, students test the explanations, revise them and apply the new knowledge to solve problems. They work in groups and are active learners. Instructors do not provide students with physical concepts or laws but create the conditions and support for learners to construct physics knowledge. A crucial resource for students is the Active Learning Guide [2]. It consists of sequences of activities that facilitate the construction of physics concepts and scientific abilities. During 2005-2006 students did not have to purchase a textbook but acquired the ALG and worked tasks in it. In the lab students designed their own experiments supported by lab write-ups and scientific abilities rubrics [3]. The former did not have detailed directions on how to conduct an experimental procedure, but had guiding questions and prompts.

During the academic year when the study was conducted the instructor in charge was enthusiastic but lacked experience. The majority of the lab and recitation TAs had never taught.

PREVIOUS WORK

Although extensive literature explores the effects of teacher attitudes on the success of reforms, students' opinions are much less investigated. Fawcett [4] suggested that students go through three different stages: comfortable dependence, anxiety and comfortable independence. He found that the change takes time and creates concern and fear in students.

Hammer found that students' expectations and attitudes help explain learners' performance in introductory physics courses [5]. UMPERG created the MPEX survey [6] to probe students' thinking about physics and learning. They found that there is a large difference between novice and expert attitudes and discovered that student expectations tend to deteriorate with instruction. Similar results were found by the

CP883, *2006 Physics Education Research Conference*, edited by L. McCullough, L. Hsu, and P. Heron

Colorado PER Group using the CLASS instrument [7].

METHOD

Rationale: Due to the nature of the posed questions, we believe that the most appropriate approach is an exploratory qualitative study. We have to identify the relevant variables triggering students' responses and study the interactions among students' perceived expectations, abilities and beliefs about physics knowledge. We need to: identify the different ways in which students' may respond to change; describe the context and conditions that may prompt various kinds of students' reactions to innovations; and point to the possible causes. This approach is called a grounded theory [8] which explains a set of observations makes predictions within certain boundaries.

In a grounded theory a theoretical model is generated through the collection and analysis of the data [8]. It is crucial to try to not anticipate possible outcomes during each phase of the study, in order to not impede the emergence of the actual model. There are three steps in the data analysis in the grounded theory: open coding, axial coding and selective coding.

Population and data collection: There were 170 students in "Physics for the Sciences" 193/194 during the academic year of 2005/2006. The population was quite heterogeneous. Some students were interested in environmental, exercise sciences, pharmacy, chemistry or medicine. Many wanted to maintain a high GPA and this contributed to their high stress level.

We used the data from interviews with nine students. We collected data during the last weeks of a two-semester course. In April/May 2006, the instructor in charge asked for volunteers. As we were especially interested in investigating the circumstances and ideas of students opposed to the reformed course, we invited two students who had previously expressed strong opinions against the course teaching and learning methods. Due to the small sample we cannot claim that the participants represent the whole class.

We conducted nine 45-minutes individual interviews. Students' responses were audio-taped and transcribed. Researchers were not instructors in the course and we conducted interviews in neutral, non-threatening spaces such as student centers and university gardens. The interviews were semi-structured, with the general aim of getting the students to talk about their experiences in the course. The researchers used the predetermined sequence of questions but tried to keep the questioning conversational. Some of the questions are listed below.

a) What do you think about physics in general?
b) How is Physics 193/194 similar to or different from other science classes that you had or you are taking?
c) What do you like the most and the least in 193/194?
d) Tell me about your experience in taking 193/194.
e) What do you think about 193/194 teaching methods?
f) Have your ideas about physics changed somehow?
g) How would you describe your performance in Physics 193/194?

The goal of open-ended questions was to access the students' perspective and to learn about students' understanding and judgments in their own terms.

Data analysis: During the open coding phase, we examined the data by breaking them into small portions and trying to identify different categories. We were guided, in part, by the language of the students describing their experiences. The categories for the textual words of the subjects are called "in vivo" codes. However we were cautious because different people may use the same worlds differently therefore it is important to capture the meaning. For example, consider the meaning of the word "learning" when used by one of the students in the interview: "I'm learning more in chemistry, but how much am I going to remember in chemistry? I would not be confident with my ability to do a chemistry problem. But I might be more confident in my way to solve a physics problem. I mean I can spit out a physics formula for you and understand what they mean, but chemistry I couldn't."

The appearing codes were compared and contrasted with one another to reduce the amount and to obtain a small number of refined categories that constituted the most important ideas of the study. The procedure continued until the categories were saturated - the analysis of the texts did not produce any more codes and many events in each category supported the codes. Axial coding followed the open coding. There, the open codes were reduced to several categories and the data in each of category were judged against that category's properties. We used a theory-generation software package named Atlas.ti [9], developed for such studies. In the final phase, selective coding, we related the core category to the other categories by creating a visual representation of the interrelationship among them (Fig.1).

FINDINGS

Our analysis yielded the following categories:

(1) *Attitude:* Positive or negative students' disposition toward the course.

(2) *Perceptions:* Students' awareness and interpretation of several relevant aspects of the course. Students' perceptions affected their attitudes.

(2a) *Course Purposes:* Students' understanding the goals of the course and the purposes of learning tasks.

(2b) *Learning:* Students' thinking of how they learn.
(2c) *Difficulty:* Students' perception of the affordability of the course goals and effort required.

(3) *Variableness*: Changes in students' perceptions and views of physics and the course.

Attitude Overall, most of the students expressed positive feelings about the course. However, we saw a wide range of responses. Some really enjoyed the course: "for a science class I feel that this is one of the better ones". Some could barely tolerate it: "I just think [the class] it's stupid". Others were content but not passionate: "this is not a bad class to be required to take". The majority appreciated some aspects and decried others or fluctuated in their support for the format and methods: "We needed to figure it out which was good and bad I mean it was frustrating but it made us really think and figure it out on our own."

Course Purposes Many students understood some of the different goals for the course: for example, the development of understanding and to a lesser degree the acquisition of scientific abilities. "I feel like I understand much more of the aspects in terms. I have trouble accepting things and I have to accept things in physics, but in this course I have to accept less of them and understand more of them". Another student said "I know like the course wants to teach us how to approach a topic scientifically like with hypothesis, prediction and everything."

However a disturbing fact was that most of the students at one or another instance mistook or ignored the objectives and had difficulty understanding the purpose of many of the tasks that they were asked to complete. For example, some of the students complained that labs and large room meetings were not always synchronized because a lab activity occasionally preceded a corresponding large room meeting about that same topic. On the contrary, labs and lectures were carefully matched, and in some cases the lab activities served as preparation for the whole class discussion. This fact was problematic for some students. Similarly most of the interviewed students complained about having to calculate the uncertainty for every quantitative result in lab: "We understood after the first lab that the uncertainties are important but we didn't need to be pounded into our heads for the next ten labs in 193 and then the entire eleven labs in 194, calculating them every single time for every single instrument." One student protested to using multiple representations "different ways of representing motion… I'm never going to use that… and I don't remember... they're like convention." These misunderstandings might have triggered negative reactions to the innovations.

Learning We found that some students believed that they learned by reading or listening to information: "I read the book and then I do practice problems". Others argued that listening to instructors or consulting a textbook was not enough; that they needed to engage in other activities to learn: "Lecture is interesting and can be fun but you are just sitting there and you're having one person tell you what you need to know and you are not grasping the concepts." Those, who thought that they needed "to figure out things on their own", were much more inclined toward the course.

Difficulties Those students who found the subject or the assignments difficult tended to reject the innovative course. Most of the students reported arriving on the first day of class thinking that physics was a very difficult subject. When they faced goals, tasks or epistemologies that were in dissonance with those in their prior science classes, their perceived difficulty of the course increased. "I came in with the mentality that physics is the devil", or "I wasn't looking forward to physics actually back in September cause thinking back to high school I just remember being very confused and like frustrated." Several felt that they lacked the aptitudes needed for physics: "some things in physics don't click with me", "physics is important, it's just not for me". In this state of mind encountering unfamiliar goals or tasks increases the level of students stress: "I was pretty surprised, I was like this isn't physics class cause normally when you first in general physics you go to lecture and you get Newton's Law right, but then here it's like today it's like observation, experiment, and everyone was like huh? It feels like this kind of too broad, I'm confused, and in the beginning I feel like oh I'm not really learning anything." Students did not expect new goals and tasks. They also found the idea about the nature of science embedded in the course was different from their previous notions. Students experienced this as an unsetting feature that attributed to physics and not to other sciences: "Biology for me is a hardcore factual thing subject, that can be understood by facts... but physics on the other hand is really like a guess and check type science. I like something with a solid answer", "I find that like biology or chemistry is more factual by the book type thing where physics can be… it changes according to the circumstances you are looking at" and "maybe physics in general isn't as straight forward as other sciences."

Variableness There was a clear trend in students' opinions. Students recalled that became significantly more positive about through the academic year: "I think it's an overall better feeling, for whatever reason it took a while to understand the goal". This transition is best represented by a student who said that if given the choice, "I'd probably pick calc based physics in this format because there are aspects of 193:194 (*number to designate the course)* that are very good and the recitations are awesome ... I would take calc

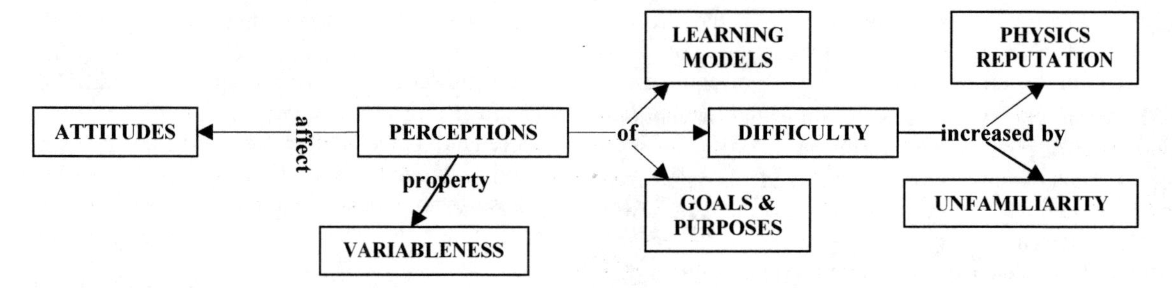

FIGURE 1. Model for students' response to the innovative physics course

based in this format because of the demonstrations and cause I think things have taken pretty good detail without the calc. With the calc, I think it would just explain stuff perfectly." She added: "I actually learned a lot in lab this semester." However, "I hated this class when it first started. I hated going to lecture, I hated going to lab." Therefore students' perceptions can change modifying their attitudes.

The findings are represented in Figure 1.

DISCUSION AND IMPLICATIONS

Our overall impression of the results of this study was quite positive. Students focused their narratives mainly on issues of learning and not on grades. Most of them expressed positive judgments about the course and its learning methods. More importantly they enjoyed the course for the reasons that education researchers had in mind when they developed the philosophy, the goals, structure and tasks in the course supporting students actively constructing meaningful knowledge. As students said: "it helps you figure it out, and that's, you know, you have your notes from lecture that you can refer to, but it's really you trying to bring it all together". Other facets of the course that students appreciated were their collaborative work and the demonstrations during large room meetings. By contrast, the majority of the interviewed students complained about aspects that were not essential to the reform. In fact major sources of dissatisfaction were the rush during the last minutes when the instructor tried to finish his lesson plan and the large amount of writing required for lab reports. Students' most frequent complaint was that no textbook was assigned. It created a feeling of uneasiness and insecurity among them. They missed something "solid" that would have supported their developing concepts.

The ideas that the individual student held about learning were an important factor affecting the easiness of how she or he embraced the reforms. Those students who believed in a transmission paradigm of knowledge were more likely to reject the *ISLE* course. They thought that they did not need all the hassle, extra effort and increase of time because they learned by listening to the lecturer and reading the book. During interviews we observed many students in mixed epistemological states, appreciating their active engagement and learning responsibility only at times or for certain aspects of the process.

Students' lack of knowledge about learning led to misunderstanding of course and assignments goals. This is particularly disturbing because the instructors made a special effort to communicate those objectives to the students in documents included in the course package. Student learning development must be an essential outcome of any course. We need to communicate this objective explicitly and implicitly, in the design of assignment and assessment. It might be helpful to students to take course on epistemology.

Many students mistrusted innovations because of malfunctions due the inexperience of the instructors or to the novelty of the approach. Therefore when implementing innovations, a special effort must be made to train TA's and support instructors in their process of adjusting to new formats.

This work was supported by the NSF grant REC059065.

REFERENCES

1. E. Etkina and A. Van Heuvelen, *Forum of the American Physical Society,* 12-14, (Spring 2004).
2. A. Van Heuvelen and E. Etkina, *The Physics Active Learning Guide,* Pearson/Addison Wesley, SF, 2006.
3. E. Etkina, A. Van Heuvelen, S. White-Brahmia, D. T. Brookes, M. Gentile, S. Murthy, D. Rosengrant, and A. Warren. Phys. Rev. ST Phys. Educ. Res. **2**, 020103, (2006).
4. G. Fawcett, *J. Lit. Research,* 30 (4), 489-514, (1998).
5. D. Hammer, *Cogn. & Inst.,* 12, 151-183, (1994).
6. E. F. Redish, J. M Saul and R. N. Steinberg, *Am. J. Phys.*, 66 (3), 212-224, (1997).
7. W.K. Adams, K.K. Perkins, N.S. Podolefsky, M. Dubson, N.D. Finkelstein, and C.E. Wieman, Phys. Rev. ST Phys. Educ. Res. 2, 010101, (2006).
8. A. Strauss and J. Corbin, "Basics of Qualitative Research: Grounded Theory Procedures and Techniques," Newbury Park, CA: Sage Publications, (1990).
9. T. Muhr. "Atlas-Ti: The Knowledge Workbench" [Computer software]. CA: Scolari Publications, (1996).

Sustaining Change: Instructor Effects in Transformed Large Lecture Courses

Steven J. Pollock, Noah D. Finkelstein

Department of Physics, University of Colorado, Boulder, CO 80309-0390

Abstract. We investigate the transfer of classroom reforms from PER faculty to more traditional physics-research faculty. This study is part of an ongoing effort to assess necessary and sufficient requirements for success with research-based course transformations. We have previously demonstrated the ability of PER faculty to replicate the success that other researchers achieve when implementing research-based reforms in large, introductory calculus-based physics courses. Here, we present new data from four implementations of Physics II, including quantitative and qualitative measures of successful transfer of courses to new faculty: validated pre/post surveys covering content, assessments of student views about physics and learning physics, and informal affective surveys. We investigate questions of sustainability, reproducibility, and instructor effect through a contextual constructivist theoretical lens, and find that replication of research-based results is possible, but a variety of factors including instructor beliefs and institutional constraints play important roles.

Keywords: course transformation, replication, course assessment
PACS: 01.40.-d, 01.40.Di,01.40.Gm

INTRODUCTION

Successful transformation of a physics course requires more than the initial implementation of research-based tools and practices. Questions of sustaining and replicating these course changes are central to the study of course transformation. At our institution, the University of Colorado at Boulder (CU), we have begun transforming our introductory calculus-based sequence with the explicit course goals of developing student conceptual understanding and supporting constructive beliefs about the nature of physics and learning physics. Once course changes are implemented, there are two main approaches to sustaining course transformations: 1) establish a dedicated set of faculty to continue the new practices, and 2) develop a mechanism for including the broader array of departmental faculty to engage in the new course structures. Our approach follows the latter strategy. There is considerable folk knowledge about the significance (or lack thereof) of instructor effects, and related issues of teacher-proof curricula, but questions about sustainability and transfer of educational reforms are not sufficiently well understood from a theoretical perspective [1,2,3]. We present new data from our ongoing implementation of Tutorials[4], Peer Instruction with personal response systems[5], and computer homework systems, in order to document the success of handoff, to demonstrate the importance of particular curricula, and to further investigate the importance of the role of the instructor.

In previous work [6,7] we investigated replication and essential features of the multiple reforms and the dynamics among them in Physics I, followed by an investigation[8] of transfer and replicability of Physics II to faculty who were also involved in PER research. Here, we present new data which follows Physics II as traditional research faculty teach in this environment. We demonstrate successful handoff of curricular reforms; students demonstrate significant conceptual gains, sustained productive beliefs about physics and favorable attitudes towards the course. However, these successful outcomes are not guaranteed or uniform. While a myriad of factors influence classroom practices and outcomes, these data suggest that there are at least four critical factors shaping the success of course transformation: institutional support, faculty beliefs and practices, curricular tools, and the students themselves.

BACKGROUND/ENVIRONMENTS

Phys 1120 is the calculus-based E&M course at CU. There are three 50-minute lectures per week. All terms studied here used ConcepTests[5] with peer discussion in class. All weekly recitations were

CP883, *2006 Physics Education Research Conference*, edited by L. McCullough, L. Hsu, and P. Heron

replaced with Washington *Tutorials*[4], including weekly online Tutorial pretests and hand-written/hand graded homework, as well as traditional end-of-chapter problems using one of two computer systems [9,10]. For the Tutorials, graduate student TAs teamed with undergraduate Learning Assistants (LAs) [7,11], all of whom attended weekly preparation sessions following the University of Washington model.

To compare courses, we collected three principal types of data. A pre/ post conceptual exam, the BEMA[12], was administered in the first and last weeks of the term. A pre/ post research-based survey on student attitudes and beliefs about learning, the CLASS[13], was administered online. Finally, informal surveys were given online to collect students' opinions about the value of course elements at the end of semester. Courses at CU are dependent on individual instructor choices; a summary of some key features and differences among the four terms of implementation are presented in Table 1. In general, this course has a "lead" instructor (who lectures) and a "secondary" instructor in charge of recitations, many course administrative details and background work.

Term:	Fa04	Sp05	Fa05	Sp06
Lead Prof:	A	B	C	D
Secondary:	B	C	(none)	E
# Sections/ # Students:	2/479	1/333	2/455	1/365
Text:	HRW[14]	RK[15]	RK	RK
Homework system:	CAPA[9]	CAPA	CAPA	MP[10]
Faculty in PER?	A & B	B	(none)	(none)
Faculty in TA Prep?	A (start) & B	B (start) & C	C & lead TA	E & lead TA

TABLE 1. Summary of administrative, curricular and instructor backgrounds in the four consecutive implementations of this reformed course.

To summarize: Prof's A and B, both involved in PER, implemented Tutorials and ConcepTests Fa04. Prof. B attempted to closely replicate the course the following term (except for switching to the Knight text [15]), helping Prof. C (a traditional instructor assigned to the course) who quickly took over the Tutorial preparations. Prof. C took over the entire class the next term, his first experience ever teaching a large lecture course. He attempted to keep all materials (including lecture notes, homeworks, etc.) essentially unchanged from the previous term. He was assisted in running Tutorial preparation sessions by a new PER grad student, who had not yet run Tutorials on her own. The fourth term, two new (non-PER) faculty participated. Lecture notes were completely rewritten to utilize PowerPoint, and the online homework system was switched to Mastering Physics [8]. The same lead TA assisted the new professor in Tutorial preparations. A "book" of materials and suggestions was prepared by Prof. A to assist in those sessions. Roughly half the TA's from any given Fall term return to TA the following spring, and more than ¾ of the LA's are new each term.

DATA

Results of the conceptual assessment are shown in figure 1. The first two implementations replicated one another closely (59% posttest, statistically indistinguishable) [8]. In Fa05 the post-test average of 50% demonstrates successful achievement (high by national standards [12]), though is a statistically significant decline from the prior semesters' scores. The final distribution from Sp06 is visibly similar to the results from the initial implementations, although closer examination reveals fewer students on the high end of the distribution, and a density of students in the lower end, resulting in the 53% posttest average..

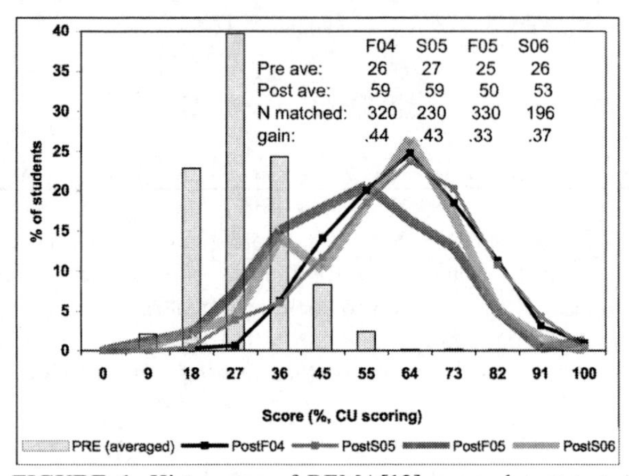

FIGURE 1. Histograms of BEMA[12] pre and post tests. (Pretests are very consistent, so averaged data are shown.) Data from the first two implementations (with PER faculty) are nearly identical, shown as solid lines; data from next two courses (with non-PER faculty) are shown with broad lines.

Student attitudes and beliefs about physics and the learning of physics were assessed with the CLASS survey[13]. A summary of the post-pre shifts in percent favorable response is shown in figure 2, for overall score and the "conceptual connections" sub-category. (Pretest scores are shown numerically to the side of the figure.) The courses mostly exhibit relatively small negative shifts (much smaller than the double-digit negative shifts generally observed in traditionally-taught courses [16]). The third semester (which scored lowest on the BEMA content survey) also shows the most negative CLASS trends.

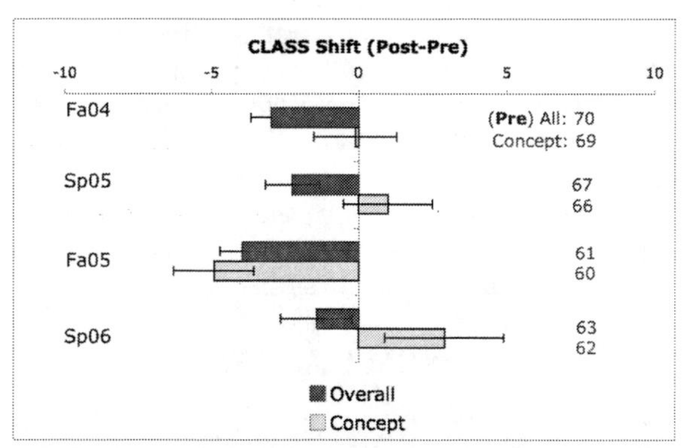

FIGURE 2. CLASS [13] shifts in pre to post favorable scores. Matched students, in overall score, and "conceptual connections" sub-category. Semesters are labeled on the left, numbers on the right indicate average pre-test scores

Fig. 3, summarizes some student opinions from the four course implementations, measured with an end-of-term online survey. Five questions surveyed students' ratings of: (1) how much they learned in lecture and (2) in tutorial, (3) how much they enjoyed lecture and (4) tutorial, and (5) overall rating of course. Responses were provided on a five-point scale. Fig. 3 shows student responses answering non-negatively, i.e. in the top 3/5 response categories. Students' affective responses are consistent across implementations, with small variations consistent with those seen in conceptual (BEMA) and CLASS scores.

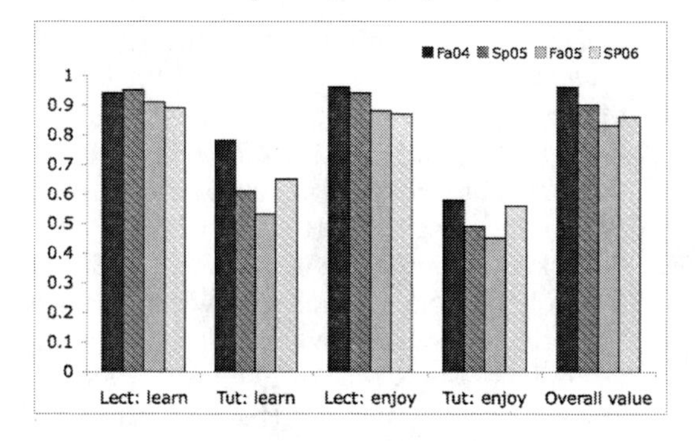

FIGURE 3. Online affective survey at the end of the terms, showing students' % neutral or positive rating of learning and enjoyment in lecture, Tutorial, and the course overall.

DISCUSSION AND ANALYSIS

To identify the factors that influence the success of these course transformations, it is helpful to compare the various conditions of implementation from a contextual constructivist perspective [7,17]. Contextual constructivism uses *frames of context* to help identify the particular factors that shape (and are shaped by) student learning. From this perspective we may identify four, roughly nested, frames of context that affect the success of the course transformation at CU: institutional level of support, faculty beliefs and background experience, curricular tools that are used in practice, and the students' background. These frames are nested in the sense that student background knowledge is evoked by tasks that are part of the curricular structure. The curricula are selected and framed by the instructors and their beliefs about the course. The instructor too is bound by institutional framing. Of course these frames are highly interacting, and neither strictly hierarchical nor mutually exclusive [17]. We compare the four implementations of the course transformation and use these frames to identify variation in application and success.

From the first implementation (Fa04) to the second (Sp05), we note that the institutional structure remains relatively constant– the number of faculty assigned to the course, the number of students per lecture section. The faculty have similar backgrounds: both have extensive experience in the introductory sequence, are involved in PER, and understand the underpinning philosophy and approaches to the curricular innovations. The course tools are very similar, with the mild exception that the textbook changed.[1] The other course elements, Tutorials, Peer Instruction, the homework system, and use of the web are very similar. Finally, the majority of the students in this course had arrived from a partnering mechanics course. In these two cases, the activities in the prior semester were similar in character, using PER-based materials. Notably however, the students in the Sp05 implementation had used the Knight workbooks [15] in recitation the prior semester, rather than Tutorials (which had been used for the Fa04 students' prior semester). Ultimately, we observe that these two courses are indistinguishable in terms of conceptual gains, shifts in students' beliefs about physics and learning physics. The one notable distinction is that students are less positive about the Tutorials. It is possible that this results from students' lack of prior experience with the Tutorials, or the new instructors slight de-emphasis of Tutorials in lecture and exams.

The second transfer of teaching responsibility occurs from Sp05 to Fa05. In many senses Fa05 is the highest fidelity form of replication from the prior semester. The tools from the prior implementation (Sp05) were used in as similar a fashion as broader constraints allowed. Lecture notes, ConcepTests, Homework sets, and Tutorials were nearly identical. Significantly, however, this implementation of the course is the only instance when a *single* faculty member is assigned to teach the entire course. This institutional constraint limits the amount of time the faculty member can spend on the course. While dedicated to the course, the work-load of this

[1] Because students generally do not read the text as assigned [19], it is unclear as to whether this would impact student achievement.

instructor is higher than any other instantiation. The institutional constraints are further exacerbated by the fact that this implementation is the first time this instructor is teaching in a large-scale environment. Nonetheless, this faculty member is committed to the endeavor, trained with the lead instructor the prior semester, and begins to adopt many of the beliefs favored by the PER community [18]. The students arrive from a course the prior semester that did not make as much use of PER materials as either of the prior implementations. Despite the limitations in institutional support, faculty background, and student prior experience, we do find measures of success that are considered significant by national standards [12]. Students post significant conceptual gains, show modest declines in their beliefs, and have generally favorable attitudes to the course and Tutorials. These successes may in part attributable to the materials and the faculty commitment.

The final transition occurs with a handoff of the course from Fa05 to Sp06. In many respects, the Sp06 incarnation is a mixed state of the initial implementations (Fa 04) and the more recent (Fa 05) implementation. The institutional support of the course transformation returns to standard level– two assigned faculty. The faculty themselves have no background in PER, nor have they trained in use of the PER tools. However, a variety of support structures (PER faculty discussing the reforms, the locally developed guide materials and a Lead TA) exist to facilitate their use of these tools. The actual curricular tools are largely the same, though lecture notes are re-written (and put into PowerPoint) and the homework system, Mastering Physics, is based on research. The students come into the course with similar backgrounds as the Fa05 cohort – less familiarity with PER tools, and specifically, no prior work with Tutorials. Overall success appears to be a mixed state of the prior implementations. Student conceptual achievement is high, falling between the first implementation and the last implementation. Interestingly, students post the most favorable beliefs in this course, which may be due to the instructors' commitments and framing [18], the new research based text and homework system, or likely, a combination of these factors. Student affect is generally favorable and again a mixture of the prior cases of implementation.

CONCLUSIONS

A comparative analysis of each of these implementations of course transformation suggests two conclusions: 1) new faculty can be brought in to transformed courses, potentially allowing for sustainable success of these reforms, and 2) there are many contributing factors affecting success, including: institutional commitment, faculty background, the tools and practices used, and the students themselves. While these findings may not be surprising, and are consistent with findings in prior research in PER, these results demonstrate some of the intricate relations of these components in real-world settings of large-scale classroom instruction in physics.

ACKNOWLEDGMENTS

Thanks to APS/AIP/AAPT Colorado PhysTEC and NSF CCLI (DUE #0410744) for grants which support our transformed classes, and the NSF STEM-teacher preparation grant that supports learning assistants. Enormous thanks to Kathy Perkins for CLASS data, the U. Washington's Physics Education Group, the CU Physics Department, and the PER at Colorado group.

REFERENCES

1. Fullan, M, (2001). The New Meaning of Educational Change. (3rd Ed. Teachers College Press: New York).
2. Henderson, C and Dancy, M. (2006). *2005 PERC Proceedings*, **818**,149.
3. Shulman L. (2000). "From Minsk to Pinsk: Why a scholarship of teaching and learning?" 1(1) 48-52
4. McDermott, L., Shaffer, P., and the PEG, (2002). *"Tutorials in Introductory Physics"*, Prentice Hall.
5. Mazur, E. (1997). *Peer Instruction,* Prentice Hall.
6. Pollock, S., (2005). *2004 PERC Proc*, **790**, 137.
7. Finkelstein, N. Pollock, S., (2005). *Phys. Rev. ST Phys. Educ. Res.* 1, 010101.
8. Pollock, S., (2006). *2005 PERC Proc.* 818, 141.
9. CAPA homework system: see www.lon-capa.org
10. Mastering Physics: www.masteringphysics.com
11. Otero, Finkelstein, McCray, Pollock, (2006). "Who is Responsible for preparing Science Teachers?" *Science* ***28**, 445*
12. Ding, L et al, (2006). *Phys Rev ST: PER,* 2, 010105, see http://www.ncsu.edu/per/TestInfo.html, and private communication with the authors.
13. Adams, et al. (2006), Phys Rev ST: PER, 2, 010101 and ref list at http://class.colorado.edu
14. Halliday, Resnick, and Walker (2001). Wiley
15. Knight, R., (2004). Pearson
16. Redish, Saul, Steinberg (1998). *Am. J. Phys.* **66** 212-224; Redish, (2003). *Teaching Physics.*
17. Finkelstein, (2005). *IJSE*, **27**,1187
18. Turpen, Finkelstein, Pollock, (2006). AAPT poster DB02-03 and *Announcer* **36**(02), 119.
19. Podolefsky, Finkelstein (2006). *Phys Tchr.* **44,** 338

Adaptation and Implementation of a Radically Reformed Introductory Physics Course for Biological Science Majors: Assessing Success and Prospects for Future Implementation

Charles De Leone[*], Robin Marion[*], and Catherine Ishikawa[†]

[*]*California State University, San Marcos*
[†]*University of California, Davis*

Abstract. The physics department at California State University San Marcos has nearly completed work on an NSF CCLI-A&I funded project to adapt and implement UC Davis' reformed introductory physics course for students in the biological sciences. As part of the project, a group of physics instructors met to discuss criteria for measuring the implementation's success and the feasibility of implementing the course at other institutions. Criteria for measuring success fell into three areas—student outcomes, institutionalization of the course, and adherence to the original course's core philosophy. This paper describes the criteria in more detail, presents data for outcomes already measured, and discusses the challenges of measuring other outcomes. Finally, the paper briefly discusses the likelihood of instructors at other institutions meeting with the same or better success at implementing the course.

Keywords: curriculum adaptation, curriculum implementation, course evaluation
PACS: 01.40.Fk, 01.40.Gm, 0.1.40.Jp.

INTRODUCTION

For the last two years, after a two year pilot project, the first author has been formally adapting UC Davis' reformed introductory physics course[1] and implementing the course at California State University San Marcos (CSUSM). As part of an NSF CCLI-A&I funded project, the authors are assessing the success of this adaptation and implementation. While student achievement outcomes provide an obvious starting point for assessment, other factors like the sustainability of the course can be considered when assessing whether implementation of a reformed course was successful. This paper discusses some possible criteria for measuring success and some challenges specific to evaluating an adopted course, as well as presenting some initial results. Many of the ideas presented were generated at a workshop held to gain additional perspective on the success of this course implementation and to explore the feasibility of further implementations at other institutions. Participants included physics instructors from the University of California system, the California State University system, and the California Community Colleges system.

Table 1 lists criteria distilled from discussion among the instructors who met. In assessing a transferred course the most fundamental question is whether or not the course benefited students. In addition to various student achievement outcomes, the group wanted to see improved communication skills and better attitudes toward physics (both affective and epistemological).

This group recognized, however, that any student gains arising from a reformed course would be lost if the course "died" or reverted, so a number of criteria relating to the institutionalization of the course were identified. One of these criteria is stable enrollment, especially if students can choose alternative sections to avoid the reformed course. Other criteria related to course stability had to do with faculty resources—the original adopter's ability to continue teaching the reformed course and the existence of other faculty willing and prepared to teach the course. Acceptance of the course—within the department, at higher administrative levels, in other departments (biology, in particular), and at other institutions—would also help ensure course permanence. Finally, idealistically, the course would have some sort of impact on the institutional culture, possibly encouraging others to teach with a more student-active format.

[1] Contact Wendell Potter for materials: whpotter@ucdavis.edu

CP883, *2006 Physics Education Research Conference*, edited by L. McCullough, L. Hsu, and P. Heron

Fidelity to the original course was the one criterion specific to an adapted course, as opposed to a new reformed course. This could include adherence to the core philosophies of the course (based in constructivist learning theory, in this case). Structuring the course so that less time was spent for lecture and more time for discussion/lab was a primary concern for the course originator present at the meeting. Others at the meeting thought that facilitation style would be important, but that a broad range would be acceptable.

Many of the criteria described above lack commonly used and accepted measurement methodologies, let alone benchmarks for success. Predicting instructor burnout or determining how fully the administration has accepted the course require creative measures. Even some of the student outcomes, like degree of cooperation and sophistication of discourse, are studied rarely enough to make choosing a method of measurement and comparing to other courses difficult. Also, deciding on an appropriate benchmark for measuring success presents a challenge. Should results be compared to the original UC Davis course, the prior CSUSM course, other traditional and reformed courses, or an established benchmark for success? For this project, the answer to the question varies for each student measure and is based more on the availability of data than on consideration of which comparison would be most appropriate. With these issues in mind, the next sections present the work that has been completed and plans for future study.

TABLE 1. Suggested criteria for assessing an adaptation/implementation, with data already collected and to be collected (in parentheses) for this project.

Criteria	Data Available
Student Outcomes	
Achievement	
Conceptual understanding	FCI, CSEM, grades
Problem solving	Observation
Reasoning skills	Observation, interviews
Communication/Interaction	
Degree of cooperation	Observation
Sophistication of discourse	Observation
Affect	Student evaluations, interviews
Attitudes/Beliefs	CLASS, observation, interviews
Institutionalization of Course	
Course permanence / stability	
Enrollment stability	Enrollment records
Adopter's enthusiasm / endurance	Personal opinion, interviews
Involvement of other faculty	(Interviews)
Acceptance of the course	(Survey, interviews)
Impact on local culture	(Survey, interviews)
Fidelity to Original Course	
Curricular materials	Course records
Structure (use of time)	Course records
Facilitation	Observation

METHODS

The course being adapted is based on a non-traditional introductory calculus-based physics course for students in the biological sciences [1]. It follows a non-traditional content sequence, beginning with energy conservation rather than kinematics, and has a reduced emphasis on mechanics. Other topics include kinetic theory, waves, pressure and flow, and electricity and magnetism. The topic areas are organized around physical models (e.g., conservation of energy and flow with dissipation) with continued explicit emphasis on the nature of these models and on how to apply the models to solve problems. No standard textbook is used, but rather a set of notes developed by the course originator at UC Davis.

As implemented at CSUSM, there is one instructor who meets with the students twice weekly, with each class meeting lasting 3 hours. This differs from the UC Davis course, which has two 2.3-hour teaching assistant led discussion/labs and one separate 80-minute lecture per week. The students work in an active learning environment with a small amount of lecture woven in. In a typical class meeting, students participate in a number of activities that are about 1 hour in length. Students work together in groups to respond to a series of prompts and then report their responses to their peers in a whole class discussion. Total lecture time is about 1 hour and 15 minutes per week and is used mainly to organize the students' ideas about phenomena they encountered in their group activities.

At CSUSM there are typically 25 to 30 students enrolled in each section of the course, the majority of whom are juniors and seniors. Many are transfer students from community colleges. Most take physics later in their academic career than is suggested by their recommended course schedule. All have either taken calculus or are taking calculus concurrently. The class is ethnically diverse, and often students are the first generation in their family to attend college. Over the course of the implementation, 60% of the students were women.

Students take quizzes every two weeks (every week for the UC Davis course). Quiz questions always emphasize applying the model to make inferences, and students are expected to back up all claims with textual or mathematical arguments. The instructor develops a rubric for each question after reading through a number of responses.

Students have completed two common tests of conceptual understanding—the Force Concept Inventory (FCI) [2] and the Conceptual Survey in Electricity and Magnetism (CSEM) [3]—and the Colorado Learning Attitudes about Science Survey

(CLASS) [4]. The FCI was given as a pre- and post-test during Spring, 2005 and 2006. The CSEM was given as pre- and post-tests in Fall, 2005. Results from the CSEM data are in the process of being analyzed, while CLASS and FCI data are available as of this writing. The CLASS survey was given in Spring 2005 and Spring 2006; Spring 2005 data is presented in this paper.

Qualitative analysis of the implementation will be the subject of a future publication, but we briefly mention the procedure used here for the benefit of others planning to assess an adopted course. Data included field observation and videotaping of 6 classes, interviews with students (four times during the year), and regular interviews with the instructor during the two-year implementation period. These data were analyzed to look for emergent themes relating to both student outcomes and the role of the instructor.

RESULTS AND DISCUSSION

Student Outcomes

Attitudes toward science and sophistication of discourse improved noticeably, while conceptual gains were present but unremarkable. Matched pre-post CLASS data (N=20) indicated a positive shift in the number of favorable responses for 30 out of 36 questions. This shift was distributed across the 7 belief categories. The average percent of favorable responses increased from 52% to 65%. No CLASS data currently exists for the UC Davis course, but compared to other reported data, the initial favorable percentage is lower and the increase is greater [5]. Because physics courses have had a reputation for causing negative shifts on attitude surveys, any positive shift points to success in the course.

Qualitative analysis of video, observation, and interview data gave further support to the attitude shifts evident in CLASS survey results. Students were more likely to work as a group, keep thinking and trying when solutions weren't obvious, focus on the big picture, and apply models than they were before. They also became better at articulating their thinking. More detail on this qualitative analysis will be presented in a future paper.

Initial FCI scores from Spring 2006 (N=26) were quite low (28% correct). The normalized gain for the class (0.3) is comparable to that seen for students in the UC Davis course and is in the middle range relative to other classes (reported in Hake, 1998) with that initial score [6]. Relative to other reformed classes, the gains are not particularly high, but considering the small amount of time given to mechanics (7 weeks) this result is not surprising. As more concept inventories are developed, validated, and used on large numbers of courses, fair assessment of this and other reformed courses that emphasize topics other than mechanics should become easier.

Future data collection efforts may include other concept inventories and analysis of quiz questions identical or very similar to ones used in the UC Davis course in the past. While differences in student populations would make comparisons difficult, this data could still be useful. A more ambitious, and possibly more revealing, instrument would be one aimed at assessing the sophistication of student discourse. Qualitative observation data from this study could serve as a springboard to development of such and instrument.

Other Outcomes

In general, the opinion of those involved in this project is that the institutionalization and fidelity criteria have been met fairly well. For these criteria, however, no standard measures of success exist. Thus, we present what information we have and our plans for gathering further information.

Course enrollment data is available for the initial two-year pilot and the two years of formal adoption. During this period enrollment has grown from 17 to 46 students. It should be noted that while this is the only course designed for students in the biological sciences, students can take the standard introductory physics sequence which is offered in a more traditional lab and lecture format. Based on end of semester evaluations, students appreciate the course. Sixteen out of 35 "strongly agreed" that the "Overall quality of the course is high" in the first semester of the course during Spring 2005 (30 out of 35 at least "agree" with this statement). In the second semester during Fall 2005, 18 out of 41 "strongly agreed" that the "Overall quality of the course is high" (35 out of 41 at least "agreed" with this statement). Further investigation into student opinions about the course is in progress. Over the course of the pilot and formal implementation, the number of students needing to repeat the course due to low grades (C- or below at CSUSM) is 10%.

No complaints from the department, university administrators, other departments, or other universities have been received. The administration has shown support by offering course release time, which was essential for the project. Since the implementation, chemistry instructors started using wipeboards in an upper division physical chemistry course after seeing them in this course, and the physics department is currently reworking standard calculus-based physics labs. As information on acceptance of the course and impacts on local culture currently consists of unsolicited anecdotal information, data in the form of brief surveys or interviews with faculty and administrators will be collected.

So far the first author is the only CSUSM faculty member who has taught the course. Having observed the difficulty many graduate student instructors have in adjusting to the changing role of the instructor in this course, we question whether an instructor "coming in cold" would be able to achieve the same success as the adopter. The adopter plans to gradually involve other tenure-track faculty in the course to ensure course longevity.

Regarding fidelity of the adaptation, the originators of the course foresee allowing some latitude for institutions adapting and implementing the course, and those at the meeting agreed that the CSUSM course fell within the acceptable range. What exactly constitutes fidelity to the course proved to be a more difficult question. A range of facilitation styles, re-ordering of topics, and use of time could be acceptable when applied within the basic philosophy of the course. At the extremes, straying from the basic philosophy would be easy to spot, but otherwise this criterion could prove difficult to measure. Other adapted courses might require more fidelity to the original course and thus this criterion might be more important and easier to measure than it is in this case.

CONCLUSIONS

The CSUSM course has had positive student outcomes, is on its way to becoming a stable presence at the university, and has inspired change in other courses. Does this mean that others could adopt the course at their institutions with similar results? At this point, "it depends," seems to be the most appropriate answer. While this adopter faced structural hurdles such as scheduling two labs weekly for each section while avoiding scheduling conflicts with other courses, he was also was uniquely qualified to implement the course. Having taught the course at UC Davis for 6 years, he was familiar with the discussion/lab activities and the rubric method of grading quizzes. He also has worked to develop a facilitation style that is challenging yet fun for students.

We doubt that an instructor without experience teaching in an active-learning setting could take written material from the UC Davis course and adapt it with the same success as the first author. The ideal method for preparing future adopters would be an apprentice model like those that have worked at UC Davis (for faculty new to the course and graduate student instructors).

For others considering assessing an adaptation and implementation project, we would like to emphasize how useful it was to involve colleagues from the education department. A non-physicist may notice strengths and weaknesses that an experienced physicist would overlook. Also, we note that long term assessment projects would be useful to support, given that some student benefits may come to fruition well after students leave the course.

ACKNOWLEDGMENTS

The authors acknowledge contributions of the workshop participants, Michael Anderson, Emily Ashbaugh, Tom Fleming, Elizabeth Gire, Chance Hoellwarth, Roger King, Pat Len, Graham Oberem, Wendell Potter, and Ed Price. We would also like to acknowledge Elizabeth Gire for her work analyzing CLASS data, and Tom Fleming for his work analyzing FCI data. Special thanks go to the course originator Wendell Potter for his support throughout this project. Lastly, the authors would like to acknowledge support for this work from NSF Grant DUE-0410991.

REFERENCES

1. De Leone, C.J., W.H. Potter and L.B. Coleman (1996). "Radically Restructured Introductory Physics Course at a Large Research University," in *The Changing Role of Physics Departments in Modern Universities, Part One*, AIP Press, p. 829.
2. Halloun, I., and D. Hestenes (1985), The initial knowledge state of college physics students, Am. J. Phys. **53**, 1–43–1055
3. Maloney, D., T. O'Kuma, C. Hieggelke and A. Van Heuvelen (2001). Surveying students' conceptual knowledge of electricity and magnetism, *Phys. Educ. Res., Am. J. Phys. Suppl.* **69** (7) S12-S23.
4. Adams,W.K., K.K Perkins, N.S. Podolefsky ,M. Dubson, N.D. Finkelstein, and C.E. Wieman, (2006) New instrument for measuring student beliefs about physics and learning physics: Colorado Learning Attitudes about Science Survey, *Phys. Rev. ST Phys. Educ. Res.* **2**, 010101
5. Perkins, K.K., W.K. Adams, N.D. Finkelstein, and C.E. Wieman, (2004). "Correlating Student Beliefs With Student Learning Using the Colorado Learning Attitudes about Science Survey" in, *PERC 2004 Proceedings,* edited by J. Marx, P. Heron, and S. Franklin, American Institute of Physics, pp. 61-64.
6. Hake, R.R., (1998) Interactive-engagement versus traditional methods: A six thousand-student survey of mechanics test data for introductory physics courses, *Am. J. Phys*. **66**, 64–74

Diffusion of Educational Innovations via Co-Teaching

Charles Henderson*, Andrea Beach†, and Michael Famiano*

*Department of Physics
†Department of Educational Leadership, Research, and Technology
Western Michigan University, Kalamazoo, MI, 49008, USA

Abstract. Physics Education Research (PER) is currently facing significant difficulties in disseminating research-based knowledge and instructional strategies to other faculty. Co-teaching is a promising and cost-effective alternative to traditional professional development that may be applicable in many situations. This paper discusses the rationale for co-teaching and our initial experience with co-teaching. A new instructor (MF) co-taught with an instructor experienced in PER-based reforms (CH). The pair worked within the course structure typically used by the experienced instructor and met regularly to discuss instructional decisions. An outsider (AB) conducted interviews and class observations with each instructor. Classroom observations show an immediate use of PER-based instructional practices by the new instructor. Interviews show a significant shift in the new instructor's beliefs about teaching and intentions towards future use of the PER-based instructional approaches.

Keywords: Dissemination, Educational Change, Higher Education
PACS: 01.40.Fk, 01.40.Jp

INTRODUCTION

Many models for promoting educational reform in the teaching of college science courses are based on the premise that an individual or group of faculty develops and tests a new instructional strategy and then disseminates it to other instructors. Dissemination typically involves transmission-oriented activities such as talks, workshops, and publications. In this paper we identify some of the difficulties with this model and describe an alternative model that may be helpful in overcoming some of these difficulties.

INSTRUCTIONAL CHANGE

Some proposed instructional innovations seek to make the current structure more efficient (surface changes) while others seek to alter core beliefs, behaviors, and structures (fundamental changes). Most reforms in science teaching call for fundamental changes [1]. This distinction between fundamental and surface changes is important because fundamental changes always face significant resistance while incremental changes often do not [2].

A major obstacle to the implementation of a fundamental instructional change is that instructors attempting to change traditional practices are already acculturated into and surrounded by a system that reflects their current practices and views [3]. Thus, they must undergo a learning (or reacculturation) process to incorporate the new skills and assumptions into their existing mental structures [3].

Even when instructors are able to successfully make internal changes they are typically still immersed in their current situation. Many aspects of this situation likely conflict with their new teaching culture. Commonalities such as large classes, content coverage expectations, classroom infrastructure, scheduling constraints, poor student preparation, and institutional reward structures all appear to favor traditional instruction. In addition, students often resist new instructional strategies.

Most instructors recognize that it is unlikely for any set of instructional materials, no matter how carefully developed, to match the constraints of their specific classroom situation. Local customization is always necessary. Because of the conflicts with existing cultures (both personal and institutional), new instruction that calls for fundamental changes is often altered by instructors during the customization process and implemented as surface changes. Although such implementation may keep some of the surface features of the innovation, it is essentially traditional instruction [4]. Henderson and Dancy have developed the term *inappropriate assimilation* to describe this type of "adoption" [5]. One unfortunate result is that instructors who think that they have adopted an

CP883, *2006 Physics Education Research Conference*, edited by L. McCullough, L. Hsu, and P. Heron

instructional strategy may incorrectly conclude that the strategy is ineffective.

Problems with Standard Dissemination

Common methods of dissemination include talks, papers, and workshops aimed at convincing individual faculty to change their instruction and giving them information and materials in support of a specific research-based strategy. While such strategies may be suitable for the promotion of incremental changes, it is unlikely that such a transmissionist approach can promote fundamental changes. This is for the same reasons that transmissionist instructional strategies for science students typically do not promote significant changes in student mental models. An additional barrier to standard transmissionist dissemination strategies is the complex nature of teaching itself. Similarly to any complex task, much of a teacher's decision-making is implicit [6]. It would be an overwhelming task for a curriculum developer to make all of the necessary implicit decisions explicit and equally overwhelming for an instructor to attempt to internalize these decisions. In contrast, apprenticeship-based models have been shown useful in the development of such tacit knowledge.

Although cost data is not commonly reported in the educational research literature, cost effectiveness of dissemination strategies is clearly an important consideration. In terms of traditional dissemination strategies, cost figures are available in an evaluation of the NSF Undergraduate Faculty Enhancement (UFE) program [7]. According to the evaluation, between 1991 and 1997, UFE made $60,963,917 in awards for faculty development workshops. This funded over 14,400 undergraduate faculty from all types of institutions to attend over 750 workshops. In telephone interviews, 40% of faculty participants said that they had made at least moderate changes in their teaching methods. Thus, each change cost ~$10,600 (not adjusted for inflation). This is a low estimate for a fundamental change since not all changes in teaching methods are fundamental changes and self-report data can be biased in terms of over-reporting changes.

CO-TEACHING

The practice of co-teaching was developed by Roth and Tobin as an alternative to the standard student teaching practice associated with most K-12 teacher preparation programs [8]. In standard student teaching, the student teacher typically first observes a number of the master teacher's classes and then takes over the class on their own. Roth argues that student teachers do not often develop the tacit knowledge necessary to be good teachers under this arrangement [8]. During co-teaching the student teacher and master teacher share responsibility for all parts of the class. Student teachers "begin to develop a feel for what is right and what causes us to do what we do at the right moment" [ref [8], p. 774]. Although we are aware that co-teaching activities have occurred at the college level at other institutions, we are not aware of any other studies that have sought to document the results of such an arrangement.

Co-Teaching Reduces Risk

New instructors are typically risk averse and afraid of making mistakes that may hurt their chances of getting tenure. Thus, any departure from traditional instruction must be made as risk-free as possible – both in terms of student satisfaction and time demands. Co-teaching, as enacted in this project, does this in two ways. First, it allows the experienced instructor to set up a course structure that is known to work in the particular context. This gives the new instructor a safe place to practice new ways of interacting in the classroom and minimizes the risks of problems arising while switching cultures. In terms of required time, during the co-teaching semester, the new instructor has the benefit of previously-used materials and only has to prepare for being in charge of the class about half the time. This leaves additional time and energy available for some of the other more reflective aspects of co-teaching.

Cost of Co-Teaching

In the co-teaching model described in this paper, the only cost is a replacement instructor to teach one class. This allows two instructors to co-teach a single class and have time for additional discussion and reflection without increasing the total amount of time spent on teaching duties. The recommended part-time rate for a 4-credit class at WMU in Fall 2005 was $2,800.

WMU CO-TEACHING PROJECT

The goal of co-teaching in the current study was to acculturate MF into research-based physics instruction as embodied in the design principles developed and enacted by the WMU Physics Teachers Education Coalition (PhysTEC) faculty. The co-teaching took place in the lecture portion of an introductory calculus-based physics course at Western Michigan University. The 4-credit course met each weekday for 50 minutes and contained about 70 students, mostly engineering majors, in a stadium-style lecture hall. CH and MF were both listed as the instructor of record for the course. There were five basic co-teaching activities.

(1) CH and MF alternate being in charge of class. Although both of the instructors were present during each class session, they alternated being "in charge" of the class on a weekly basis. The person in charge would typically preside over whole-class discussions or presentations. Much of the class time was spent by students working in assigned small groups. During this time both instructors would circulate around the lecture hall and interact with groups. The first draft of weekly quizzes or exams were developed by the instructor in charge.

(2) Weekly meetings between CH and MF. Each Friday, CH and MF met for approximately 1 hour. During that time they talked about how things went during the past week and any difficulties that arose. The instructor in charge of the following week would then present his initial plans, which were discussed. In addition to this weekly scheduled meeting, CH and MF frequently had shorter discussions about the course at other times.

(3) Course structure set up by CH to support PhysTEC design principles. The research-based course structure was specifically designed so that much of class time would be used having students working together in small groups while discussing important physics ideas.

(4) MF had access to materials used by CH in previous offerings of the course. At the beginning of the semester, CH gave MF a CD with electronic copies of all the course activities and assignments used in the previous semester. MF typically used, with minor modifications, about half of these and developed the other half of the course activities himself.

(5) MF teaches course on his own during subsequent semester. MF taught the same class on his own during the following semester -- Spring 2006.

DATA COLLECTION AND ANALYSIS

A faculty member from the college of education (AB) conducted individual interviews with CH and MF at the beginning, middle, and end of the co-teaching semester. Interviews focused on thoughts about how the course was going, general beliefs about teaching and learning, and perceptions about co-teaching. A final interview was conducted with MF at the end of Spring 2006 once he had taught the course on his own. AB also observed class sessions at the beginning, middle, and end of the co-teaching semester.

CH and AB independently analyzed the interview transcripts and observation field notes looking for three things: 1) evidence related to MF's instructional practices; 2) evidence related to MF's beliefs about teaching and learning; and 3) evidence related to MF's intentions towards future instruction. There was a large degree of agreement between the two analyses and all disagreement was resolved through discussion.

RESULTS

Teaching Practices: During the observations by AB, both CH and MF were judged to be working with similar effectiveness within the interactive PhysTEC class structure. She did notice some more subtle differences, though. For example, in her first observation of MF she writes "*MF was somewhat more structured than I saw CH . . . MF presented concepts and then problems that exemplified them. Less of having students generate concepts.*"

Although noticeable, these differences were considered by AB to be minor compared to the large difference between the co-taught course and traditional science courses. There no noticeable changes in MF's instructional practices during the semester. These observations suggest that the scaffolding provided by the course structure was effective, right from the start, in helping MF teach in a non-traditional way.

Without this structure it is likely that MF would have put much more emphasis on facts and principles lecturing. The likelihood of this was confirmed by MF during the first interview "*[If I were teaching the course by myself] I would probably treat it more like a lecture.*" (MF1#222-223)

Beliefs About Teaching and Learning: MF's beliefs about how students learn appeared to be consistent throughout the semester and largely aligned with the beliefs behind the PhysTEC course structure. Thus, even though MF envisioned his teaching as some variation of a traditional lecture, he did not think that students would get a lot out of such a lecture. "*A student sitting in a lecture listening to you is going to do what most students do, and that is, fall asleep or walk away and not learn something.*" (MF1#128-130). This disconnect between beliefs about learning and instructional practices appears to be common in all types of faculty [9].

Although MF's beliefs about student learning were consistent throughout the semester, his beliefs about teaching appeared to change. He was initially concerned that the PhysTEC course structure was too much of a departure from the lecture method. This changed by the middle of the semester. "*It really seems to be a good method.*" (MF2#88) His largest initial concern appeared to be student resistance to such an interactive class structure. Thus he did not envision such methods being successful until he experienced the students being engaged.

Intentions Towards Future Instruction: As MF's beliefs about teaching changed, his intentions towards future instruction also began to change. MF was initially skeptical about the course design where much of the class time was used having students work together on problems. By the mid-term interview, he was beginning to become comfortable with the course design, but was still largely non-committal about his future instruction. *"You know, it [the co-teaching experience] taught me something that I am going to adopt aspects of in future courses."* (MF2#196-197) By the end of semester, though, he seemed to have shifted his perception to be very favorable towards the course structure. "*My class [next semester] is going to be very similar to what we did last semester. . . .It's going to be almost identical.*" (MF3#272-273)

Even though at the end of the Fall co-teaching experience, MF indicated that his Spring 2006 course would be "almost identical" to the co-taught course, he later decided to make some changes to the course structure. Although the Spring 2006 was well within the PhysTEC structure, all movement was towards a more traditional course. In addition, almost all changes were made in order to reduce the instructor time required or to reduce perceived student dissatisfaction. For example, in Spring 2006, MF decided to change the written homework from a group to an individual assignment. This was largely based on his perception that students did not like the group homework assignments. In contrast, although quiz corrections were quite popular with students, MF decided not to use quiz corrections in Spring 2006 due primarily to the extra time required for grading.

The final interview at the conclusion of the Spring 2006 semester revealed that MF was unhappy with many of the changes that he made and planned to go back to a course structure more closely aligned to the Fall 2005 course. He indicated that his direct experience with co-teaching followed by teaching alone convinced him that the course elements were important enough in promoting student learning that they were worth extra time and possible student dissatisfaction.

CONCLUSIONS

Co-teaching appears to have been successful in creating a fundamental instructional change -- more closely aligning MF's beliefs and instructional practices with the WMU PhystTEC design principles.

Co-teaching is a cost-effective model that shows significant promise for the dissemination of instructional innovations. It appears to be able to provide a much larger change for much less money than common workshop models since the only cost is a replacement to teach one course so two instructors can team up without increasing their workload. Co-teaching is effective because it immerses an instructor in the new instructional context and provides scaffolding and modeling to ensure success. An interesting question for future study would be to examine the effectiveness of co-teaching in promoting instructional change in experienced faculty.

Of course, co-teaching is only appropriate when there is a teacher available who is experienced in teaching the target course in a research-consistent manner. Thus, co-teaching cannot solve all dissemination problems. Yet, when the conditions are right, it should not be overlooked. With the expanding presence of PER researchers/teachers in physics departments, it could have a significant impact.

ACKNOWLEDGEMENTS

This project was supported, in part, by the Physics Teacher Education Coalition (PhysTEC), which is funded by the National Science Foundation.

REFERENCES

1. J. Handelsman, D. Ebert-May, R. Beichner, P. Bruns, A. Chang, R. DeHaan, J. Gentile, S. Lauffer, J. Stewart, S. M. Tilghman and W. B. Wood, "EDUCATION: Scientific Teaching," *Science*. **304**, 521-522 (2004).
2. R. F. Elmore, "Getting to Scale with Good Educational Practice," *Harvard Ed. Review*. **66** (1), 1-26 (1996).
3. M. Fullan, *The New Meaning of Educational Change*, New York: Teachers College Press, 2001.
4. J. P. Spillane, *Standards Deviation: How Schools Misunderstand Educational Policy*, Cambridge, MA: Harvard University Press, 2004.
5. C. Henderson and M. Dancy, "When One Instructor's Interactive Classroom Activity is Another's Lecture: Communication Difficulties Between Faculty and Educational Researchers," (Paper presented at the AAPT Winter Meeting, Albuquerque, NM, 2005).
6. D. C. Berliner, "Ways of thinking about students and classrooms by more and less experienced teachers," in *Exploring Teachers' Thinking*, edited by J. Calderhead, London: Cassell Publishing, 1987.
7. C. Marder, J. McCullough and S. Perakis, *Evaluation of the National Science Foundation's Undergraduate Faculty Enhancement (UFE) Program*, SRI International, 2001.
8. W.-M. Roth, D. Masciotra and N. Boyd, "Becoming-in-the-classroom: A case study of teacher development through coteaching," *Teaching and Teacher Education*. **15**, 771-784 (1999).
9. L. Norton, J. T. E. Richardson, J. Hartley, S. Newstead and J. Mayes, "Teachers' beliefs and intentions concerning teaching in higher education," *Higher Education*. **50**, 537-571 (2005).

Assessing the Effectiveness of a Computer Simulation in Introductory Undergraduate Environments

C.J. Keller, N.D. Finkelstein, K.K. Perkins, and S.J. Pollock

University of Colorado at Boulder
Department of Physics, Campus Box 390 Boulder, CO 80309

Abstract. We present studies documenting the effectiveness of using a computer simulation, specifically the Circuit Construction Kit (CCK) developed as part of the Physics Education Technology Project (PhET) [1, 2], in two environments: an interactive college lecture and an inquiry-based laboratory. In the first study conducted in lecture, we compared students viewing CCK to viewing a traditional demonstration during *Peer Instruction* [3]. Students viewing CCK had a 47% larger relative gain (11% absolute gain) on measures of conceptual understanding compared to traditional demonstrations. These results led us to study the impact of the simulation's explicit representation for visualizing current flow in a laboratory environment, where we removed this feature for a subset of students. Students using CCK with or without the explicit visualization of current performed similarly to each other on common exam questions. Although the majority of students in both groups favored the use of CCK over real circuit equipment, the students who used CCK *without* the explicit current model favored the simulation more than the other group.

Keywords: Computer Simulation, PhET, Peer Instruction, Laboratory, Electric Circuits.
PACS: 01.40.Fk, 01.50.H-, 01.50.Lc

INTRODUCTION

Our previous studies on the use of simulations in college environments have investigated a simulation's impact on conceptual understanding in labs and recitations. In interactive recitations using *Tutorials in Introductory Physics* [4], we demonstrated that students perform similarly on measures of conceptual understanding using either a simulation or real equipment [5]. However, in a traditional laboratory, students using a simulation outperformed their counterparts who used real equipment on a conceptual survey of the material and in coordinated, hands-on tasks [6]. We follow these lines of inquiry by studying the use of a simulation in an interactive lecture that utilizes *Peer Instruction* [3] and an inquiry-based laboratory environment to study the effects of an explicit visual model of current flow within this simulation on students' conceptual understanding and their attitudes towards the simulation.

A computer simulation, known as the Circuit Construction Kit (CCK), was introduced into these environments to investigate its impact. In the study conducted in lecture, we observe significant improvements on concept test performance in the domain of DC circuits by students who view a demonstration using CCK compared to those who view an equivalent physical demonstration and associated chalkboard explanation. In the study conducted in an inquiry-based laboratory, we observe no difference in exam performance between students who used versions of CCK with or without the explicit model for current. The majority of students in both groups favored the use of CCK over their experience with real equipment. Finally, students who used CCK *without* the explicit current representation rated the simulation more favorably than students using CCK *with* this representation.

PHET SIMULATIONS

The simulation used in these studies was developed and tested by the Physics Education Technology (PhET) project [1, 2]. The PhET project has developed approximately 60 freely downloadable physics, chemistry, and mathematics simulations that include most topics covered in a typical introductory physics sequence. The simulation, CCK, (Fig. 1) allows students to build simple DC circuits using batteries, wires, resistors, light bulbs, and switches. The simulation utilizes Kirchhoff's laws to accurately model current and voltage for circuits created by the user. A virtual workplace is provided where users can place components, connect them together, and

CP883, *2006 Physics Education Research Conference*, edited by L. McCullough, L. Hsu, and P. Heron

measure current and voltage using virtual ammeters and a voltmeter. Additionally, CCK provides the user an explicit visual representation of current flow by representing electrons as small spheres dots that obey current conservation. As part of this study, we explore what happens when this visual current model is not present.

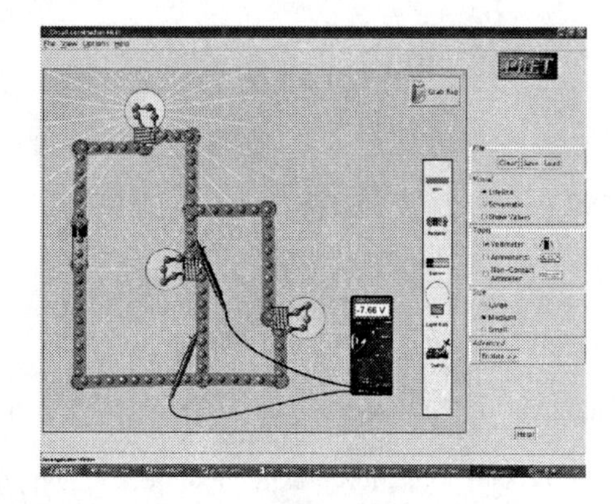

FIGURE 1. Screen shot of CCK *with* the visual representation of current. Modified version of CCK *without* current representation does not show spheres in circuit elements.

EFFECT OF *PEER INSTRUCTION*

Classroom Environment: The first study took place in Fall 2004 in a large, interactive, calculus-based, introductory physics course at the University of Colorado at Boulder. This course was the second semester in a two-semester sequence intended mainly for engineering and physics majors and consisted of 360 enrolled students. Topics include electricity, magnetism, waves, and lenses. The course was divided into two nearly identical 50-minute lectures, each meeting three times per week with the same instructor. One lecture was held at 10:00am ('10am,' N~180), and the other was held at 12:00pm ('noon,' N~180). *Peer Instruction* was implemented in these lectures along with clickers. Students met weekly for a 50-minute, TA-led recitation, during which students worked on *Tutorials* in small groups. For a complete description of this course, see [7].

Procedure & Data Collection: When a concept test was initially given, students were instructed to not discuss with their peers and answer the question independently. After students responded, the instructor did not present the students' aggregate responses to the entire class. The instructor would perform a demonstration or lecture, and students were allowed to discuss the question with each other. Students then responded individually a second time to the same question. The results of students' responses were then revealed to the class and the instructor would present the answer, if necessary.

Data were collected on 5 concept tests given in this format over the term on different topics. For 2 of the 5 questions, CCK was shown to students in the 10am lecture, and an equivalent explanation or physical demonstration was performed in the noon lecture. As a control, the remaining 3 questions were carried out in a similar manner in both lectures.

Results: Because students answered the same concept test twice, pre and post discussion results are available. In Fig. 2, the students in the 10am lecture have a statistically larger absolute gain (pre to post) than the noon lecture on the 2 questions where CCK was shown (p=0.002, two-tailed z-test). The average of these questions for students who saw the simulation improved from 59.8% to 92.1%, while students who did not see the simulation improved from 61.2% to 83.1%. The remaining 3 questions demonstrate that the two lectures are improving similarly during the same instruction (p=0.54).

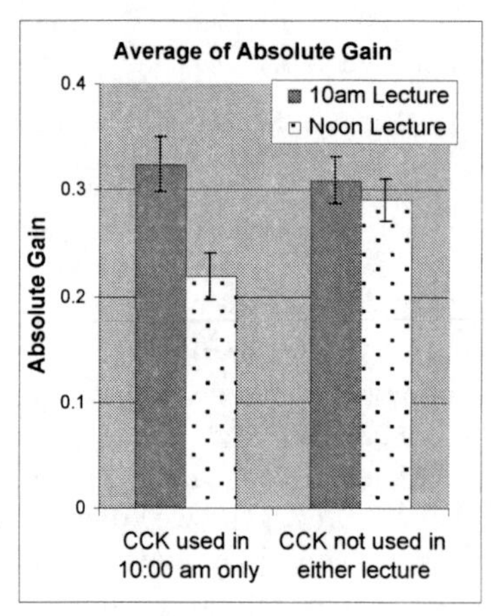

FIGURE 2. Absolute gain for 5 concept tests. CCK was only used in the 10am lecture for data on left side.

Discussion: Although a more exhaustive study is necessary to make substantial claims, these results suggest that a simulation can potentially spur more productive discussion than viewing real demonstrations during *Peer Instruction*. One possible explanation for these results is that CCK's explicit visual representation for current provides a more productive feature upon which students may base their discussions.

EFFECT OF VISUALIZATION

Classroom Environment: This course was the second semester of a two semester sequence in algebra-based introductory physics at the University of Colorado at Boulder held Fall 2005 (N~160). Topics covered in this course included E&M and modern physics. This course has a coupled two-hour laboratory section that met weekly, consisting of 15 to 30 students each. Four teaching assistants were assigned to seven laboratory sections. Students completed 10 inquiry-based labs during this 15-week course, working in groups of 2 to 5 students. During weeks where a laboratory was not assigned, a discussion section was held instead.

Procedure and Data Collection: The 3rd and 4th labs of the semester covered voltage and current in DC circuits, respectively. During the 3rd lab, *all* students in the course used real equipment to complete the lab. During the 4th lab (referred to as the 'CCK lab'), all students used CCK and two groups were formed—one group of students used CCK *with* the explicit current visualization ('Current group,' N=65, 3 laboratory sections), and the second group used CCK *without* the current visualization ('No Current group,' N=90, 4 laboratory sections). Both groups completed the same laboratory on DC circuits. Students were given no specific instructions on how to use the simulation, nor did any have prior formal experience using CCK.

Data assessing student conceptual performance were collected from three exams over the course of the semester. A common midterm exam was given during the 9th week of instruction, approximately 3 weeks after the intervention. This exam consisted of 20 multiple-choice questions, 7 of which were related to DC circuits. The BEMA exam[1] [8] was given to all students only during the last (15th) week of instruction. The BEMA was *not* given during the first week of instruction[2]. During the 16th week, a common final exam was given consisting of 30 multiple-choice questions, 2 of which were related to DC circuits.

Online prelabs contained Likert-scale questions (1 to 5; *not at all* to *very* useful/enjoyable) probing students' attitudes towards the perceived usefulness and enjoyment of the prior week's laboratory. Two additional questions were added to compare the usefulness and enjoyment of real circuit equipment with CCK after students completed the 3rd and 4th labs.

Results: The average of all 20 questions on the midterm exam for both groups is statistically similar, as are the averages of all 7 questions related to DC circuits and all 13 questions on other topics (Table 1).

The average of all 30 questions on the final exam for both groups is statistically similar (p=0.5), with the Current group averaging 57.4% (N=60) and the No Current group averaging 56.5% (N=78). The averages of the 2 questions on DC circuits are also statistically similar, and the averages for the Current and No Current groups are 75.0% and 73.7% (p=0.8).

TABLE 1. Results of midterm and BEMA exam.

	Questions (p-value)†	% correct [standard error]	
		Current Group, N=60	**No Current Group, N=81**
Midterm	All (p=0.2)	60.3 [1.4]	62.8 [1.2]
	DC Circuit (p=0.7)	66.1 [2.3]	67.4 [2.0]
	Non DC-Circuit (p=0.2)	57.2 [1.8]	60.3 [1.5]
		N=54	**N=68**
BEMA	All (p=0.6)	37.9 [1.2]	37.0 [1.1]
	DC Circuit (p=0.9)	37.6 [2.5]	37.8 [2.2]
	Non-DC Circuit (p=0.5)	38.0 [1.4]	36.7 [1.2]

†p-values calculated by a two-tailed z-test.

The averages for post-BEMA questions in aggregate appear in the bottom of Table 1. All three categories are statistically similar.

Overall, we observe more favorable attitudes towards the CCK lab compared to all other labs during the term (Fig. 3). The average student rating regarding the usefulness of the CCK lab is more positive for *both* groups than the average of the remaining labs. This trend is even more pronounced for the enjoyment of the labs. Interestingly, we do observe differences between the two groups in their attitudes towards the CCK lab, with the No Current group having a more positive attitude towards the usefulness of the CCK lab than the Current group ($p<0.002$). We observe a similar difference on the CCK lab for the enjoyment of the labs (p=0.01). The differences between the two groups on all other individual labs are statistically similar ($p>0.05$).

The two additional questions asking students to compare CCK to real equipment also demonstrate favorable attitudes towards CCK. The fraction of students that rated the simulation as better than real equipment in terms of utility was 77.6% and 88.2% for the Current and No Current groups (p=0.08), respectively. The average ratings of comparative utility (on a scale from 1 to 5, *much less* to *way more* useful/enjoyable) are 4.13 for the Current group and 4.41 for the No Current group, (p=0.05). The question asking students to compare their enjoyment of the simulation to real equipment also demonstrates favorable attitudes towards CCK, where 56.1% and 79.3% of the students gave a favorable response (Current and No Current group, respectively; $p<0.002$). The average for the Current group (on a scale from 1 to 5) is 3.68, while the No Current group has an average of 4.13 (p=0.008).

[1] This exam also included questions 10, 11, and 12 from the ECCE [9]. Additionally, questions 9, 12, 18, 28, and 29 of the BEMA are not included because this material was not covered in the course.

[2] Scores tend to be predictably low on the pre BEMA exam.

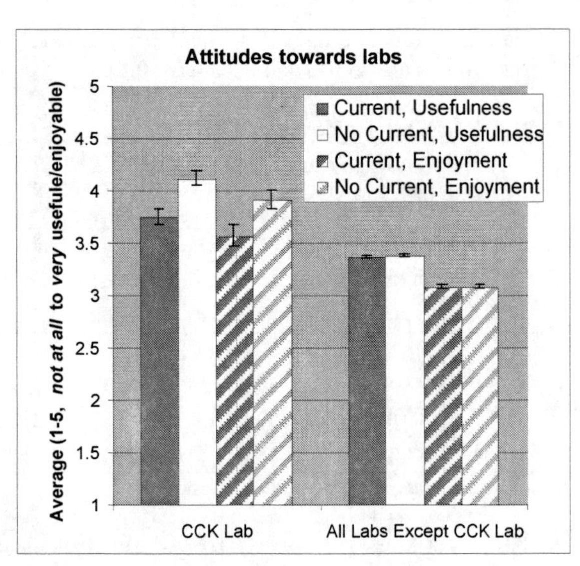

FIGURE 3. Average response from students on two questions regarding the usefulness and enjoyment of labs.

Discussion: No differences in conceptual understanding of DC circuits were observed between students who did and did not see the current visualization when using CCK to complete a laboratory. However, we do observe differences between students' attitudes towards the two different versions. Students who used CCK *without* the current model had a more positive attitude towards the perceived usefulness and enjoyment of the simulation. We hypothesize that this surprising result stems from how well matched the version of CCK was to the lab task. Perhaps predictions the students must make about circuits during the lab causes the lab to be unchallenging for these students who used a tool that explicitly shows current. Future work will include interviews to expand the hypothesis and design of curriculum to exploit the affordances of CCK.

Despite these differences, it should be noted that *both* groups had more favorable attitudes towards CCK than real equipment. The majority of students felt that using CCK was more useful and more enjoyable than their previous experience with real equipment.

CONCLUSION

This paper presents two studies documenting the effect of a computer simulation, known as CCK, in two different classroom environments. In the first study conducted in lecture, we observe significant improvements on concept test performance in the domain of DC circuits by students who view a demonstration using CCK compared to those who view a physical counterpart. In the following study conducted in an inquiry-based laboratory, we observe no difference in exam performance on questions relating to DC circuits between students who used CCK either with or without the explicit model for current. Despite the absence of a difference on conceptual understanding between the two groups, we do find that students who used CCK *without* the explicit current model rated the simulation more favorably than students using CCK *with* this model. We believe this to be related to how well matched the tool is to the task.

It should be noted that the majority of students in both groups favored the use of CCK over their experience with real equipment.

ACKNOWLEDGMENTS

This work is supported by NSF-DUE #0442841 and #0410744, NSF REC #0448176, the Kavli Operating Institute, Hewlett Foundation, Colorado PhysTEC, APS, AIP, and AAPT. The authors would like to express their gratitude to the PhET Lead Programmer, Ron LeMaster, and the creator of CCK, Sam Reid. Many thanks to Carl Wieman and the rest of the Physics Education Research Group at the University of Colorado at Boulder.

REFERENCES

1. K. Perkins, W. Adams, M. Dubson, N. Finkelstein, S. Reid, C. Wieman, R. LeMaster. "PhET: Interactive Simulations for Teaching and Learning Physics," *Physics Teacher* **44**, 1 (2006).
2. Physics Education Technology Project, http://phet.colorado.edu
3. E. Mazur, *Peer Instruction*. Prentice Hall, New Jersey, 1997.
4. L.C. McDermott, P.S. Schaffer. Prentice Hall, New Jersey. 2002.
5. C.J. Keller, N.D. Finkelstein, K.K. Perkins, and S.J. Pollock. *PERC 2005*, Salt Lake City, Utah.
6. N.D. Finkelstein, W.K. Adams, C.J. Keller, P.B. Kohl, K.K. Perkins, N.S. Podolefsky, S. Reid, R. LeMaster. Phys. Rev. ST Phys. Educ. Res. 1, 010103 (2005).
7. N.D. Finkelstein, S.J. Pollock, Phys. Rev. ST Phys. Educ. Res. **1**, 010101 (2005).
8. L. Ding, R. Chabay, B. Sherwood, R. Beichner, Phys. Rev. ST Phys. Educ. Res. **2**, 010105 (2006).
9. The ECCE (Thornton & Sokoloff) is part of Workshop Physics: http://physics.dickinson.edu/~wp_web/wp_resources/wp_assessment.html

Characterization of Instructor and Student Use of Ubiquitous Presenter, a Presentation System Enabling Spontaneity and Digital Archiving

Edward Price*, Roshni Malani†, and Beth Simon†

*Department of Physics
California State University, San Marcos

†Computer Science and Engineering Department
University of California, San Diego

Abstract. Ubiquitous Presenter (UP)* is a digital presentation system that allows an instructor with a Tablet PC to spontaneously modify prepared slides, while automatically archiving the inked slides on the web. For two introductory physics classes, we examine the types of slides instructors prepare and the ways in which they add ink to the slides. Modes of usage include: using ink to explicitly link multiple representations; making prepared figures dynamic by animating them with ink; and preparing slides with sparse text or figures, then adding extensive annotations during class. In addition, through an analysis of surveys and of web server logs, we examine student reaction to the system, as well as how often and in what ways students' utilize archived material. In general, students find the system valuable and frequently review the presentations online.

*http://up.ucsd.edu/about/

INTRODUCTION

Electronic lecturing possesses strengths and limitations. Computer-based digital projection allows the use of high quality pictures or diagrams, the incorporation of simulations, applications, or web materials, and can facilitate sharing of prepared content by instructors or web-based publication. In contrast, board-based lecturing possesses a naturally controlled pacing and allows extemporaneous presentation of material. Most PER-based curricula de-emphasize lecturing, yet lecturing remains widespread; furthermore, many alternate uses of class time require a shared presentation space. As a result, the features of presentation systems and the ways they are used are important for traditional and PER-based activities. The idea of affordances – those uses to which a tool is naturally suited – is useful for comparing presentation methods.[1] Following Norman, we use 'perceived affordances' to mean "the perceived and actual properties of the thing... that determine just how the thing could possibly be used."[2] While the chalkboard affords impromptu presentations, digital presentation systems such as PowerPoint do not. Table 1 compares the perceived affordances of these presentation modes.

The Tablet PC is an augmented laptop computer with a stylus that can be used to "write" on the screen. Ubiquitous Presenter (UP) is a Tablet PC-based system developed at the University of California, San Diego, based on Classroom Presenter.[3-5] UP allows faculty to write on ("ink") and augment prepared digital material (slides, pictures, etc.) in real time in class. Ink is automatically archived stroke by stroke and can be reviewed synchronously via a web interface. Thus UP uniquely combines affordances of both digital and board-based presentation, as shown in Table 1. The system also supports in-class interaction by students with web-enabled devices – though that aspect of the system is beyond the scope of this paper.[6]

In this paper, we will explore the use of UP's inking and student review features through the study of two introductory physics classes. We characterize ways in which the two instructors capitalized on UP's affordances to use ink in combination with prepared materials during lecture, and analyze specific uses that are enabled by UP's affordances: linking multiple representations, filling in templates or "sparse slides", and adding dynamic elements to prepared material. We do not claim that these techniques could not be

TABLE 1. Perceived affordances of different presentation systems.

Perceived affordance	Board	Digital	UP
Spontaneous changes	Yes	No	Yes
Archiving	No	Yes	Yes
Including prepared material	No	Yes	Yes
Natural pacing	Yes	No	Yes

CP883, *2006 Physics Education Research Conference*, edited by L. McCullough, L. Hsu, and P. Heron

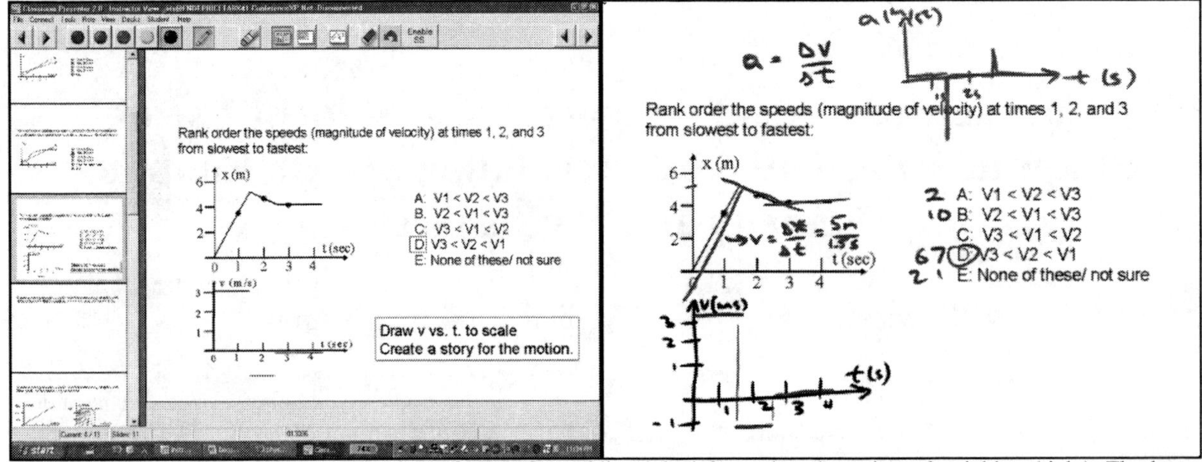

FIGURE 1. Screenshot of instructor's view of UP before adding ink (left), and student view after inking (right). The lower graph and text in the instructor's view are instructor objects, and are not visible to the students.

accomplished using traditional presentation systems; rather, that they are naturally and easily employed given UP's affordances. We also provide preliminary analysis of student use of the system through self-reported student surveys and web statistics on actual system use.

INSTRUCTOR USE

In this section, we explore the ways in which two physics instructors used UP in their introductory courses. We focus on characterizing patterns of use common to both instructors, and relate those uses to the affordances described above. Course A (57 students, taught by EP) was a semester long course for physical science and mathematics students at a public regional university; course B (180 students) was a quarter long course for life science students at a public research university. Both instructors were using UP for the first or second term, were trained and supported in their use of the system, and described themselves as comfortable with the system.

During class, instructors can add ink of several colors, erase, undo, and create extra blank space. Figure 1 shows a fully inked slide, in this case, an in-class question with solution. An individual "ink stroke" is captured as a placement of the Tablet pen on the screen until the moment of lift from the screen. Writing a single word may account for several strokes. Drawing of diagrams almost always happens in many strokes. In Course A, 40 lectures had inked slides, out of 42 class meetings. On average, lectures had 6.9 inked slides, with 31.7 ink strokes. In course B, 24 lectures had inked slides, out of 29 class meetings. Representing an alternate use, lectures in course B had more slides (12.8 on average) with less ink (19.6 strokes on average), as compared to course A.

Figure 2A shows an example of linking multiple representations. In this slide, the instructor has used lines to link pictorial representations to mathematical representations. Variables are thereby explicitly and

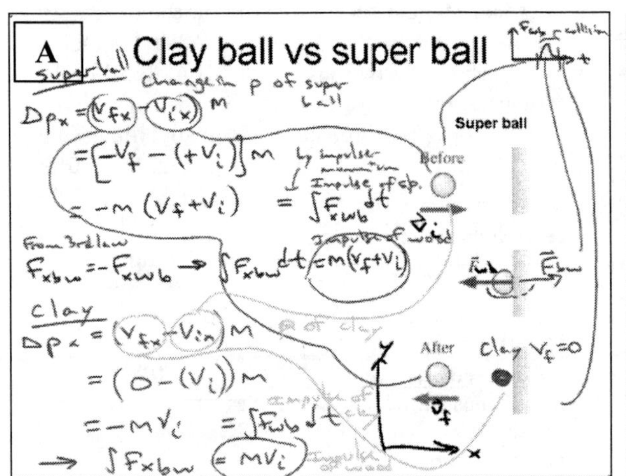

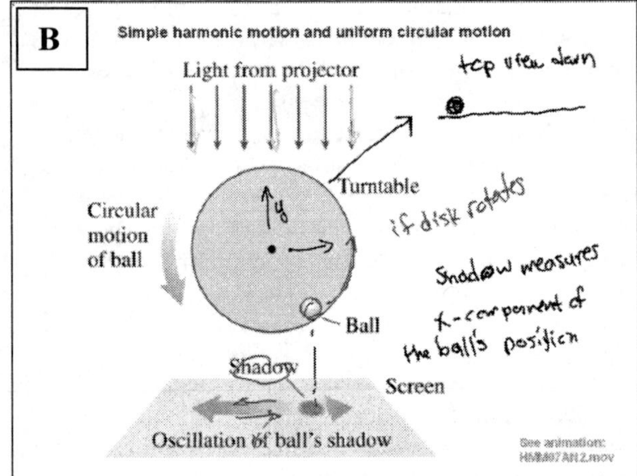

FIGURE 2. Fully inked slides showing use of ink to link representations (A) and make a figure dynamic (B).

graphically connected to the physical situations they represent. Instructors also linked pictorial and graphical representations, written problem statements and mathematical representations, and variables and expressions substituted for them. This pattern of usage rests on the affordance of spontaneous inking: instructors can easily add or erase a line in a different color. While an instructor writing on the board or using PowerPoint might connect representations with a gesture, instructors writing on a TabletPC running UP readily used ink for this purpose, thereby creating a graphical link.

Next, we describe two patterns of use that rely on mixing prepared materials with ink added in class. The instructors often extensively inked sparse slides – slides with little prepared material. The sparse prepared material included outlines (filled out in class); rhetorical, motivational, or transitional statements (leading to related analysis or derivations); problem statements (worked as examples in class); and graph or figure templates (drawn on in class, *e.g.* to show data collected in class). Thus, the prepared material may act as an anchor, a prompt, or a workspace, while the ink is added dynamically, so that the presentation is "created" in front of students. Instructors are frequently frustrated with the "canned" feel of PowerPoint lectures and the limitations of animation that must be prepared in advance. Because UP affords spontaneous inking, instructors can capture the "live" feel of a chalkboard lecture. Indeed, we find that slides are often inked extensively, to an extent that would be overwhelming if presented all at once instead of built up incrementally.

In contrast to extensively inking sparse slides, both instructors also used prepared figures extensively, often drawn from materials provided by the textbook publisher. Prepared figures often served as a focal point for inking, and instructors often used ink to make features of the figure dynamic, as shown in Figure 2. In this example, the instructor has traced over the light rays, drawn a line showing the motion of the ball, and indicated the path of the moving shadow. The figure is thus transformed from a static picture to a dynamic animation, a technique that is particularly suited to describing physical systems.

STUDENT USE

Students experienced the system during class and were also able to view slides online after class. They could view slides without the ink added during class (uninked), fully inked (the default), or at any intermediate stage. Students could therefore "replay" the lecture. We assessed student use of the system in two ways: surveying students and analyzing web server logs of actual user access. We consider both data sources in forming a comparative picture of courses A and B.

Survey results

A self-report style survey reveals students' perceived benefits from use of the system. Response rates were 74% (course A) and 54% (course B) of enrolled students. Not all students responded to all questions. Table 2 indicates the number of students who reported that the system had a very positive or slightly positive impact. The first two rows report on the overall impact; the last four report on the impact on student learning experience based on specific instructor usage modes.

Though differences in reported positive impact exist between the courses, a majority of students felt the system had a positive impact. For both courses, students respond more positively to questions about specific pedagogical impacts (such as drawing a diagram) than to general questions on the impact of the system on attention and understanding. Differences between the courses may reflect differences in instructors' use of the system and student populations.

Students were also asked to characterize their use of the web to review instructor ink after class: given five options, select all that apply. The options were "review of class within a few days", "solve homework problems", "review for a test or a quiz", "because I was not present for class physically", and "other". The most common reply in both courses (43% for A, 74% for B) was to review for a test or a quiz. In both courses, server hits increased just before tests and quizzes. The next most common response in course A was to solve homework problems; in course B, to review a class within a few days. In both courses, approximately 1/4 of the students reported at least one use of the inked slides because they were not present for class physically. These usage patterns reflect the course requirements: homework was graded for credit

TABLE 2. Students reporting positive impacts.

Topic	Course A	Course B
Attention to lecture	81% (34/42)	53% (52/98)
Understanding of lecture materials	71% (30/42)	62% (60/97)
Gave answers to student questions	86% (36/42)	54% (51/95)
Explained a concept	83% (35/42)	73% (71/97)
Drew a diagram, or picture	93% (39/42)	77% (74/96)
Used pen colors	81% (34/42)	80% (77/96)

in course A but not course B, and tests were biweekly in course B but approximately monthly in course A.

Web results

The UP web server records detailed information about when students access the online notes, including which lectures, which slides, which version of ink on a slide, etc. Furthermore, the instructor's actions of inking in class and changing slides are also recorded.

In course A, 93% of students created a user account and at least viewed one lecture slide, while 83% of course B students did the same. The number of slides in course A is 631, and 3 of the 53 (5.7%) students had at least that many hits – meaning they potentially looked every slide in the course, assuming they just looked at the "final inked" versions of each slide. The number of slides in course B is 473, and 62 of the 149 (41.6%) students had at least that many hits. There are many possible explanations for the dramatic differences in student access rates in the two courses, including student motivation, time for studying, study habits, and perceived value of reviewing the slides.

Students are overwhelming more likely to only look at the final inked version of a given slide, rather than to employ the ink replay feature that would allow them to review any process revealed by the incremental inking on the slide. In course A, 8 out of 53 (15.1%) of users had 10% or more of their server hits on "progressive" inked slides (not a completely uninked or "final" inked slide). In course B, the students viewed relatively more progressive inked slides: 41 out of 149 students (27.5%) exceeded the 10% threshold. This is an unexpected finding; the developers and instructors anticipated that the ability to view the lectures notes in progress would be valuable to and utilized by students.

Given that most students primarily viewed fully inked versions of the slides, the number of hits (or traffic) that the students generated can be analyzed in comparison to the total number of slides available in their course. Table 3 shows the number of students who viewed less than 50%, between 50-100%, between 100-200%, and more than 200% of the total number of slides available. Compared to students in course A, more students in course B accessed a greater percentage of available slides, a difference that may be related to differences between the students' goals, motivation levels, extra-curricular commitments, and perceived value of this kind of review.

TABLE 3. Student review of slides on the web. Number who viewed certain % of available slides

% Slides viewed	Course A	Course B
< 50%	77% (41/53)	44% (66/149)
50% – 100%	17% (9/53)	14% (21/149)
100% - 200%	6% (3/53)	20% (30/149)
> 200%	0% (0/53)	21% (32/149)

CONCLUSIONS

Tablet PCs can support novel instructor uses due to a combination of affordances not found in other presentation methods, specifically the ability to include prepared electronic materials and add spontaneous ink. In a review of two instructors' inked lectures from introductory physics classes, we note common uses of ink to link multiple representations, fill in a template or sparse slide, and animate prepared diagrams or pictures. Additionally, the Ubiquitous Presenter system's automated archiving of lecture materials provides an interesting new resource for students by making all instructor content available after class. In general, students report a positive perceived impact on their learning and understanding. Students' reported reasons for reviewing lecture material online are consistent with course structure. Though many students reviewed material online, surprisingly few students replayed the "process" of any given slide from lecture.

ACKNOWLEDGMENTS

We would like to thank Barbara Jones and Shane Walker for providing access to their lecture materials and for supporting the survey of their students. Additionally, this work was made possible, in part, by grants from Microsoft Research and Hewlett Packard.

REFERENCES

1. Gibson, J. J.. *The Ecological Approach to Visual Perception.* Boston: Houghton Mifflin (1979).
2. Norman, Donald A. *The Design of Everyday Things.* New York, Doubleday (1988).
3. Wilkerson, M., Griswold, W., Simon, B. in *SIGCSE Technical Symposium on CS Ed.*, Feb. 2005
4. Anderson, R. Anderson, R., Hoyer, C., Price, C., Su, J., Videon, F., Wolfman, S. Computers and Graphics, **29** 480 (2005)
5. Anderson, R., Anderson, R., Hoyer, C., Wolfman, S. in *Proceedings of CHI (Conference on Human Factors in Computing Systems)*. pp. 567-574 (2004)
6. Examples are at http://physics.csusm.edu/eprice

Impact of a Classroom Interaction System on Student Learning*

Joseph Beuckman, N. Sanjay Rebello and Dean Zollman

Department of Physics, 116 Cardwell Hall, Kansas State University, Manhattan, KS 66506-2601

Abstract. We have developed and implemented a Web-based wireless classroom interaction system in a large-enrollment introductory physics lecture class that uses HP handheld computers (PDAs) to facilitate real-time two-way student interaction with the instructor. Our system is ahead of other "clicker" based PRS (Personal Response System) that is limited to multiple-choice questions. Our system allows for a variety of questions. It also allows for adaptive questioning and two-way communication that provides real-time feedback to the instructor. We have seen improved performance on course assessments through use of PDAs compared to PRS in the same class. We have also shown that students who use PDAs more often in class are more likely to perform better in the course.

Keywords: Physics education research, curricula, teaching methods, assessment, technology
PACS: 01.40.Fk

INTRODUCTION

Educational research (e.g. [1]) has converged on the conclusion that students learn best when they actively construct their own knowledge. However, the structure of most large-enrollment lecture classes discourages active engagement. When an instructor in a college lecture class asks questions, typically only a few students respond.

Recently, many faculty members teaching large-enrollment physics classes have begun using "clickers" to pose multiple-choice questions and increase student interaction. But these systems, though robust limit the nature of interaction and feedback to the instructor. Also, they do not replicate the kinds of open-ended questions that students have to answer on other course assessments.

We believe that wireless mobile technology such as HP IPAQ Pocket PCs, also called handhelds or PDAs (Personal Digital Assistants) offers a better solution. Through appropriately designed Web-based software we have greatly expanded the question types and improved the richness of interaction. This solution allows us to create a real-time adaptive classroom interaction system rather than merely a classroom response system.

In this study we compare students learning in two classes – one that used a one-way, multiple-choice-based PRS (Personal Response System) with those using PDAs.

RESEARCH QUESTIONS

The following research questions were framed to examine the impact of HP PDAs used in conjunction with reformed pedagogy on student learning:

Research Question 1: Did students' grades in the course improve with PDAs relative to PRS?

Research Question 2: Did students who used the PDAs in class more frequently perform better than those who used them less frequently?

LITERATURE REVIEW

An excellent review of classroom response systems and underlying pedagogy is provided by Judson & Sawada. [2] This review was extremely helpful in redesigning the pedagogy for the course. Judson & Sawada point out that it was not merely the technology, but rather the use of appropriate pedagogy that resulted in improved learning. Indeed, they warned that "an electronic response system does not come pre-packaged in an interactive learning environment."

The overarching pedagogical principles that guided our approaches are elucidated by Hake [3] who demonstrated that students in interactive learning

* This work is supported in part by HP Technology for Teaching Grant

CP883, *2006 Physics Education Research Conference*, edited by L. McCullough, L. Hsu, and P. Heron

environments performed better on conceptual learning assessments than those in traditional instructional environments. In particular, we adapted Mazur's strategy of "Peer Instruction" [4] in our classroom. Students were asked a question over the system and asked to first respond individually. Next they were asked to discuss their responses with their neighbor and finally they were asked to respond again. Crouch and Mazur [5] found that this technique greatly improves student performance on the assessment and fosters interactive and collaborative learning.

TECHNOLOGY & PEDAGOGY

Since 1995 the Kansas State University Physics Department has implemented classroom response systems in its introductory physics classes. A review of the evolution of these systems has been presented earlier.[6] Since 2001 we have been using the infrared "clicker" system called PRS (Personal Response System). The PRS system allowed for multiple-choice questions. In 2004 we obtained 40 HP IPAQ Pocket PCs (PDAs) and have since developed and implemented a Web-based classroom interaction system. The "K-State InClass" software allows for several different question types. These include: multiple-choice questions, short-answer questions and ranking task questions. It also allows for sequenced adaptive questions where the system automatically asks students a follow-up question based on their responses to a question. Additionally, the students can also send a question or comment to the instructor during class that the instructor can receive on her/his computer. This option was rarely used unless requested by the instructor.

Both the PRS and PDA systems were implemented in a class taken by about 90 students. Almost all of these students are elementary education majors. Over 95% of them are women. The instructor teaching this class is very familiar with research-based pedagogy and uses Mazur's Peer Instruction during lecture. Typically about four or five questions were asked during each class period. The instructor responded to feedback provided based on students' responses.

METHODOLOGY

To compare the PRS and PDA systems, we used data collected from the students when the PRS system was used in Fall 2003 (N=64) and later when the PDA system was used in Fall 2005 (N=87). The instructor in both years was the same and there was no statistically significant difference in the student population in these two years based on GPAs, majors in college or gender breakdown. The course content and overall pedagogical strategy did not vary significantly between these two years. Importantly, the course assessments – test and exam questions did not vary significantly. Also, the final course grades were assigned based on a fixed scale which was identical between the two years. The student population in these two courses was also statistically similar in terms of their SAT scores, etc.

Thus for the purposes of our study the main differences between the two semesters was that in one year the PRS system was used and in the other year the PDA system was used in class. The only other significant difference was the number of students enrolled in the class in these two semesters.

Our sources of data for this study included the following:

- Student course grades when the PRS system was used in Fall 2003 (N = 64)
- Student course grades when the PDA system was used in Fall 2005 (N= 87)
- Student data logs as they responded to questions posed by the instructor using the PDA system in Fall 2005 (N=87).

RESULTS & DISCUSSION

Course Grades: PRS vs. PDA

We compared student grades and GPA in the two semesters. The results are shown in Figure 1.

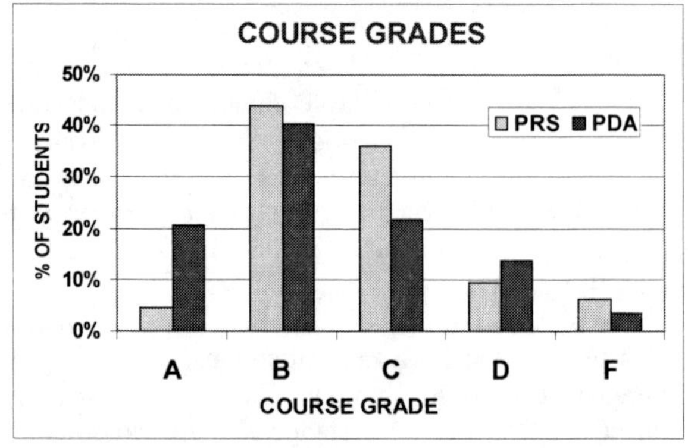

FIGURE 1: Course grades for PRS and PDA semesters

When the average course GPA for the PRS students was compared with the PDA students, we find that there is a statistically significantly higher course GPA for PDAs (2.61) vs. PRS (2.31). A t-test showed a statistical significance at the $p<0.07$ level for a two-tailed test. We also find that there is a significant difference in the grade distribution between the two semesters, with a significantly higher percentage of the PDA students scoring an 'A' or 'B' in the course compared with the PRS students.

Relationship Between Course Performance and PDA Use

The previous result indicates that the overall course performance of the PDA group was superior to the PRS group. However, we were also interested in investigating whether the higher scoring students in the PDA group were in fact using the PDA technology more often. In other words, was there a correlation between use of PDAs in the classroom and student scores?

We recorded PDA usage by each student in the class based on how many questions the students responded to over the course of the semester. Students were not provided any incentive to respond to questions, thus their participation using the technology was purely voluntary. This was also the case for the PRS system.

A Pearson correlation analysis was conducted between the PDA usage and the overall score on course exams and tests combined. The results showed a weak correlation between PDA usage R=0.37, which was significant at the F=0.0003 level for N=86.

Further, we decided to compare the PDA usage for students in different grade bands. The results are shown in Figure 2 below. The bars indicate the mean PDA usage for each grade and the error bars indicate the standard deviation in the mean PDA usage.

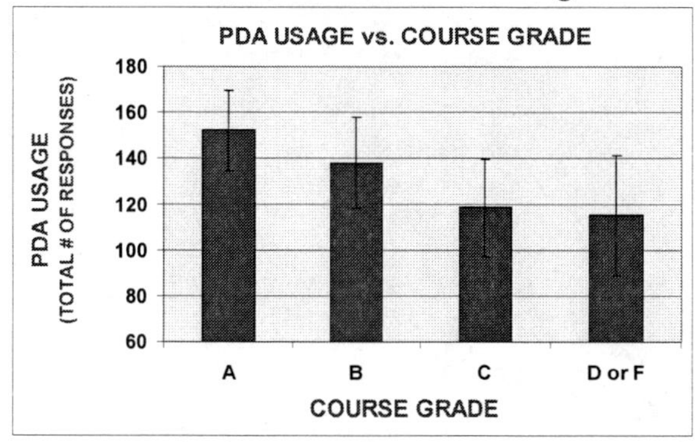

FIGURE 2: PDA usage for different course grades

An ANOVA for the PDA usage in different grade bands shown in Figure 2 yeilded a p-value = 0.030 (with F = 3.12 and $F_{Critical}$ = 2.71). Thus students who score a higher grade are significantly more likely to be using PDAs. The between-group degrees of freedom is three (3), while the within group degrees of freedom is 83 for the ANOVA.

We also compared the course GPA for the students who were LOW, MEDIUM and HIGH users of PDAs in the classroom, where LOW is defined as students who responded to fewer than one-third of the questions posed, MEDIUM defined as students who responded to between one-third and two-third of the questions posed and HIGH defined as those who responded to more than two-thirds of the questions posed by the instructor on the PDA. The results are shown in Figure 3 below.

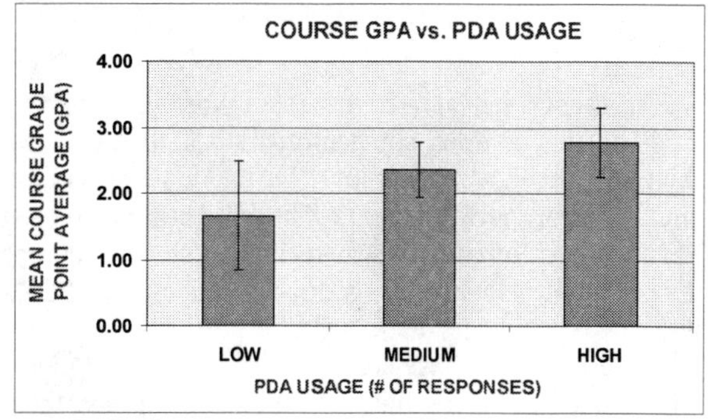

FIGURE 3: Course GPA for different usage levels

We conducted an ANOVA for the PDA usage in three groups (LOW, MEDIUM, HIGH) and found that there was a statistically significant difference in average GPA of the three groups in Figure 3 with a p-value = 0.027 (with F = 3.76 and $F_{Critical}$ = 3.11). Thus students who use the PDAs more often are more likely to perform better on the course.

Summary

There is a statistically significant improvement in overall course grades after the PDA-based system was implemented. Before the PDA-based classroom interaction system was implemented (i.e. with the PRS system) about one-half of the students secured an A or a B in class. In 2005, when the PDA-based system was implemented nearly two-thirds of the students secured an A or a B in the class. Both of these courses in successive years were taught by the same instructor, covering the same content and using very similar and partly identical course assessments.

In the class in which the PDA-based system was implemented, students who used the system more frequently in class secured higher course grades than those who used the system less frequently. Conversely, there was also a statistically significant difference between students getting an A, B, C or D/F in terms of their usage of the PDA-based system in class. Students who secured an A used the system more often than students who got a D or F. This indicates that students who used the system in class more frequently are likely to get higher grades and also that those who secure higher grades used the system in class more frequently.

Consistent with the result described above, a weak but statistically significant correlation was observed between PDA usage in the classroom and students' mean performance score on course assessments. Correlation does not imply causality, so these results do not imply that students' grades will improve merely by making them use the PDA-based classroom response system.

LIMITATIONS OF STUDY

As mentioned above, the main limitation of this study is that it does not demonstrate a direct causal relationship between use of PDAs in the classroom and superior performance. The students who used PDAs more often performed better on the course assessments. However, this does not mean that PDAs were responsible for their superior performance. One might argue that these were in fact "better" students who did several other things in the class that led them to perform better on the course assessments for reasons unrelated to the PDAs.

The only way to establish causality in this kind of study is to examine the ways in which students used the PDAs more carefully and investigate whether the ways in which students used the PDAs were indeed responsible for superior performance. Such a study would be a natural extension of the present work.

CONCLUSIONS

We sought to address the following research questions in this study:

Research Question 1: Did students' grades in the course improve with PDAs relative to PRS?

Yes, we did find a statistically significant improvement in course grade distribution between the semester in which we had not used the PDA-based system and the one in which we did use the system for similar students and identical content and instruction.

Research Question 2: Did students who used the PDAs in class more frequently perform better than those who used them less frequently?

Yes, we did find that more frequent users secured higher course grades and conversely that students who secured higher course grades had used the system more frequently in the class.

In spite of the promising results above, as explained in the previous section, the results of this study must be viewed with caution because correlation does not imply causality. More research needs to be completed to understand the ways in which students use PDAs so that we can identify classroom practices of using PDAs that are most beneficial to student learning.

REFERENCES

1. R. K. Sawyer, *The Cambridge Handbook of The Learning Sciences*. (Cambridge University Press, New York, NY, 2006).
2. E. Judson and D. Sawada, Journal of Computers in Mathematics and Science Teaching **21** (2), 167 (2002).
3. R. R. Hake, American Journal of Physics **66** (1), 64 (1998).
4. Eric Mazur, *Peer Instruction: A User's Manual*. (Prentice-Hall, Upper Saddle River, NJ, 1997).
5. C. H. Crouch and E. Mazur, American Journal of Physics **69** (9), 970 (2001).
6. D. A. Zollman and N. S. Rebello, presented at the 2005 Winter Meeting of the American Association of Physics Teachers, Albuquerque, NM, 2005 (unpublished).

Reliability, Compliance and Security of Web-based Pre/Post-testing

Scott Bonham

Department of Physics and Astronomy, Western Kentucky University
1906 College Heights Blvd. #11077, Bowling Green, KY, 42101-1077

Abstract. Pre/post testing is an important tool for improving science education. Standard in-class administration has drawbacks such as 'lost' class time and converting data into electronic format. These are not issues for unproctored web-based administration, but there are concerns about assessment validity, compliance rates, and instrument security. A preliminary investigation compared astronomy students taking pre/post tests on paper to those taking the same tests over the web. The assessments included the Epistemological Beliefs Assessment for Physical Science and a conceptual assessment developed for this study. Preliminary results on validity show no significant difference on scores or on most individual questions. Compliance rates were similar between web and paper on the pretest and much better for web on the posttest. Remote monitoring of student activity during the assessments recorded no clear indication of any copying, printing or saving of questions, and no widespread use of the web to search for answers.

Keywords: pretest, posttest, computer assessment.
PACS: 01.40.Fk, 01.50.ht, 01.50.Kw

INTRODUCTION

Pre/post testing has become a very important tool in Physics Education Research. The Force Concept Inventory[1] (FCI) has clearly exposed shortcomings in traditional instruction,[2] transformed individuals[3] and entire departments[4] and spurred the growth of the Physics Education Research community. It has been used to compare the effectiveness of different curricular approaches,[5] evaluate the use of technology,[6] look at gender fairness in instructional methods,[7] and answer many other research questions. The success of the FCI has inspired the development of a wide range of other assessment tools, covering topics including mechanics,[8] electricity and magnetism,[9] DC circuits,[10] astronomy[11], problem-solving,[12] uncertainty,[13] and beliefs about the nature of science.[14]

In spite of their utility in revealing the state of our students and the effectiveness of instruction, these assessments are not as widely used as is desirable. An important reason is the time and effort involved in administering them during class and processing the forms afterwards. An alternative is to have students complete the assessments outside of class using a web-based system. This saves instructional time, reduces errors and puts the data directly into electronic format. With the appropriate tools built in, it can quickly and easily make feedback from the assessment available to instructors. While there are clear advantages, there are also important questions about the validity, compliance and security in this approach. The purpose of this study is to address those.

Research Questions

1. Will administration in an unproctored web environment affect how students respond? Changing administration modes could affect the reliability of items, and students could use resources not available in the classroom.

2. Will administration in an unproctored web environment affect compliance rates? Those who show up to class on the given day and those who comply with instructions to do something outside of class are not necessarily similar groups.

3. Will administration in an unproctored web environment compromise the security of the test? If an assessment is on the web, there is no way to stop students from keeping copies of questions in some format, which would compromise the validity of the instrument.

These research questions are listed in order of increasing seriousness and decreasing amounts of information in scholarly literature. Research on computer testing shows that in most cases, computer-

CP883, *2006 Physics Education Research Conference*, edited by L. McCullough, L. Hsu, and P. Heron

administered tests are equivalent or even superior to paper-administered ones[15]. Most studies show that scores are usually equivalent and that people tend to finish computerized tests quicker and find the experience more enjoyable (or at least less disagreeable). However, most of these studies have compared proctored paper-and-pencil tests with proctored computer assessments, where the students come into a computer testing center and work under supervision. In the study most similar to this one[16] half of a class took the FCI in class and the Maryland Physics Expectation survey (MPEX)[17] on line, with the mode reversed for the other half of the class. No difference was found in the scores on the FCI for the two groups (MPEX scores were not reported), though the other questions above were not addressed. The web administration system used was rudimentary, in that students were not required to log in and all questions were displayed on a single page, requiring students to scroll up and down. No post-testing and no follow-up studies were reported.

SETTING

A preliminary study was carried out using in two pairs of introductory astronomy class sections taught at Western Kentucky University the spring of 2006, one pair in the Astronomy of the Solar System course and in the Astronomy of the Stellar Systems course. Table 1 summarizes the classes in this study. Instructors A and B are highly experienced instructors, and Instructor C is a newer instructor. Classes 1 and 2 were taught in a lecture hall, while classes 3 and 4 were in one of the department's SCALE-UP style classrooms[18] and included hands-on activities along with lecture.

Table 1: Summary of courses involved in the study. N is number of students on the roster at the end of the semester.

Class	Course	Instructor	N	Paper	Web
1	Solar	A	71	ASTRO	EBAPS
2	Solar	A	64	EBAPS	ASTRO
3	Stellar	B	40	ASTRO	EBAPS
4	Stellar	C	39	EBAPS	ASTRO

Two assessments were used with each class. One is a twenty-four item multiple choice astronomy concept assessment developed for this study,[19] which will be referred to as the ASTRO. This was a quickly developed assessment with no effort to check the reliability or validity that was used as a "stand-in" to avoid the risk of compromising a well-designed astronomy concept assessment. The other assessment was the Epistemological Beliefs Assessment for Physical Science (EBAPS).[20] This is a thirty item assessment with five subscales: *Structure of scientific knowledge*, *Nature of knowing and learning*, *Real-life applicability*, *Evolving knowledge* and *Source of ability to learn*. Because of the 'opinion' nature of the questions it was felt that security was less of an issue with this assessment, and permission was given by the author for web use.[21]

In each of the pair of classes, one section took the WKU-ASTRO assessment in class using paper and pencil while the other section took the EBAPS assessment in class. Students were then asked to take the other assessment on-line outside of class during a specified period (a little over a week) and class credit was given to those who complied. Both assessments were given pre/post in classes 1 and 2, while the paper assessment was not given at the end of the semester in classes 3 and 4.

SYSTEM

The web-based assessment system was developed by the author using PHP and MySQL. Students log onto the system using their school username and student identification number. They then receive an instruction page, where they are asked to complete the exam at one sitting, not use other resources, and not print, save or copy material. They are then presented with the currently available assessment for their class. Questions are delivered one at a time and by default the system automatically advances to the next question once the student selects one of the choices. There are controls to go back to previous questions, skip ahead to following questions, and to mark a particular question to return to latter. Once the student has gone through all of the questions, the system will return to any skipped questions and marked questions before finishing the assessment.

On the posttest, a script was included to monitor any behavior that could potentially compromise the integrity of the test, such as searching for answers on the web, copying material to another application, saving a page and printing. JavaScript and Java were used to capture events (actions) that could signify such activities and send information about them to be recorded on the server. Potential copying and searching for answers on the web was identified using the focus gained and focus lost events of the browser, which happen every time the browser becomes and ceases to be the active window on the screen. On Microsoft Internet Explorer (MSIE) the contents of the clipboard were also recorded to compare with question text. Printing was monitored on MSIE by events signaling the start and ending of printing, and on non-MSIE browsers by a small, virtually invisible embedded Java Applet. Finally, loading of pages was monitored to detect any that had not been recently sent by the server, which would indicate a saved page being reloaded. It should be noted that there is innocent activity that could trigger many of these events and that these methods are not entirely fool-proof. It is possible for individual events to escape detection, but unlikely that wide-spread violations could.

RESULTS

Validity

First the scores on the two exams were compared between those completed in the traditional format and those on the web. Table 2 summarizes the two-tailed t-tests results and shows no significant differences on the ASTRO and all five EBAPS scales. A question-by-question comparison using the chi-squared test for independence between groups identified two questions (9 and 19) that demonstrated differences significant at the α=0.05 level for all courses combined but not comparing class 1 with class 2 and class 3 with class 4. One would expect that on a 30 item inventory, on average between one and two items will be by random chance found significant at the α=0.05 level. On the ASTRO, there was no difference in the over-all scores, but an item-by-item comparison identified four questions with significant differences at α=0.02 or less dealing with brightness of stars, nature of shooting stars and weightlessness on the space station. This set included a question with an awkwardly worded correct option and one where the first distracter was mostly correct.

Table 2: Summary of two-tailed *t*-tests comparing scores on the ASTRO assessment and five subscales on the EBAPS.

	M_{web}	M_{paper}	t	p
ASTRO	8.50	8.57	-0.12	0.90
Structure sci. knowledge	2.11	2.11	0.64	0.95
Nature knowing/learning	2.50	2.49	0.16	0.87
Real-life applicability	2.58	2.48	1.2	0.23
Evolving knowledge	2.50	2.48	0.12	0.91
Source of ability	3.02	2.88	1.4	0.16

Compliance

The compliance rates for the four classes are summarized in Table 3. This table lists the enrollment fraction of students on the roster at the end of the semester that took each of the assessments. The pretest compliance rate for the web is nearly as high as the paper test for all except class 4. In that class the instructor encouraged students to participate but did not give any course credit on the pretest. Extra credit was given on the post test, leading to an increase in compliance. It may also be observed that the posttest compliance rates on the web are nearly as high as the pretest rates, but that the posttest on paper is much lower than the pretest for classes 1 and 2. The paper posttests were given during the last class session, late afternoon on the last class day of the semester. The difference in compliance was mainly among freshman, suggesting that the web and paper groups are not equivalent.

Table 3: Compliance rates for each class and assessment. Total posttest paper is for classes 1 and 2 only.

	Class 1	Class 2	Class 3	Class 4	Total
Enrollment	72	63	40	39	214
pretest paper	90%	84%	88%	92%	88%
pretest web	86%	83%	83%	54%	79%
posttest paper	58%	60%	N/A	N/A	59%
posttest web	81%	71%	75%	69%	75%

Security

During the post-test, on-line student activity was monitored as previously described, recording events that could indicate printing, saving, copying or other activities. The start print event in MSIE browsers and Java print() in others would indicate a student printing a page, but no events were recorded out of the 160 students who participated. Saving of pages was monitored by listening for page load events not paired to a page serve event. Again, no events were recorded up to and past the time of the student final exams.

Possible copying and searching for answers were monitored by focus gained/lost events, and on MSIE the contents of the clipboard were also recorded. No question text or any other clipboard contents were recorded. (78% of the students used MSIE.) Time during which the browser is not the active window can also be a signal of copying or searching for answers, as well as innocent activities like checking email or closing pop-up ads. Many thousands of focus lost events were generated, but most of these were caused by a page being submitted. Eliminating all events between submission and page load, those within a second of page load, and those without a paired focus gained event left 86 focus lost events that appear to be actual instances where some other window or menu became the active window for a period of time. Figure 1 provides a histogram of the time gap between the lost focus and corresponding gained focus event for these instances, except for a single one on the EBAPS with gap of 737 seconds. Excluding that event, the median gap on the EBAPS and WKU-ASTRO were 16 and 10 seconds, respectively. Of the 33 students where an instance was recorded, almost half involved only a single instance, and over 80% had less than three instances. Nine out of twelve of the instances from the student with the most occurred on the first question, and the next highest, with 11 events, involved only 8 different questions.

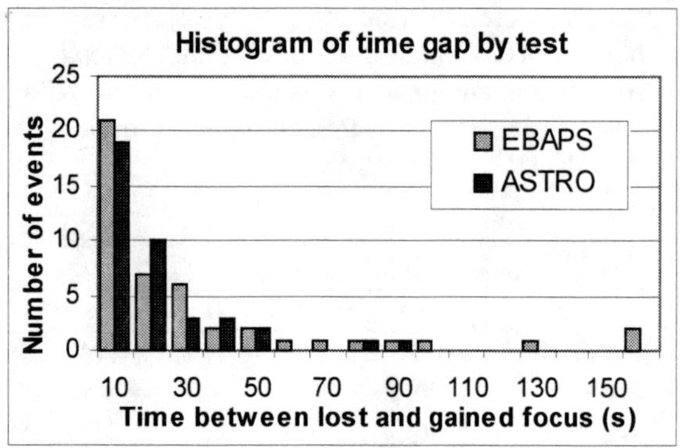

FIGURE 1. Histogram of elapsed time between focus lost and focus gained events. Median gap is 10 s for ASTRO and 16 s for EBAPS.

DISCUSSION

The results from this preliminary study are promising for using unproctored web assessment, but further study is needed. The difference in how the assessment was administered did not change response patterns on most items and there were no significant differences on the over-all scores. Furthermore, the items that showed strong format effects were all on the relatively poorly developed ASTRO assessment. The compliance rates at the beginning of the semester were nearly the same for all but one section. The instructor in that class did not provide any course credit for the web pretest, illustrating the importance of providing at least a minimal amount of class credit to ensure good compliance. Compliance rates were significantly better for the web on the posttest because of the number of students who did not come to class at the end of the semester. It appears that compliance rates on the web can be as good or even better than those using class time when some class credit is provided and the instructor makes expectations clear. This study implemented methods to monitor security of the tests, and found no signals that any of the material was being printed, copied or saved, or that there was extensive searching for answers on the web. While this method is not perfect or foolproof, it is clear that most if not all of the students complied with the request to not copy, print, save or use other resources for the questions. These preliminary results suggest that web-based pre/post assessments may provide an acceptable alternative to paper ones, though validity needs to be established for individual instruments, procedures to encourage participation need to be attended to, and security needs to be considered. Future work will include repeating the study with a larger population, using additional instruments, and estimating failure rates of the monitoring system to detect suspicious activity.

REFERENCES

1 David Hestenes, Malcolm Wells, and Gregg Swackhamer, Physics Teacher **30** (3), 141 (1992).

2 Ibrahim Abou Halloun and David Hestenes, American Journal of Physics **53** (11), 1043 (1985); Richard Hake, American Journal of Physics **66** (1), 64 (1998).

3 Eric Mazur, *Peer Instruction.* (Prentice-Hall, Upper Saddle River, NJ, 1997); Catherine H. Crouch and Eric Mazur, Amercan Journal of Physics **69**, 970 (2001).

4 Tim Stelzer and Gary Gladding, in *Forum on Education of The American Physical Society* (2001), Vol. 2002.

5 Jeff Saul, University of Maryland, 1998.

6 S.W. Bonham, D.L. Deardorff, and R.J. Beichner, Journal of Research in Science Teaching **40**, 1050 (2003).

7 Mercedes Lorenzo, Catherine H. Crouch, and Eric Mazur, Amercan Journal of Physics **74** (2), 118 (2006).

8 Ronald K. Thorton and David R. Sokoloff, American Journal of Physics **66** (4), 338 (1998).

9 Dave Maloney, Alan Van Heuvelen, Tom O'Kuma et al., (Joliet Junior College, Joliet, Illinois, 1999).

10 Paula Vetter Engelhardt and Robert J. Beichner, Amercan Journal of Physics **72** (1), 98 (2004).

11 Beth Hufnagel, The Astronomy Education Review **1** (1), 47 (2002).

12 David Hestenes and Malcolm Wells, Physics Teacher **30** (3), 159 (1992).

13 Duane Lee Deardorff, North Carolina State University, 2001.

14 Edward F. Redish, Jeffery M. Saul, and Richard N. Steinberg, American Journal of Physics **66** (3), 212 (1998); Ibrahim Halloun, International Conference on Undergraduate Physics Education (1996); W. K. Adams, K. K. Perkins, N. Podolefsky et al., Phys. Rev ST: Phys. Educ. Res. **2** (1), 010101 (2006).

15 David Zandvliet and Pierce Farragher, Journal of Research on Computing in Education **29** (4), 423 (1997); Allan C. Bugbee, Journal of Research on Computing in Education **28** (3), 282 (1996).

16 Dan MacIsaac, Rebecca Pollard Cole, David M. Cole et al., Electronic Journal of Science Education **6** (3) (2001).

17 Edward F. Redish, Richard N. Steinberg, and Jeffery M. Saul, Amercan Journal of Physics **66**, 212 (1998).

18 R. Beichner, J. Saul, D. Abbott et al., in *PER-Based Reform in University Physics*, edited by E. F. Redish and P. J. Cooney (American Association of Physics Teachers, College Park, MD, in press).

19 The effort of Richard Gelderman and Micheal Carini in developing this assessment are gratefully acknowledged.

20 Andy Elby, American Journal of Physics **69** (7), S54 (2001).

21 Andy Elby, personal communication.

Investigation and Evaluation Of A Physics Tutorial Center

Kristin N. Walker and Melissa H. Dancy

Department of Physics, University of North Carolina at Charlotte, Charlotte, NC, 28223, USA

Abstract. The Physics Resource Center (PRC) is a tutorial center offering supplemental instruction to all introductory level physics students. Physics students were surveyed to investigate and evaluate the center's effectiveness in meeting their needs. Survey results and suggestions offered by the students are reported. Findings include the dependence of PRC attendance on factors including gender, course performance, and social engagement.

Keywords: Tutorial Center, Tutoring
PACS: 01.40.Fk, 01.40.-i

INTRODUCTION

Supplemental academic programs such as peer tutoring and study skill courses have been shown to enhance the academic behavior, skills and performance of college students.[1-3] Benefits of tutorial programs like the Physics Resource Center include: the tutor and tutee learn to organize and extend their knowledge, students gain a better understanding of course expectations, content information can be retained and recreated by recalling tutor session discussions, and it can help relieve some of the strain on professors teaching large, multi-ability classes.[3,4] Successful tutoring programs contribute to positive student attitudes toward tutoring, effective interactions between the tutor and tutee, and motivation for the students to follow through and continue attending.[5,6] In this article, we report on a specific tutorial program and offer recommendations applicable to similar programs elsewhere.

PHYSICS RESOURCE CENTER

The Physics Resource Center (PRC) is a typical tutorial program, operated by the physics department at UNC-Charlotte, open to all students in the first and second semester algebra and calculus-based physics courses. Tutorial sessions in the PRC are held 15 hours each week and are led by a graduate student and possibly a physics professor holding his/her office hours. Students voluntarily attend the PRC sessions with questions about homework, concepts, and upcoming exams. Since these sessions are open to all physics courses, during one session there are often students from several different classes asking a wide range of questions. Sessions are held in a traditional classroom with bolted-down seats, which unintentionally encourage a more traditional lecture style of tutoring when several students attend.

During an average week approximately 45 out of 750 students enrolled in nine introductory physics courses attend the PRC.

Although low attendance allows more one-on-one tutoring to those that attend, it is unfortunate that more students are not attending to get additional help. The purpose of this research is to better understand why in a given week only 6% of physics students take advantage of this program and to collect information that would help the PRC better meet the needs of the students.

METHODOLOGY

A survey was designed to investigate: What types of students attend the PRC, why do so few students attend, and how the students rate the usefulness of the PRC. A web-based survey was made available to all physics students during week 13 out of a 16 week semester. Completion of the survey was voluntary and confidential. A total of 190 out of 750 introductory physics students completed the survey,

Based on their experience with the PRC, students that participated in the survey were directed into one of four tracks which will be identified as "no hear" (NH), "no attend" (NA), "no return" (NR), and "attend" (A) throughout this paper as described in Table 1. Each student's survey contained tailored questions based on their track about their PRC

CP883, *2006 Physics Education Research Conference*, edited by L. McCullough, L. Hsu, and P. Heron

experience and opportunities for open-ended comments and suggestions to improve the PRC.

TABLE 1. Track Descriptions		
Track	**Description**	**% of Respondents**
No Hear (NH)	Students claimed to have never heard of the PRC	12
No Attend (NA)	Students knew about PRC but never attended	30
No Return (NR)	Students attended a PRC session but never returned	15
Attend (A)	Students attended the PRC on regular basis	43

STUDENT PROFILE

Out of the 192 students that completed the survey, 89% were from the algebra-based classes where extra credit was offered.

Gender Differences in Tutorial Attendance. Like most science and engineering disciplines there were larger percentages of males than females both in the survey response sample and in the courses at large. Table 2 shows the number of males and females in each survey track. Results from a chi-squared test showed no significant statistical difference between the percentage of males and females in each track; both males and females are similar when it comes to PRC attendance.

TABLE 2. Gender distribution in each track. No statistical gender difference (χ^2=1.42)

	NH	NA	NR	A	Total
# of Male	13	33	20	51	117
% Male	11	28	17	44	
# of Female	10	24	10	31	75
% Female	13	32	13	41	

Course Performance and Tutorial Attendance. To gain an understanding of what type of students are in each track, students were asked to provide their exam scores. Grades analyzed were from the 154 students in the algebra-based classes that provided their first two or three exam scores. Within each track the normal distributions of exam grades recorded from all algebra-based classes combined were calculated and plotted in Figure 1.

A t-test was performed on each pair of means at a 5% significance level. A significant difference ($t > 1.64$) was found between tracks "no attend" and "no return", and between tracks "no attend" and "attend" with t values of 1.73 and 3.98 respectively.

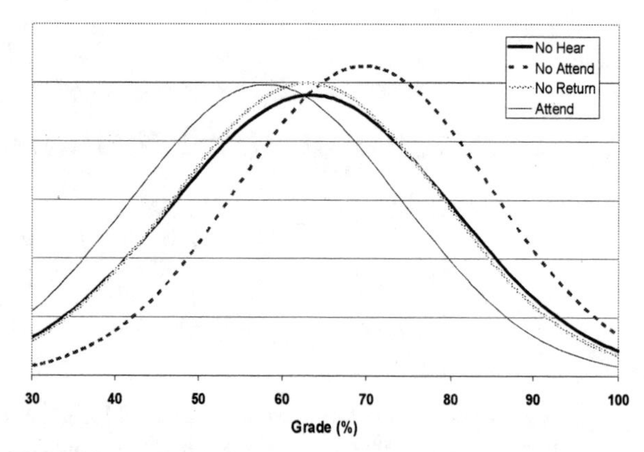

FIGURE 1. Total student grade distribution in each track for all algebra-based classes combined. Regular attendees perform less well on exams.

Previous results do not take into account the differences between classes. To create a more controlled group the grades of a single first semester algebra-based class were analyzed. Out of this class 41 students provided their exam scores for the first two exams. Normal distributions for each track were calculated and plotted in Figure 2.

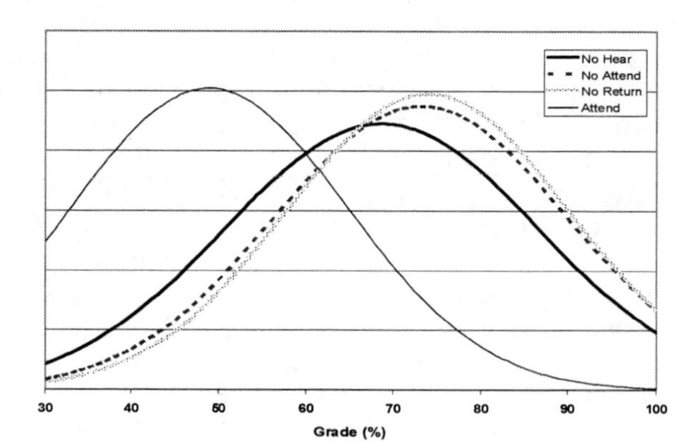

FIGURE 2. Student Grade Distribution in each track for a single algebra-based class

The statistical t-test at 5% significance was again performed on these class scores showing statistical difference between tracks NH and A, NA and A, and tracks NR and A with t-values of 2.35, 3.58, and 2.91 respectively. Both sets of exam grades show a large difference between those that have never attended and those who attend regularly with the regular attendees not doing as well on exams. We believe this difference is due to those who need the help the most being more likely to attend, rather than any adverse effects of tutorial attendance.

Attitudes and Tutorial Attendance. To better understand their attitudes, students were asked to agree or disagree statements about tutoring and physics and statements from the Colorado Learning Attitudes about Science Survey[7]. For each statement the students' responses were scored using Equation 1 which weights the percentages of each response on a scale from -200 to 200 where a score of 200 is a result of 100% of the students strongly agreeing with the statement. Students' survey response scores from each track are listed in Table 3. It appears that students generally view tutoring as valuable, including those who never attended the tutorial center. There are no statistical significant differences between tracks preventing a comprehensive investigation of the relationship between student attitudes and attendance. Research has been done to try to answer why students that need the most help tend to avoid seeking help[8].

$$\text{Score} = 2*(\%\text{ Strongly Agree}) + 1*(\%\text{ Agree}) + 0*(\%\text{ Neutral}) - 1*(\%\text{ Disagree}) - 2*(\%\text{ Strongly Disagree}) \quad (1)$$

TABLE 3. Attitude questions and results – Similar responses across tracks

Question	NH	NA	NR	A
Tutoring is a positive solution to get help	102	130	130	140
I am not satisfied until I understand why something works the way it does	87	122	109	122
I do not spend more than five minutes stuck on a physics problem before giving up or seeking help from someone else	-82	-87	-107	-105
I can usually figure out a way to solve physics problems	14	37	17	26

Social Engagement and Tutorial Attendance. Students were asked if they attended the PRC in groups or by themselves. Results in Table 4 indicate that those that attend solo are less likely to return and that many regular attendees come with others.

TABLE 4. How students generally attended the PRC

% Students	NR	A
By themselves	80	57
With one or two other people	17	36
With several people	0	5

Because our data is correlational, we can not be certain if attending with friends increases attendance or if people who are likely to attend regularly are also likely to study socially. Based on our observations in the tutorial center itself, we suspect that coming with others does encourage repeat visits. For example, tutorial instructors note that when students come in groups of two or more they work together, teach each other, and explain concepts to each other. They appear to be less singled-out and to smile and laugh more when they come with friends. This likely improves the quality of their experience in the center and increases the likelihood of a repeat visit.

EVALUATION OF THE PHYSICS RESOURCE CENTER

Students were asked why they do or don't attend or return to PRC. Students that never heard of PRC before were asked why they would or wouldn't attend. Table 5 and 6 show the percentages of the students' responses. Blank spaces represent answer options that weren't offered to the corresponding track, not zeros.

TABLE 5. Top reasons students attend the PRC

% Students	NH	NR	A
To check homework answers	52	10	39
Get help where I got stuck on homework	74	67	83
Help with setting up the problem	70	37	53
To have concepts explained	78	50	41
To have confusing class or textbook examples explained	70	27	31

TABLE 6. Top reasons students didn't attend the PRC

% Students	NH	NA	NR
Time Conflict	91	67	57
Can usually figure out the homework myself	26	43	
Work on homework with classmates	4	29	
Isn't beneficial	13	10	37
Find help elsewhere	0	0	23
Too many students from other classes			37
Questions were not relevant to me			33
Was more confused after leaving			23

Students that have attended PRC at some point were asked if they agree or disagree that the PRC was helpful in solving problems, explaining concepts, and test preparation. Percentages from tracks "no return" and "attend" that agree the PRC was helpful in the three areas are presented in Table 7.

TABLE 7. Percentages of those agreeing that the PRC was helpful in the following areas

	No Return	Attend
Problem-Solving	54 %	89 %
Conceptual Understanding	44 %	76 %
Test Preparation	27 %	66 %

RECOMMENDATIONS FROM STUDENTS

The fact that only 6% of the current physics students attend the PRC in a given week indicates the need for change. Students were asked to suggest ways to improve the PRC to better meet their needs. Table 8 lists the most common suggestions made and the number of students that referred to this improvement.

TABLE 8. Students' top suggestions on how to improve the PRC

	Number of references
More times/better times	52
More specific to class/ more structure	20
Class and PRC need to coincide	15
More instructors/ smaller groups	14
Better test preparation	12
Additional examples and study material	11

Offer convenient times for the students. Time conflict was the main reason students gave for not attending the PRC. Holding sessions in the evenings or right before or after class to better fit their schedules will provide more opportunities for students to attend.

Split the sessions based on course content. Making PRC more specific appears to be an important improvement. By offering a more structured program to the needs and goals of each individual class, PRC instructors can be better prepared and offer more guidance to the students. In addition, making the sessions more course specific is more likely to encourage student interaction in the center, which appears likely to encourage repeat visits as indication by the data on returns and social interaction.

Greater communication between professors and PRC instructors. This will help in the understanding of each other's goals and expectations for the class. Students will benefit when class and the PRC coincide in the material being covered on a weekly basis and possibly allow more material to be covered.

Environment that promotes small group work available at all times. Smaller groups and one-on-ones working at a table instead of a lecture classroom is beneficial in meeting the specific needs of the students. This could be accomplished by have more PRC tutors and holding the PRC in a room with tables that is open at all times allowing students to come in and work together with or without a PRC tutor present.

Create a library of additional physics resource materials. It was suggested that the Physics Resource Center have a library providing students with additional materials such as additional examples, review sheets, old exams, and study material besides homework problems for those students that need additional resources besides lectures and homework. Make this library available to students at all times.

CONCLUSION

In addition to the recommendations from the students themselves, we add two additional recommendations based on the data collected as well as our own observations in the center.

Get the students through the door. Promote the PRC and find a way to get students to attend early in the semester while motivation to do well is still strong. Over 40% of the students had either never heard of the center or had never attended it. Those who had attended generally had positive experiences. It is important to get students to attend at least once. Once a student has attended they know exactly what the center has to offer and are more likely to return if the center offers a valuable service for them.

Encourage students to attend with friends and to interact with other students while at the center. Students are more likely to attend when they know that they aren't alone in their physics struggle. When students attend with other classmates it gives them the chance to work together, learn from each other, and possibly meet others and form a study group.

REFERENCES

1. D. M. Lidren and S.E. Meier, "The effects of minimal and maximal peer tutoring systems on the academic performance of college students," *Psychological Record* **41**(1), 69-78 (1991)
2. D. S. Bender, "Effects of study skills programs on the academic performance of college students," *J. of College Reading and Learning* **31** (2), 209-216 (2001)
3. C. Singh, "Impact of peer interaction on conceptual test performance," *American Journal of Physics* **73** (2), 446-451 (2005, May).
4. M. A. Powell, "Academic tutoring and mentoring: A literature review," California Research Bureau, California State Library, CRB-97-011 (1997, Oct)
5. N. Henderson, M. Sami Fadali, J. Johnson, "An Investigation of First-Year Engineering Students' Attitude toward Peer-Tutoring," *32nd ASEE/IEEE Frontiers in Education Conference,* F3B-1-3 (2002)
6. M. T. C. Chi, "Constructing Self-Explanations and Scaffolded Explanations in Tutoring," *Applied Cognitive Psychology* **10**, S33-S49 (1996)
7. W. K. Adams, K. K. Perkins, N. S. Podolefsky, M. Dubson, N. D. Finkelstein, and C. E. Wieman, "New instrument for measuring student beliefs about physics and learning physics: The Colorado Learning Attitudes about Science Survey," Phys. Rev. ST Phys. Educ. Res. 010101, 1-14 (2006)
8. A. M. Ryan, P. R. Pintrich, and C. Midgley, "Avoiding Seeking Help in the Classroom: Who and Why?," *Educational Psychology Review* **13**, 93-114 (2001

Cultivating Problem Solving Skills via a New Problem Categorization Scheme

Kathleen A. Harper*, Richard J. Freuler∓, and John T. Demel∓

*Department of Physics, The Ohio State University, 191 W Woodruff Ave, Columbus, OH 43210
∓First-Year Engineering Program, The Ohio State University, Room 244 Hitchcock Hall, 2070 Neil Ave, Columbus, OH 43210

Abstract. When one looks at STEM disciplines as a whole, the need for effective problem solving skills is a commonality. However, studies indicate that the bulk of students who graduate from problem-solving intensive programs display little increase in their problem solving abilities. Also, there is little evidence for transfer of general skills from one subject area to another. Furthermore, the types of problems typically encountered in introductory STEM courses do not often cultivate the skills students will need when solving "real-world" problems. Initial efforts to develop and implement an interdisciplinary problem categorization matrix as a tool for instructional design are described. The matrix, which is independent of content, shows promise as a means for promoting useful problem-solving discussion among faculty, designing problem-solving intensive courses, and instructing students in developing real-world problem solving skills.

Keywords: interdisciplinary problem solving, STEM, problem categorization, teaching problem solving
PACS: 01.40.Fk, 01.40.gb, 89.20.Kk

INTRODUCTION

When one looks at the STEM (science, technology, engineering, and mathematics) disciplines as a whole, one aspect that pervades all subdisciplines is the need for effective problem solving skills. Indeed, one might argue that teaching college students to problem-solve may be more important than teaching them specific content knowledge, given that it is nearly impossible to predict what specific content today's graduates will need to know even 10 years from now.[1-6] However, studies indicate that the bulk of students who graduate from problem-solving intensive programs display little increase in their problem-solving abilities, even after solving thousands of problems.[7] Further, there is little evidence for transfer of general skills from one subject area to another, from one course to another, or even between different topics in the same course.[8] It has been argued that in order for students to achieve cognitive growth, they must solve a variety of types of problems, according to some taxonomy, such as Bloom's.[9,10] This paper describes some initial interdisciplinary efforts to improve student problem solving skills via a new method of problem categorization.

CONTEXT OF THE WORK

The Fundamentals of Engineering for Honors (FEH) program at Ohio State has included some coordination of topics in physics, engineering, and mathematics since 1997 in an effort to 1) help students have appropriate background for each course and 2) assist students in making connections between the different subject areas. One common element in all three disciplines is problem solving, but until recently there had not been much discussion of this prevalent aspect of STEM education in the coordination efforts. It was postulated that the interdisciplinary nature of the FEH program would make it a useful setting to successfully impact the issues referred to in the introduction..

Interviews of FEH team members from a variety of disciplines indicated that there are few commonalities in the way different instructors describe problems.[11] Without a common framework it is difficult for instructors to describe problems from their courses to each other. Further, given that novices have a difficult time seeing commonalities within one discipline area, let alone across disciplines, without a common framework it should not be surprising that little transfer occurs from one discipline to another. If instructors are assigning problems with common characteristics, it would be useful to know and make this connection explicit to the students. If students begin to see some relationships between problem

CP883, *2006 Physics Education Research Conference*, edited by L. McCullough, L. Hsu, and P. Heron

solving in different courses, they might become more adept at problem solving. Likewise, if the instructors help students see that certain strategies and skills tend to be successful in approaching certain types of problems,it might have an impact.

DEVELOPMENT OF THE PROBLEM CATEGORIZATION SCHEME

A large body of work details the differences between how novices and experts solve problems. Of particular relevance here is that novices tend to categorize problems according to their surface features where experts categorize problems based on the concepts needed to solve the problem. [12] However, experienced instructors also know that students run into difficulty when encountering problems that go beyond "plug 'n chug," and require them to apply real-world skills such as approximating, making assumptions, estimating, or doing additional research to solve the problem. Part of this may be due to lack of practice with these skills; part of this may be due to a lack of recognition of when to appropriately apply them.

Although experts categorize in a deeper way than novices, the categorization is largely dependent upon the content knowledge that the experts have. A person can be an expert in one field and a novice in another, and so, if he or she is asked to categorize problems in two very different fields, may apply different strategies to each. Therefore, teaching a categorization scheme based upon content might be expected to have weaknesses because: 1) the instructor's content knowledge is broader and better organized than the students; [13] and, 2) any skills learned that are highly connected to the content would not be expected to be transferable. Therefore, non-content aspects of problem solving should be expected to be more transferable.

The categorization scheme that has been developed is a two-dimensional matrix, where one axis indicates the nature of the solution (no possible solution, exactly one correct solution, or multiple correct solutions) and the other describes the nature of the given information (insufficient information, exactly sufficient information, or excess information). This scheme is indeed content independent. Additionally, typical STEM employees will encounter problems that fit in most, if not all, of these boxes in their work.

As an example of how this matrix works, consider Figure 1. Each cell of the categorization matrix contains a list of skills that might be appropriate to employ when solving a problem of this nature, regardless of its content.

As an example of how problems that are similar both in terms of basic content area and presentation fit into different blocks of the matrix, consider Figure 4 (at the end of the paper), which shows a set of nine similar yet different statics problems, each in the appropriate cell.

The point of this is that problems which share a number of common characteristics can be quite different in the manner in which they are successfully approached. Most problems typically encountered in introductory courses will fit into one of these matrix cells, although some problems might be better thought of as "mixed states" (e.g., a problem with some excess information, but other needed information lacking). The point is *not* that any problem situation can be modified to fit in all the blocks. In fact, it appears that many standard textbook problems can be rather easily modified to fit in about six blocks, but do not readily lend themselves to the entire matrix.

Nature of Provided Information		Nature of the Answer: None	Nature of the Answer: One	Nature of the Answer: Two or more
	Insufficient	Analyze Approximate Assess/Evaluate Assume Estimate Model Research Verify	Analyze Approximate Assess/Evaluate Assume Estimate Model Research Verify	Analyze Approximate Assess/Evaluate Assume Estimate Model Optimize Research Verify
	Exactly Sufficient	Analyze Assess/Evaluate Model Verify	Analyze Assess/Evaluate Model Verify	Analyze Assess/Evaluate Model Optimize Verify
	Excess	Analyze Assess/Evaluate Discriminate Filter Model Sort Verify	Analyze Assess/Evaluate Discriminate Filter Model Sort Verify	Analyze Assess/Evaluate Discriminate Filter Model Optimize Sort Verify

FIGURE 1. General Problem Categorization Matrix with Associated Skills

A look at chapters of two traditional texts (one in physics, one in mechanical engineering) indicates that the vast majority of problems are in the center square of the matrix, having exactly enough information and one correct answer. Figure 2 summarizes these findings.

One might think that the emergence of more "reformed" or research-based textbooks might better address the need for a variety of problem types. Although these books often do a better job of addressing conceptual understanding, the problem distribution in these books is not much different than their traditional counterparts, as shown in Figure 3.

		Answers		
		None	One	> One
Information	Insuff		HRW 1.5%	
	Exact	HRW <1% BJ 2%	HRW 97% BJ 89%	HRW <1% BJ 9%
	Excess			

HRW - Halliday, Resnick & Walker, Physics, 6th Ed
Ch 2 - Kinematics, Ch 4 - Kinematics 2 & 3 D
(All end of chapter problems)
BJ - Beer & Johnson, Statics, 6th Edition
Ch 3 - Rigid Bodies, Ch 6 - Truss Analysis
(114 problems)

FIGURE 2. Problem Distribution in Traditional STEM Texts

		Answers		
		None	One	> One
Information	Insuff		K <1% H 2%	K <1%
	Exact		K 89% ST 86% H 86%	K 9% ST 13% H 22%
	Excess			

K - Knight, Physics, 1st Ed
Ch 2 - Kinematics, Ch 4 - Motion in a plane
(All end of chapter problems)
ST - Sheppard & Tongue, Statics, 1st Ed
Ch 6 - Rigid Bodies, Ch 9 - Truss Analysis
(100 problems)
H - Hughes-Hallet, Calculus, 1st Edition
Ch 2 - The Derivative, Ch 5 - Using the Derivative

FIGURE 3. Problem Distribution in Reform STEM Texts

ONGOING WORK

This summer (2006), the project team is in the process of soliciting a variety of problems from faculty to see if they fit into the matrix to determine if further modifications are needed. Also, a sample of STEM faculty are reviewing the categorization scheme to see if 1) they agree with the descriptions, 2) they can add more terms to the matrix, and 3) they can think of any problems that do not fit in the matrix. Further tweaking may be necessary. Eventually, all basic problems in the FEH sequence should be categorized according to this or a similar scheme.

At the same time, a variety of problems are being developed to fill in the sparsely occupied boxes of the matrix, particularly for engineering graphics and physics mechanics, since these are two of the FEH courses that will be taught in the fall.

The next step will be to engage the FEH faculty in discussions to determine as a staff a strategy for utilizing this categorization as an instructional tool. Included in this work will be syllabus development, lecture modification, and problem selection. At the same time, a third axis will be added to the matrix, further categorizing problems utilizing a taxonomy such as Bloom's. This third classification will again assist the instructors in developing an approach to improve problem solving instruction, both within individual courses and also program-wide. As strategies are developed, the research staff will design an assessment plan for their implementations.

CONCLUSIONS

The problem categorization matrix, while still in its early stages, shows promise as a useful tool in guiding faculty problem-solving discussions. Using the categorization scheme to do a preliminary analysis of introductory textbooks has revealed that their problems usually do not require application of the real-world problem solving skills that STEM majors need; this could be one reason for the poor development of such skills cited in the literature. Development of additional problems to better cultivate these skills and utilization of the new categorization scheme by an interdisciplinary group of faculty may have a more substantial impact. Further development and research will show whether or not this is the case.

ACKNOWLEDGMENTS

We are grateful to the FEH staff, as well as the problem-solving researchers nationwide, who have served as sounding boards for our ideas.

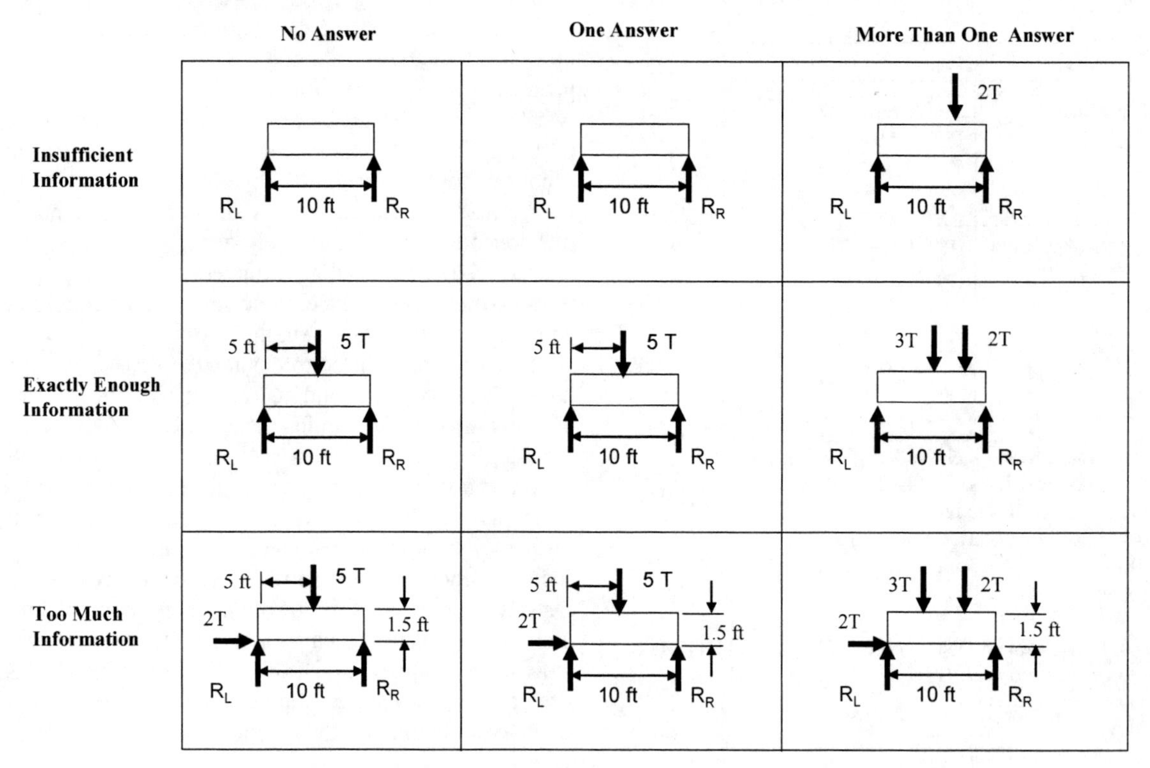

FIGURE 4. Example Statics Problems in the Problem Categorization Matrix
Note that the problem statement at the top of each column pertains to each problem in that column.

REFERENCES

[1] H. A. Simon, in *Problem Solving and Education: Issues in Teaching and Research*, edited by D. T. Tuma and F. Reif, Hillsdale, NJ: Lawrence Erlbaum Associates, 1980.

[2] H. F. O'Neil, *Computers in Human Behavior* **15**, 255-268 (1999).

[3] J. Schacter et al., *Computers in Human Behavior* **15**, 403-418 (1999).

[4] A. P. Carneval et al., *Workplace Basics: The Essential Skills Employers Want* San Francisco, CA: Jossey-Bass , 1990.

[5] D. Rosil, "What are Masters Doing?" College Park, MD: AIP Statistics Division, 1996.

[6] See, for instance, the accreditation criteria of ABET.

[7] D. Woods, *J. Coll. Sci. Teach* **23**, 157-158 (1994) and **12**, 446-448 (1983).

[8] *How People Learn*, edited by J. Bransford et al., Washington, DC: National Academy Press, 1999.

[9] R. M. Felder, and R. Brent, *Journal of Engineering Education* **93**, 269-277(2004).

[10] R. M. Felder and R. Brent, *Journal of Engineering Education* **93**, 278-291 (2004).

[11] J. Demel et al., in *Proceedings of the ASEE 2005 Annual Conference*, ASEE, 2005.

[12] M. T. H. Chi et al *Cognitive Science* **5** , 121-152 (1981).

[13] J. H. Larkin., *Engineering Education* **70**, 285-288 (1979).

Comparing Explicit and Implicit Teaching of Multiple Representation Use in Physics Problem Solving

P. Kohl,[I] D. Rosengrant,[II] and N. Finkelstein[I]

[I]*Department of Physics, University of Colorado, Campus Box 390, Boulder, CO 80309*
[II]*Rutgers, The State University of NJ, GSE 10 Seminary Place, New Brunswick NJ 08901*

Abstract. There exist both explicit and implicit approaches to teaching students how to solve physics problems involving multiple representations. In the former, students are taught explicit problem-solving approaches, such as lists of steps, and these approaches are emphasized throughout the course. In the latter, good problem-solving strategies are modeled for students by the instructor and homework and exams present problems that require multiple representation use, but students are rarely told explicitly to take a given approach. We report on comparative study of these two approaches; students at Rutgers University receive explicit instruction, while students from the University of Colorado receive implicit instruction. Students in each course solve five common electrostatics problems of varying difficulty. We compare student performances and their use of pictures and free-body diagrams. We also compare the instructional environments, looking at teaching approaches and the frequency of multiple-representation use in lectures and exams. We find that students learning via implicit instruction do slightly better and use multiple representations more often on the shorter problems, but that students learning via explicit instruction are more likely to generate correct free-body diagrams on the hardest problem.

INTRODUCTION

Instructors and researchers in PER have long argued that students can benefit from solving problems that require the use of more than one representation (equations and bar diagrams, for example)[1-5]. These problems are said to require a more complete understanding of the underlying physics than traditional 'plug and chug' problems[1,2]. There have been several studies in which students are taught explicit steps for solving problems that use multiple representations [3,4,5]. Another, less-studied approach is to model good problem-solving techniques for students implicitly, without teaching explicit steps. Arguments can be made in favor of either the explicit or implicit approaches. For example, an explicit approach gives students an easy-to-follow 'checklist,' though it might also result in dependence on algorithms executed with little understanding. We are unaware of any studies directly comparing explicit and implicit approaches to teaching multiple representation problem solving, and we wish to know whether PER-informed implicit teaching of these skills can be comparable to explicit approaches.

In this study, we consider two introductory algebra-based physics courses taught at large state universities. In one of these (at Rutgers, the State University of New Jersey), the professor taught an explicit set of multiple representation problem-solving steps for physics problems involving forces[6]. These steps were then emphasized many times over the course of the two-semester sequence. The course at the University of Colorado included many similar multiple representation problems, but did not teach a specific heuristic. Rather, the professor modeled good problem-solving techniques in lecture and in solutions to homeworks and exams.[1]

We have two goals in this study. First, we wish to look for differences in performance on multiple-representation problems in the two classes, and for differences in how often and how successfully students in each course used multiple representations in their solutions. Second, we wish to compare the instructional approaches of the two courses as well as the density of representation use in each course, with the intention of relating that to student performance. We can identify three major results so far. First, students from both courses, despite the differences in implementation, use multiple representations in their solutions frequently as compared to students from traditional courses. Second, student performance in each course is very similar. Third, students from the "implicit" course at CU use more representations more often on shorter problems, while students from the "explicit" course at Rutgers are much more likely to draw a complete and correct free-body diagram (FBD) on the most difficult problem.

[1] The differences in multiple representation use are not the only differences between the two courses. Rather, they are the differences we choose to focus on during this study of student use and success with multiple representations.

CP883, *2006 Physics Education Research Conference*, edited by L. McCullough, L. Hsu, and P. Heron

	Prob. 1	Prob. 2	Prob. 3	Prob. 4
Rutgers	0.38 (235)	0.56	0.32	0.38 (155)
CU	0.38 (314)	0.56	0.43	0.40 (269)

TABLE 1. (Above) Performance of Rutgers and CU students on the recitation questions. The difference on problem 3 is significant (p=0.008, two-tailed binomial test). The numbers in parentheses are the sample sizes for problems 1-3 and problem 4.

TABLE 2. (Right) Top: Fraction of students drawing pictures with their problem solutions. Bottom: Average number of forces correctly identified with solutions. Statistical significances are all two-tailed binomial tests.

Pictures	Prob. 1	Prob. 2	Prob. 3	Prob. 4
Rutgers	0.89 (235)	0.89	0.03	0.73 (155)
CU	0.92 (314)	0.91	0.13	0.90 (269)
p-value	X	X	0.0001	<0.0001

Forces	Prob. 1	Prob. 2	Prob. 3	Prob. 4
Rutgers	0.59 (235)	0.52	0.03	1.71 (155)
CU	0.80 (314)	0.71	0.11	1.69 (269)
p-value	0.008	0.05	0.0002	X

We consider these results to be consistent with the course environments, as we argue in the discussion.

METHODS

The study involved second-semester large-lecture algebra-based physics courses from CU and Rutgers, taught in the spring of 2006. Both schools are large public universities, with similar SAT/ACT scores for incoming students. The instructors involved had also taught the first semester of the sequence. Both courses can be described as reformed in nature. A more complete course description can be found in the Data section. Students in each course received a set of four electrostatics problems in recitation that either required calculation of a force or specified forces in the problems. An example problem is included as a note[7]. These problems were ungraded, though students received credit for significant effort.

Students were also given a more challenging problem, intended to be very difficult to solve without an FBD. This problem[8] was given with multiple-choice answers on an exam in the Rutgers course, and was given as a free-response recitation quiz in the CU course.

To complement these problems, we analyzed the representational content of the exams in each course. The procedure (described in [9]) involved quantifying the fraction of lecture time and exam points associated with verbal, mathematical, graphical, and pictorial representations, as well as noting how often multiple representations were used together. We will reserve lecture analysis for a long-format paper, where we will establish that both lectures were rich in representation and multiple representation use.

DATA

We divide our data into four sections. First, we will provide a short description of the problem-solving heuristic taught in the Rutgers course. Second, we shall consider student performance and representation use on the four recitation problems. Third, we shall examine performance and representation use on the harder exam/quiz problem. Finally, we characterize the representational content of the Rutgers and CU exams. All data include only the students who took both semesters of the physics sequence from the instructor in question.

Course Descriptions. The Rutgers course uses the ISLE curriculum, which is inquiry-based and spends considerable time on the use of multiple representations[10]. In addition, they use the Active Learning Guide workbook in lecture and in recitation, which includes many tasks designed to teach multiple representation use[6]. The instructor teaches students an explicit problem-solving heuristic with five main steps: Picture and Translate, Simplify, Represent Physically, Represent Mathematically, and Evaluate. Students also learn an explicit procedure for constructing free-body diagrams. The CU course includes reforms such as personal response systems (clickers) and PER-based labs and recitation activities[11] (including PhET computer simulations[12]). It also includes substantial multiple representation use in lecture and in homework and exam tasks, but little *explicit* instruction in multiple representation use is given.

Recitation Problems. Table 1 shows the fraction of Rutgers and CU students answering each of the four recitation questions correctly. The number in parentheses indicates the sample size. Student performance was comparable on problems 1, 2, and 4. On problem 3, 43%of the CU students answered correctly, compared to 32% of the Rutgers students. This difference is statistically significant at a p=0.008 level using a two-tailed binomial proportion test. Problem 3 is unique in that it was the only problem that provided a picture and free-body diagram along with the problem statement.

The top of Table 2 shows the fraction of students that drew a picture with their problem solutions. Since problem 3 included a picture, for that problem we show

Exam	Correct	1 Force	2 Forces	3 Forces
Rutgers	0.56 (283)	0.09	0.22	0.51
CU	0.29 (280)	0.23	0.31	0.32

TABLE 3. (above) Fraction of students answering exam/quiz problem correct, and fraction correctly identifying 1, 2, or 3 (out of 3) forces with a free-body diagram. The fractions identifying all three forces correctly are different with $p<0.0001$.

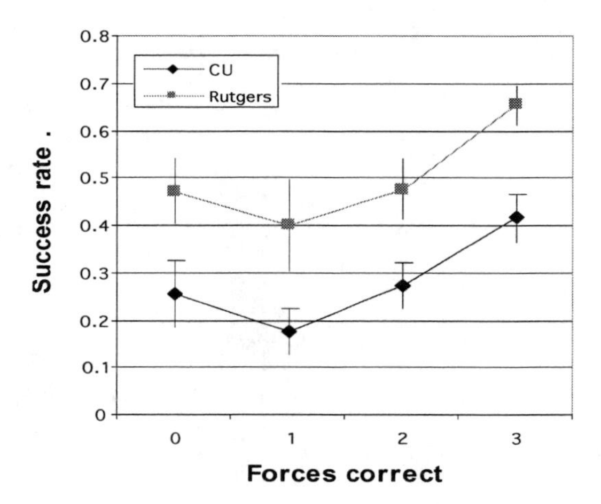

FIGURE 1. (right) Student performance on exam/quiz problems as a function of number of forces correctly identified. Note that the Rutgers problem was multiple choice, so absolute performances are not directly comparable.

the fraction of students that re-drew their own picture. Picture use is very high on problems 1, 2, and 4, with an overall average of 87% of problem solutions including a picture. CU students drew a picture significantly more often for problem 4.

The bottom of Table 2 shows the average number of correct forces that students identified on their solutions. Since problem 3 included a free-body diagram and only a few students re-drew an FBD, the average number of correct forces is very low, and we instead show the fraction of students re-drawing any forces. We see that students from CU drew more correct forces on average for problems 1and 2. Note that problem 4 (and only problem 4) asked students to include a complete FBD with their answer, which in that case would include two forces.

Exam/Quiz Problem. Table 3 shows the fraction of students answering the exam/quiz problem correctly at Rutgers and at CU. We also see the fraction of the class that correctly identified one, two, or three forces in their free-body diagrams. An FBD was not required as part of the answer, but the problem was extremely difficult to complete otherwise. As noted previously, the Rutgers problem was given in a multiple-choice format, while the CU problem was given as free-response, so direct performance comparisons are difficult. Since guessing on a five-choice exam should result in a 0.20 success rate, we expect that the CU score would be boosted by that much (or more, since the choices constrain the order of magnitude of the answer) given a multiple-choice format. With that in mind, the CU and Rutgers performances are probably similar.

In Figure 1, we plot problem success versus number of forces correctly identified for both courses. We see the same trend in both courses: Students who identify no forces correctly do as well (possibly better) than those who identify only one. Those that identify two do as well as those who identify none, and those who identify all three do significantly better than all others. The error bars are large, though overlap is reduced if one averages both data sets together (not shown).

Exam content. In Figure 2, we show the fraction of the exams devoted to different representations and multiple representations. We average over all of the exams from the *first* semester of this physics sequence, as the study took place before any exams in the second semester, and so only those from the first semester could influence these students. We see that the CU exams generally used more representations more often, as compared to the more mathematical exams at Rutgers.

DISCUSSION AND CONCLUSION

We find that the CU group, which had received implicit problem solving instruction, was somewhat more likely to use multiple representations in their solutions to short problems. However, students in the Rutgers group (who received explicit training in multiple-representation problem solving) were much more likely to generate complete and correct free-body diagrams on the most difficult problem (51% vs 32%, as in Table 3). In situations where the performances could reasonably be compared, they were very similar.

Perhaps the most striking result is that students in both courses used multiple representations quite often in their problem solutions (80 or 90 percent of the time). While there were significant differences in implementation, both courses were PER-based and had representation-rich lecture components. For comparison, Van Heuvelen observes much less frequent multiple representation use in traditional courses[3].

Also notable is the observation that CU exams were richer in representations. We have hypothesized previously[9] that representation-rich exams might foster improved representational competence (as such exams hold students accountable for developing broad representational skills), and it is possible that these richer

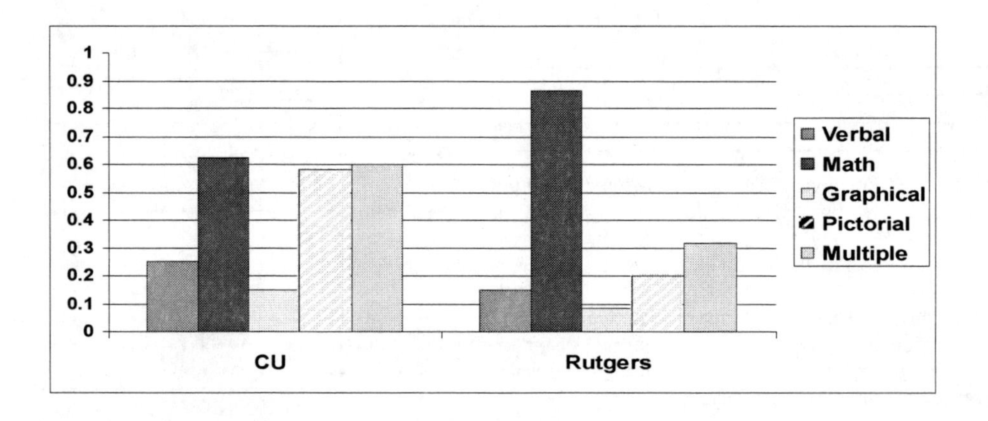

FIGURE 2. Averaged fraction of first-semester exam problems containing listed representations. "Multiple" category refers to the use of two or more representations in one problem.

exams contributed to the CU students' frequent multiple-representation use on the shorter problems, despite the less explicit problem-solving instruction. More detailed analysis of the course environments is required, and will be reserved for a long-format paper.

We have also seen that the problem-solving algorithm taught to the Rutgers group is detailed and would involve considerably more effort than a standard "plug 'n chug" approach. This suggests to us an explanation of the differences we observed in representation use. If the students learn to use free-body diagrams and pictures in a specific way, and that way is lengthy, students may be inclined to only use multiple representations when the problem is difficult enough to make the extra effort clearly worth it. Thus, we see the Rutgers students using FBDs much more often on the challenging exam/quiz problem, and using pictures and FBDs less often on the simpler problems. Interestingly, the CU students do not significantly outperform Rutgers students on these shorter problems, raising the question of whether multiple representation use should be valued independent of performance.

It is not clear from these data whether the explicit approach to teaching problem solving was superior to the implicit approach, or vice versa. Indeed, such a judgment may depend on the goals of the instructor. If teaching a specific heuristic were to result different representation use on different tasks, the instructor would need to decide whether that tradeoff was desirable. Without more specific criteria, it appears to us that both appear to have been successful in fostering multiple representation use in solving problems. It is possible that elements of each course could be usefully combined, as well.

ACKNOWLEDGEMENTS

This work was supported in part by an NSF Graduate Fellowship. Special thanks to the PER groups at CU and Rutgers.

REFERENCES

1. Larkin, J.H. (1983). The role of problem representation in physics. In *Mental Models,* D. Gentner. and A. Stevens. (eds.). Hillsdale, NJ: Lawrence Erlbaum, pp. 75-99, 1983.
2. Reif, F. (1995). Understanding and teaching important scientific thought processes. *Am. J. Phys.* 63, 17-32.
3. Van Heuvelen, A. (1991). Learning to think like a physicist: A review of research-based instructional strategies. *Am. J. Phys*, 59(10). 891-897.
4. Van Heuvelen, A. and Zou, X. (2001). Multiple reps. of work-energy processes. *Am. J. Phys*, 69(2). 184-194.
5. Dufresne, R. J., Gerace, W. J., and Leonard, W. J. (1997). Solving physics problems with multiple representations. *The Physics Teacher*, 35, 270-275.
6. Van Heuvelen, A. and Etkina, E. (2006) *Active Learning Guide*. Addison Wesley, San Francisco, CA.
7. A small (100 g) metal ball with +2.0 μC of charge is sitting on a flat frictionless surface. A second identical ball with -1.0 μC of charge is 3.0 cm to the left of the first ball. What are the magnitudes and directions of the forces that we would have to apply to each ball to keep them 3.0 cm apart?
8. A small metal ball with $Q_1 = +2.0$ μC of charge hangs at the end of a vertical string. A second, identical ball with $Q_2 = -2$ μC of charge hangs at the end of a vertical string. The tops of the strings are brought near each other, and the strings reach an equilibrium orientation (no longer vertical) when the balls are a distance d = 3.0 cm apart. If the force of the Earth on each ball is $F_1 = F_2 = 30$ N, what is the force T of the string on each ball?
9. Kohl, P.B and Finkelstein, N.D. (2006) The effect of instructional environment on physics students' representational skills. *Phys. Rev. ST: PER*, 2, 1, 010102.
10. Etkina, E. & Van Heuvelen, A. (2001). Investigative Science Learning Environment: Using the processes of science and cognitive strategies to learn physics. *PERC Proceedings 2001.* Rochester, NY, 17-21.
11. Finkelstein, N.D. and Munsat, T. http://www.colorado.edu/physics/EducationIssues/lab_revisions/
12. http://phet.colorado.edu

An Overview of Recent Research on Multiple Representations

David Rosengrant, Eugenia Etkina and Alan Van Heuvelen

Rutgers, The State University of New Jersey
GSE, 10 Seminary Place, New Brunswick NJ, 08904

Abstract. In this paper we focus on some of the recent findings of the physics education research community in the area of multiple representations. The overlying trend with the research is how multiple representations help students learn concepts and skills and assist them in problem solving. Two trends developed from the latter are: how students use multiple representations when solving problems and how different representational formats affect student performance in problem solving. We show how our work relates to these trends and provide the reader with an overall synopsis of the findings related to the advantages and disadvantages of multiple representations for learning physics.

Keywords: Multiple Representations, Free-Body Diagrams, Student Problem Solving.
PACS: 01.40.Fk

INTRODUCTION

A representation is something that symbolizes or stands for objects and or processes. Examples in physics include words, pictures, diagrams, graphs, computer simulations, mathematical equations, etc. Some representations are more concrete (for example, sketches, motion diagrams and free-body diagrams) and serve as referents for more abstract concepts like acceleration and Newton's second law—they help student understanding. Mathematical representations are needed for quantitative problem solving. More concrete representations can be used to help apply basic concepts mathematically. For example, students can learn to use free-body diagrams to construct Newton's second law in component form as an aide in problem solving. Consequently, many educators recommend the use of multiple representations (MRs) to help students learn and to solve problems [1-7].

This manuscript describes recent multiple representations studies by the physics education research community [1-17] including our own work in this field.

RECENT TRENDS

In this section we provide an outline of the recent trends in multiple representation research [2003-2005] in the PER community. These trends form a logical sequence. The sequence begins with the major question of whether using MRs helps students learn concepts and learn to better solve problems. Concerning problem solving, what instructional innovations actually help students use MRs while solving problems? And if they do use them, then how do they use them to help solve the problem?

A separate line of research relates to problem posing – how does the representation in which the problem is posed affect student performance and their decision to use another type of representation when solving the problem?

Table 1 compiles the studies used for this paper into an easy reference for those who wish to read the full articles. The numbers in the table correspond to the references in the manuscript.

TABLE 1. References to multiple representation studies.

Research Trend			
MRs help students learn concepts and solve problems	6, 7		
	Do students use them?	**Do they help students**	**What MRs do students choose?**
Use of MR to solve a problem	5,8,11,12	5,8,11,12	12,13
Use of MR to pose a problem		13,14,15,16, 17	13

Next, we describe the details of the different trends.

CP883, *2006 Physics Education Research Conference*, edited by L. McCullough, L. Hsu, and P. Heron

Multiple Representations Help Students Learn Concepts and Solve Problems

Hinrichs [6] describes how using a system schema (object of interest is circled, objects that are interacting with it are circled and then connected to it via labeled arrows) helped his students learn dynamics. He used the system schema as part of a sequence of representations (problem text, sketch, system schema, free body diagram, and finally equations) to solve a problem. He compared classes where he used system schemas with classes where he did not use schemas. The 28 students who learned to use system schemas increased from 1.1 ± 1.0 questions out of 4 questions correct on Newton third law FCI pretest questions to 3.7 ± 0.8 on the post-test. The 31 students who did not learn to use a system schema scored a 1.2 ± 1.0 on the same 4 questions on the pre-test and 2.8 ± 1.2 on the post-test. The author reports that the system schema had a significant effect on student learning.

Finkelstein et al. [7] used computer simulations to aide students in learning DC circuits. The simulations provide visual representations of concepts such as current flow and Kirchoff's laws. They found that students who learned concepts related to DC circuits via computer simulations and never built a real circuit performed significantly better on 3 exam questions that related to DC circuits than students who learned using real circuits. They also found that the former students could build and explain real circuits faster (14 minutes compared to 17.7). The authors report that this is a significant difference. There was no significant difference on non-circuit questions. The visual simulation representations had a significant effect on understanding and problem solving.

Use of Multiple Representations and Problem Solving

Several studies investigated whether the use of multiple representations in courses affected student problem solving. De Leone and Gire [8] studied how many representations students in a reformed course used when solving open-ended problems on quizzes and tests. They analyzed student's work on 5 problems and found that 31 of 37 students used 4 or more representations total (they called them high MR users).

Our group [5,11,12] investigated students' use of one representation – a free body diagram (FBD) – when solving multiple choice problems in mechanics and static electricity. The experimental group used the *ISLE* curriculum in which multiple representations are central to students' learning [9]. We found that on average 58% of the students drew an FBD while solving a multiple choice problem even though they knew that no credit was given for the diagrams. Only 15% of students in traditional settings use FBDs to help them solve problems [10].

The research by Rosengrant, Etkina and Van Heuvelen and by DeLeone and Gire shows that if students learn physics in an environment that emphasized the use of multiple representations, students will use them to help solve problems. Does the use of these different representations improve problem-solving performance?

DeLeone and Gire [8] found that those students who successfully solved 3 or more of the 5 coded problems were all high MR users. They drew a picture, an extended force diagram, an energy system diagram, or plotted a graph. On 4 out of those 5 problems students who used representation other than mathematical had a higher success rate than those who did not. De Leone and Gire did not assess the quality of the representations that student constructed.

We investigated a similar question, but took our analysis one step further [5,11]. We related student success on a problem with not only the presence, but the quality of the representation—in our case a free body diagram. Table 2 contains the average results from our two year study of 245 students answering several multiple-choice problems in different conceptual areas. The first column states the quality of the free body diagram assessed by a specially designed rubric on a scale of 0-3. The second column shows the number of students who correctly solved a problem with that quality of FBD divided by the total number of students who drew an FBD of the same quality. The last column is the percentage of the previous column. The average percentage of correct answers for all problems was 60% (a measure of the difficulty of the problems). The results suggest that FBDs are most beneficial to students if they are constructed correctly. If a student constructs an incorrect free body diagram, then they actually have a lower chance to correctly solve the problem then if they had no free-body diagram.

TABLE 2. Average of Two Year Study [11].

Quality of FBD	Number of students with correct answer divided by total number with same quality diagram	%
Correct (3)	251/295	85
Needs improvement (2)	261/370	71
Inadequate (1)	69/181	38
None (0)	304/619	49

In the PERC 2005 proceedings paper we reported on a qualitative study (6 students) [12] using think aloud interviews that investigated what representations students chose to help them solve problems involving forces and why they constructed the representations. We found that all students, even those who could not solve the problem, drew a picture for the problem situation but only those who were in the reformed *ISLE* course constructed free body diagrams to help them solve the problem. The high achieving students in the sample used the representations not only to help them solve the problem but also to evaluate their work.

In new and unpublished work, we have also found that the way the problem is posed can affect whether students will use an FBD to solve it. We used several multiple-choice exams with 245 students where a total of 12 problems that involved forces (mechanics or electrostatics) were selected for the study. On the exam sheets, many students drew FBDs for some problems and few for others. Was there any pattern in their choices? To answer this question we grouped these 12 problems into three categories based on the number of students who constructed an FBD to help them solve that problem [Table 3]. We placed a problem in a 'Low' category if fewer than 50% of the students constructed an FBD to solve it; in a 'Medium' if between 50 and 60% constructed an FBD; and in the 'High' category if more then 60% constructed an FBD.

Though all of the problems were multiple-choice in design, there were features that were different across the problems. Some problems had a picture in the text and some did not. Some were more difficult than others. Some problems asked students directly to determine a force and some did not. Finally, some problems were in mechanics and some in electrostatics. How did these differences contribute to students' decisions to draw an FBD? Table 3 shows the relationship between these factors and the questions in each group.

TABLE 3. Possible relationships between type of problem and how likely students are to construct a diagram for it.

Factors	Low	Medium	High
# of problems	4	3	5
Picture present	2 with 2 w/o	2 with 1 w/o	0 with 5 w/o
Average Success Rate	52.75%	65.7%	63.8%
Problem asks for a Force	2 No 2 Yes	1 No 2 Yes	0 No 5 Yes
Type of Problem	2 Mech. 2 Elect.	1 Mech. 2 Elect.	4 Mech. 1 Elect.

We correlated our results from Table 3 and found that the highest correlation between the percentage of students who constructed an FBD and an influencing factor was if the problem asked for a force [Pearson correlation coefficient 0.502]. The next highest correlation was if a picture was present [correlation coefficient -0.479]. The negative correlation implies that if a picture is present, a student was less likely to construct an FBD. The difficulty of the problem and whether the problem was in mechanics or electrostatics did not have a big effect on student choices [correlations of 0.395 and 0.278]. None of the correlations was statistically significant though the top two factors were close to being significant. This is not surprising since the sample size for the number of problems in the study was very small [N=12].

This result suggests that when the instructor supplements the text of the problem with the picture, the students are less likely to construct a free body diagram to solve the problem. One explanation can be because the provided picture helps the students understand the problem situation and thus they think that they do not need to draw an FBD. Also, if the problem asks students to solve for a force, they are more likely to construct an FBD to help them solve that problem. One possible explanation for this is that the word force in the problem statement triggers a "FBD schema." Problems involving similar concepts but asking for acceleration do not trigger this schema. However, there needs to be more research in this area before we can verify these trends.

Use of Multiple Representation to Pose A Problem

This area of research investigates the relationship between student success and the representational format in which a problem is posed. The first question relates to student choices of the problem format: if they are given this choice, what will they choose? Kohl and Finkelstein [13] found that more students prefer the problem statement to be represented with a picture than with words, graphs or mathematical equations. However, this does not necessarily make them more successful in solving the problem.

For example, on a question in wave optics students who chose a pictorial format did significantly better then the control group. However in atomic physics the students who chose a pictorial format did significantly worse then students in the control group. There was no clear pattern what format made the problem more difficult. However, in their second study [16,17] they found that students who learned physics with the instructor who used lots of representations were less affected by the representational format of the problem. Therefore if we want our students to be able to reason flexibly, it appears that the use of multiple representations when they are learning new material helps.

Dancy and Biechner [14] used computer animations for some questions on the pre-test FCI in an experimental group and traditional questions in a control group. They found significant differences on 6 of the questions between the two groups. On 3 questions, the animations group performed significantly better, while on 3 other questions the control grouped performed better. After conducting interviews they found that for the problems including motion, the animations clarified the problem statement and helped students make answer choices more consistent with their understanding (and not necessarily correct). They concluded that a simulation format is especially beneficial for those students who have reading comprehension problems. Animations in the problem statement lead to a more accurate assessment of student reasoning because students have a better understanding of the intent of the questions.

Meltzer's study [15] compared students' responses to a variety of isomorphic physics problems posed in different ways: in words, with a vector diagram, with a circuit diagram, etc. He found that "student performance of very similar problems posed in different representations might yield strikingly different results" (p. 473). The same student can answer a Newton's third law question posed in words correctly and choose an incorrect answer to the very same question posed with a picture with vectors. He also found that females were particularly harmed by the non-verbal representations of the problem statements.

DISCUSSION

The general consensus of the described work is that representations are important for student learning. They assist students in acquisition of knowledge and in problem solving. We can say that using high quality multiple representations while solving a problem is a sufficient condition for success but it is not a necessary condition. Students use representations to help them understand the problem situation and to evaluate the results. Representations in problem statements can have different effects on student performance and on their choice to use other representations. For some problems a computer animation can clarify the situation for the students and help them display their real reasoning. Students who learn the material in an environment that uses more representations are less affected by the representational format of the problem statement. Another finding is that certain words in the problem statement may trigger the use of particular representations, though more research must be done to verify this finding.

There has been a recent growth of research in multiple representations. This growth is expanding rapidly with many opportunities for future researchers. They can: focus on what factors influence students to construct representations, replicate studies with other representations or investigate the quality of representations students construct.

ACKNOWLEDGMENTS

This work was supported in part by NSF grants DUE 0241078 and DUE 0336713. Special thanks go out to the entire physics education research group at Rutgers, The State University of New Jersey and to the PERC Proceedings reviewers.

REFERENCES

1. R. Dufresne, W. Gerace and W. Leonard, *Phys. Teach.*, **35**, 270-275 (1997).
2. J.I. Heller and F. Reif, Cognit. Instruct. **1**, 177-216 (1984).
3. D. Hestenes, In Modeling Methodology for Physics Teachers. E. Redish & J. Rigden (Eds.), AIP Part II. p. 935-957, (1997).
4. A. Van Heuvelen and X. Zou, Am. J. Phys., **69**, 184-194 (2001).
5. D. Rosengrant, A. Van Heuvelen, and E. Etkina, In edited by J. Marx , P. Heron, and S. Franklin, 2004 Physics Education Research Conference Proceedings, Sacramento, CA, 2004, 177-180.
6. B. Hinrichs, In edited by J. Marx , P. Heron, and S. Franklin, 2004 Physics Education Research Conference Proceedings, Sacramento, CA, 2004, 117-120.
7. N. Finkelstein, W. Adams, C. Keller, P. Kohl, K. Perkins, N. Podolefsky, S. Reid, and R. LeMaster, *Phys. Rev. ST Phys. Educ. Res.* 1, 010103, (2005).
8. C. DeLeone and E. Gire, edited by P. Heron, L. McCullough and J. Marx, 2005 Physics Education Research Conference Proceedings, Salt Lake City, UT, 2005, 45-48.
9. E. Etkina and A. Van Heuvelen, 2001 Physics Education Research Conference Proceedings, Rochester, NY, 2001, pp 17-21.
10. A. Van Huevelen, Am. J. Phys., **59**, 891-897 (1991).
11. D. Rosengrant, E. Etkina, and A. Van Heuvelen, National Association for Research in Science Teaching 2006 Proceedings, San Francisco, CA (2006).
12. D. Rosengrant, A. Van Heuvelen, and E. Etkina, edited by P. Heron, L. McCullough and J. Marx, 2005 Physics Education Research Conference Proceedings, Salt Lake City, UT, 2005, 49-52.
13. P. Kohl and N. Finkelstein, *Phys. Rev. ST Phys. Educ. Res.* 1, 010104, (2005).
14. M. Dancy and R. Beichner, *Phys. Rev. ST Phys. Educ. Res.* 2, 010104, (2006).
15. D. Meltzer, Am. J. Phys., **73**, 463-478 (2005).
16. P. Kohl and N. Finkelstein *Phys. Rev. ST Phys. Educ. Res.* 1, 010102, (2006).
17. P. Kohl and N. Finkelstein *Phys. Rev. ST Phys. Educ. Res.* 1, 010104, (2006).

Same to Us, Different to Them: Numeric Computation versus Symbolic Representation

Eugene Torigoe and Gary Gladding

Department of Physics, University of Illinois Urbana-Champaign, Urbana, IL, 61801

Abstract. Data from nearly 900 students was used to measure differences in performance on numeric and symbolic questions. Symbolic versions of two numeric kinematics questions were created by replacing numeric values with symbolic variables. The mean score on one of the numeric questions was 50% higher than the analogous symbolic question. An analysis of the written work revealed that the primary identifiable error when working on the symbolic problems was a confusion of the meaning of the variables. The paper concludes with a discussion of possible theoretical explanations and plans for future follow-up studies.

Keywords: physics education research, kinematics, symbolic algebra.
PACS: 01.40.Fk. 01.30.Cc.

INTRODUCTION

Algebra is an essential skill for students in physics. For some students the lack of algebraic facility acts as a barrier to success in physics. Studies from the mathematics education research community have shown that many students learning algebra have difficulties understanding the meaning of symbolic variables and dealing with unevaluated expressions [1]. We performed a study to measure the effect of such algebraic difficulties in the context of a physics exam.

We modified exam questions to see the effect of numeric computational and symbolic representational cues on student performance. We hypothesized that changing the numeric quantities in the problem to symbolic variables would change the way students thought about and approached the solution to the problems. Specifically, we thought that students would have difficulties with problems that required them to plug an unevaluated symbolic expression into another symbolic equation.

METHODS

The subjects of the study were students who were enrolled in the calculus-based introductory mechanics course, Physics 211, at the University of Illinois, Urbana-Champaign in the spring 2006 semester. Students in Physics 211 take three multiple-choice midterm exams and a cumulative final. While numeric questions are the most common type of question, symbolic questions are not uncommon. There were 894 students who completed one of the two randomly administered versions of the final exam.

We placed a pair of kinematics questions dealing with cars on each version of the final exam. On final exam 1, numeric values were used and on final exam 2, symbolic variables were used. The questions are shown in Figure 1.

Minor modifications were made to discourage cheating. The order of the problems and the order of the choices were reversed from one final to the other. The same choices are present in both versions of each problem, except final 1 question 7 and its partner final 2 question 6 where two choices do not agree. Both choices, however, were selected by less than one percent of the students.

We also placed a pair of Newton's 2nd law problems on each version of the final exam whose solutions involved intermediate variables that cancelled out of the final equation. The version on final 1 supplied numeric values for the intermediate variables and the version on final 2 supplied symbols for the intermediate variables. The analysis for these questions is still ongoing and will not be discussed in this paper.

RESULTS

As we expected, the mean score of the questions that used symbols were lower than the questions that did not. Table 1 lists the mean and the standard error for each question.

CP883, *2006 Physics Education Research Conference*, edited by L. McCullough, L. Hsu, and P. Heron

Final 1

A bank robber is racing towards the state line at a constant speed of 30 m/s and passes a cop who is parked 1000 m from the state line. At the moment the robber passes him, the cop accelerates from rest with a constant acceleration to try to catch the robber before he reaches the state line.

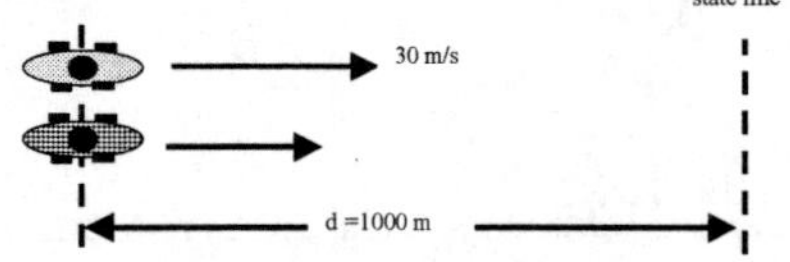

6. What is the minimum acceleration that the cop must have in order to catch the robber before he reaches the state line? (Ignore any reaction time.)

a. 0.23 m/s^2 (< 1%) b. 0.45 m/s^2 (7%) c. 0.90 m/s^2 (6%)
d. 1.80 m/s^2 (85%) e. 3.60 m/s^2 (< 1%)
(More than 1 choice: 2%)

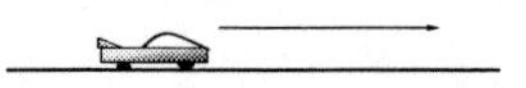

7. A car can go from 0 to 60 m/s in 8 seconds. At what distance, d, from the start (at rest) is the car traveling 30 m/s? (Assume a constant acceleration.)

a. 30 m (< 1%) **b. 60 m (94%)**
c. 90 m (<1%)(No corresponding choice)
d. 120 m (2%) e. 240 m (2%)
(More than 1 choice: 0%)

Final 2

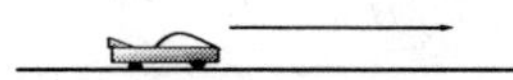

6. A car can go from 0 to a speed v_1 in t seconds. At what distance, d, from the start (at rest) is the car traveling ($v_1 / 2$)? (Assume constant acceleration.)

a. $d = v_1 t$ (< 1%) (No corresponding choice)

b. $d = \frac{v_1 t}{2}$ (25%) c. $d = \frac{v_1 t}{4}$ (27%)

d. $d = \frac{v_1 t}{8}$ **(45%)** e. $d = \frac{v_1 t}{16}$ (< 1%)

(More than 1 choice: 3%)

7. A bank robber is racing towards the state line at a constant speed of ***v*** and passes a cop who is parked ***d*** from the state line. At the moment the robber passes him, the cop accelerates from rest with a constant acceleration to try to catch the robber before he reaches the state line.

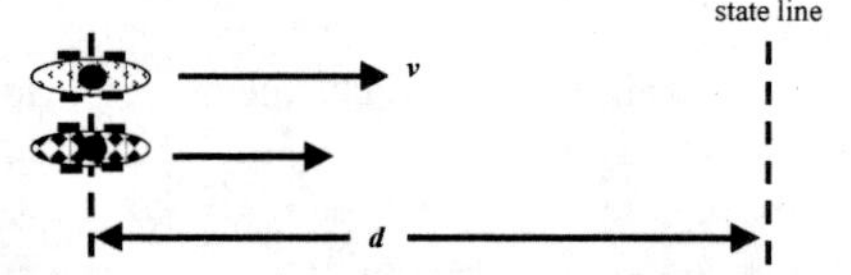

What is the minimum acceleration that the cop must have in order to catch the robber before he reaches the state line? (ignore any reaction time).

a. $a_{min} = (2v)^2/d$ (3%) **b. $a_{min} = 2v^2/d$ (57%)**
c. $a_{min} = v^2/d$ (6%) d. $a_{min} = v^2/(2d)$ (29%)
e. $a_{min} = v^2/(4d)$ (< 1%)
(More than 1 choice: 4%)

FIGURE 1. Questions placed on the final exam. Popularity of each choice is shown in parentheses and the correct choice is in bold.

TABLE 1. Exam Question Mean and Standard Error

	Final 1 (Numeric, N=453)	**Final 2 (Symbolic, N=441)**
"Bank robber"	85.0% ± 1.7%	57.4% ± 2.4%
"A car can go"	94.5% ± 1.1%	44.7% ± 2.4%

To test the equivalence of the groups we compared the mean midterm exam grade for each group. The mean midterm exam grade and standard error for the students who were given final exam 1 was 72.2% ± 0.6%; for the students who were given final exam 2 the mean midterm exam grade and standard error was 71.9% ± 0.6%. According to this measure the two groups are virtually indistinguishable.

Analysis of Written Work

At the conclusion of the exam, all students were required to turn in all exam materials. The exam packets were collected to keep the exam secure for future use and to resolve student concerns, but were not graded. Students, therefore, were not required to show any work in these packets.

We coded a sample of student written work (89 students). Table 2 demonstrates that the mean score for each of the questions for our sample was consistent with that of the overall population. The coding scheme developed and the numbers of students making each error are shown in Table 3 and Table 4.

TABLE 2. Exam Question Mean and Standard Error for the Sample of 89 Students.

	Final 1 (Numeric, N=43)	**Final 2 (Symbolic, N=46)**
"Bank robber"	83.7% ± 5.6%	60.9% ± 7.2%
"A car can go"	93.0% ± 3.9%	37.0 ± 7.1%

Even though the students were not graded based on their written solutions; a great majority of the exams from our sample showed work. In most cases it was clear what the student was doing, but there were some

TABLE 3. Coding of student work for the "A car can go" problem

Final 2 Question 6 (A car can go) Codes	Number of Students
Correct	17
Using $v_1/2 = at$ and $(v_1/2)^2 = 2ad$. Here t is misunderstood to mean the time it takes for the car to reach a speed $v_1/2$. They get the incorrect result $d = v_1t/4$.	4
Using $v_1 = at$ and $d = (1/2)at^2$. Here t in the second equation in misinterpreted and they are actually finding the distance when the car reaches a speed v_1 rather than the distance when it reaches a speed $v_1/2$. They get the incorrect result $d = v_1t/2$.	1
Using $v_1/2 = at$ and $d = (1/2)at^2$. Here t in both equations is misinterpreted. See errors 1 and 2. They get the incorrect result $d = v_1t/4$.	5
Use $(v_1/2)(t) = d$ or use $(v_1/2)(t/2) = d$. They use the equations for constant velocity. In the first they also misinterpret the meaning of the variable t. They get the choices $d = v_1t / 2$ and $d = v_1t / 4$ respectively.	2
An algebra error	3
Indeterminate	9
No work	3
Other	2

TABLE 4. Coding of student work for the "Bank robber" problem

Final 2 Question 7 (Bank robber) Codes	Number of Students
Correct	28
Using $v^2 = 2ad$. Here the v is used as if it were the cop's final velocity when he reaches the bank robber, but in the statement of the problem it is actually the bank robber's speed. They get the choice $a = v^2/(2d)$.	11
Using $t = d/v$ and $v = at$. The v in the second equation is used as if the cop's final velocity is the velocity of the bank robber. They get the choice $a = v^2/d$.	2
An algebra error	1
Indeterminate	2
No work	1
Other	1

cases in which the student's method was indeterminate. We also found that the main difficulty was not related to their inability to combine symbolic equations. The main difficulty seemed to be that when working symbolically the students often confused the meaning of the variables. They would confuse the meaning of the variable "t" in the "A car can go" problem and the meaning of the variable "v" in the "bank robber" problem.

DISCUSSION

Even though no one student received both the numeric and symbolic version of a particular question, the data suggests that 30% to 50% of the students would be able to solve a numeric version correctly but not an analogous symbolic problem. Even though we hypothesized that the students would have difficulty combining multiple symbolic equations, no evidence was found to support this hypothesis. Instead many students successfully combined symbolic equations, but did so with variable confusions.

The fact that the mean score on the 2nd question was higher than the 1st question on each final could be an indication of an effect due to the order of the questions. But because we find that the mean score on the symbolic versions are the lower than the numeric versions across exams, the order effect is small compared to the numeric/symbolic effect.

From an expert perspective the solutions to the numeric and symbolic questions are identical. The types of errors made, however, make it clear that the questions are not identical to the students. In the sample of written work we analyzed there were more students who made variable confusion errors on the symbolic questions than the total number of students who incorrectly answered the numeric questions.

Written Work and Theoretical Descriptions

The analysis of the students' written work on the symbolic version of the "bank robber" problem showed that 11/18 of the students who marked an incorrect choice also used the equation $v^2=2ad$. This error appears to be a matching of the variable "v" given in the problem with the "v_f" in the general equation "$v_f^2 = v_o^2 + 2a\Delta x$" on the equation sheet. When they make this error they are confusing the velocity of the bank robber with the final velocity of the police officer. This may be especially attractive to students because this solution only requires a single equation in order to obtain one of the choices while the correct solution requires two equations coupled by the time.

The resources framework proposed by David Hammer and colleagues [2] seems to describe certain aspects of our data. The resources framework posits knowledge pieces whose activation depends on the way the situation/task is framed. The framing determines what resources are activated and the types of activities (games) that are appropriate. The differences we observed between numeric and symbolic problems may be described as a difference in the way many students frame the two types of problems. Especially in the symbolic "bank robber" problem it appears that many students may have framed the problem as a "symbol matching" activity. It appears, on the other hand, that similar students framed the numeric version of the same problem as an activity involving the use of general equations and numeric quantities in a meaningful way.

One might imagine students framing the symbolic problems as a conceptual exercise that could be solved without the necessity of combining equations. But because the majority of the exams showed attempts to combine equations, we do not believe that this interpretation is a major factor in the results we found.

Similar theoretical frameworks have been described in the mathematics education research literature. Sfard [3] posited the existence of a duality of the procedural and the structural perspectives of mathematical concepts. Symbolic representations exemplify the dual nature of mathematical concepts. Symbolic equations can be thought of procedurally as a computation, as well as structurally as a statement of a relationship. Sfard's framework differs from the resources framework in that during the development of a mathematical concept the operational conception must precede the structural conception. In the resources framework no hierarchy of resources exists.

The analysis of the students' written work on the "A car can go" problem is less clear. The main identifiable error can by seen by combining the 2nd and 4th rows of the Table 3. 9/29 students who chose an incorrect choice used the equation "$(v_1/2)=at$" to determine the acceleration of the car. The general equation, "$v_f = v_o + at$" is appropriate, but they confuse the variable "t" in the equation with time to reach the velocity "$(v_1/2)$". To form this incorrect equation they combined information from two different sentences in the problem. When deciding what to substitute for "v_f" in the general equation only "v_1" and "$(v_1/2)$" have the correct units for the replacement. They may be influenced to choose "$(v_1/2)$" because that is the quantity of focus in the question.

Unlike the "bank robber" problem, in the "A car can go" analysis of the students' written work we found that 9/29 of the students who made the incorrect choice showed work that was indeterminate. These students either showed partial equations, many conflicting equations, or only a few sparse equations.

Cognitive load theory [4] can also be used to describe the data. This theory focuses on how cognitive resources are used. The cognitive demands on novices are much higher than on experts because novices must frequently store information in their working memory which can become quickly overloaded, while an expert can rely on pieces of compiled knowledge to minimize the strain on their working memory. It may be that many students have great difficulty simultaneously attending to the meaning of variables, the meaning of the symbolic expression, and the symbolic manipulations. The common incorrect equation "$(v_1/2)=at$" from the symbolic "A car can go" problem may be a result of trying to coordinate different pieces of information from different sentences along with the meaning of the equation on the equation sheet. This theory also seems to explain why there are so many students whose written work is indeterminate. Students whose working memory is overloaded would be expected to produce multiple and contradictory equations in the process of coordinating different information.

FUTURE WORK

In future experiments we would like to further study the differences between numeric and symbolic problems. In one proposed experiment, we plan on replacing the symbols used with uncommon symbols (e.g. b, c, q) to see if by increasing the barrier for symbol matching students would be more likely to treat these uncommon symbols in a way similar to numeric quantities. In another proposed experiment, we plan on replacing the confused variable with a numeric value, while keeping the other variables symbolic. In this variation the students would still have to setup an equation and would still have to carry an unevaluated expression from one equation to the next.

These experiments would help us identify what factors influence how students frame numeric and symbolic problems. These experiments would also allow us to determine the effect of varying the amount of symbolic information that the students have to coordinate, and thus the effects of cognitive load.

REFERENCES

1. Kieran, C., "The Learning and Teaching of School Algebra" in *Handbook of Research on Mathematics Learning and Teaching*, edited by D. Grouws, New York: Macmillan, 1992, pp. 390-419.
2. Hammer, D., Elby, A., Scherr, R., & Redish, E., "Resource, Framing, and Transfer" in *Transfer of learning from a modern multidisciplinary perspective*, edited by J. Mestre, Greenwich, CT: Information Age Publishing, 2005, pp. 89-119.
3. Sfard, A., *Educ. Studies in Mathematics* **22**, 1-36 (1991)
4. Sweller, J., *Learning and Instruction* **4**, 295-312 (1994)

Student (Mis)application of Partial Differentiation to Material Properties

Brandon R. Bucy,[1] John R. Thompson,[1,2] and Donald B. Mountcastle[1]

[1]*Department of Physics and Astronomy and* [2]*Center for Science and Mathematics Education Research*
The University of Maine, Orono, ME

Abstract. Students in upper-level undergraduate thermodynamics courses were asked about the relationship between the complementary partial derivatives of the isothermal compressibility and the thermal expansivity of a substance. Both these material properties can be expressed with first partial derivatives of the system volume. Several of the responses implied difficulty with the notion of variables held fixed in a partial derivative. Specifically, when asked to find the partial derivative of one of these quantities with respect to a variable that was initially held fixed, a common response was that this (mixed second) partial derivative must be zero. We have previously reported other related difficulties in the context of the Maxwell relations, indicating persistent confusion applying partial differentiation to state functions. We present results from student homework and examination questions and briefly discuss an instructional strategy to address these issues.

Keywords: Thermal physics, mathematics, partial differentiation, material properties, thermal expansivity, isothermal compressibility, Maxwell relations, upper-level, physics education research.
PACS: 01.40.-d, 01.40.Fk, 05.70.Ce, 65.40.De, 65.40.Gr.

INTRODUCTION

At the University of Maine (UMaine), we are currently engaged in a research project to explore student understanding of thermal physics concepts for the purposes of improving instruction. Research on student learning of thermal physics concepts in university physics courses, particularly beyond the introductory level, is rare. However, a growing body of research presents clear evidence that university students display a number of difficulties in learning many introductory and advanced thermal physics concepts [1-5].

Mathematics is a primary representation that can be used to articulate relationships among variables in physics. Mathematical facility allows a fuller understanding of empirical results, while more robust mathematical ability allows for the extension of physical concepts beyond a basic qualitative comprehension. As the physics becomes more advanced, so does the prerequisite mathematics. In thermal physics, there are topics that require specific mathematical concepts for a complete understanding of the physics. We have recently shown that thermal physics students have difficulties with regard to these mathematics concepts in addition to the above mentioned difficulties with physics concepts [5]. We have designed and administered questions to students in an upper-level thermodynamics course in the context of exploring the relationship between the physical properties of the thermal expansivity *(β)* and the isothermal compressibility *(κ)* of a system. These questions probe student understanding of the mathematics that underlie these physical properties, particularly multivariable calculus and partial differentiation.

We present results from a survey of two semesters of UMaine's *Physical Thermodynamics* course, taught in Fall 2004 and 2005 (by DBM). This course deals primarily with classical thermodynamics, covering the first 11 chapters of Carter's textbook [6] along with supplemental material. A separate statistical mechanics course is offered in the spring semester. Instruction included lecture, class discussions, and demonstrations; homework assignments included standard problems and instructor-designed conceptual questions. The instructor emphasized explicit connections between physical processes and relevant mathematical models to a greater extent than is common in typical textbooks. The homework was graded and returned with comments, and a detailed answer key was supplied to students. Data were

CP883, *2006 Physics Education Research Conference*, edited by L. McCullough, L. Hsu, and P. Heron

obtained from the fourteen students taking the courses: two juniors, eleven seniors, and one physics grad student; eleven physics majors, one math major, and one marine sciences major. All students had completed the prerequisite third semester of calculus, which includes multivariable differential calculus. All students but one had additionally completed one or more courses in ordinary and partial differential equations.

INSTRUMENT AND RESULTS

We focus here on student responses to a written question dealing with the relation between complementary partial derivatives of the isothermal compressibility and the thermal expansivity of a substance. The *"β–κ"* question was administered to students twice during the semester: as part of a homework assignment after instruction on state functions and partial derivatives, and again after the homework was graded with instructor comments and returned with an answer key, in a slightly modified form as part of a graded examination (Figure 1).

(a) Show that in general $\left(\frac{\partial \beta}{\partial P}\right)_T + \left(\frac{\partial \kappa}{\partial T}\right)_P = 0.$

(b) With the usual definitions of isothermal compressibility *(κ)* and thermal expansivity *(β)*, for any substance where both are continuous, show how these two derivatives are related:

$$\left(\frac{\partial \kappa}{\partial T}\right)_P \text{ and } \left(\frac{\partial \beta}{\partial P}\right)_T .$$

FIGURE 1. "*β-κ* question" asked to students on (a) homework and (b) a midterm examination. Based on Problem 2-9 in Carter's text [6].

The thermal expansivity of a thermodynamic system is related to the partial derivative of the system volume with respect to temperature at a fixed pressure: $\beta \equiv \frac{1}{V}\left(\frac{\partial V}{\partial T}\right)_P$. The isothermal compressibility is related to the partial derivative of the system volume with respect to pressure at a fixed temperature: $\kappa \equiv -\frac{1}{V}\left(\frac{\partial V}{\partial P}\right)_T$. Physically, β describes the response of the system volume to a change in temperature while κ describes the response of the system volume to a change in pressure. By convention, the negative sign is included in the definition of κ recognizing that the volume of a system always decreases with increased pressure. Division by volume makes the properties of β and κ *intensive*, that is, a material property of a substance, independent of the sample size.

In order to answer the question, students must first take the requested derivatives of β and κ. Application of the product rule and the chain rule results in the following expressions:

$$\left(\frac{\partial \beta}{\partial P}\right)_T = -\frac{1}{V^2}\left(\frac{\partial V}{\partial P}\right)_T\left(\frac{\partial V}{\partial T}\right)_P + \frac{1}{V}\frac{\partial^2 V}{\partial P \partial T} \quad (1)$$

$$\left(\frac{\partial \kappa}{\partial T}\right)_P = \frac{1}{V^2}\left(\frac{\partial V}{\partial T}\right)_P\left(\frac{\partial V}{\partial P}\right)_T - \frac{1}{V}\frac{\partial^2 V}{\partial T \partial P} \quad (2)$$

It is easy to see that the first terms in equations (1) and (2) are exact opposites, containing products of first partials, while the second terms contain complementary mixed second partials – one taken with respect to pressure then temperature, the other vice-versa. It turns out that for any function for which the second partial derivatives are defined and continuous, the mixed second partials are identical, regardless of the order of differentiation. This relationship is known as Clairaut's Theorem, or as "the equality of mixed second partials." Applying Clairaut's Theorem in this case allows one to see that the complementary second partial derivatives of β and κ are identically opposite.[1]

TABLE 1. Student responses to β-κ question.

Category of Student Response (Non-exclusive)	# Student Responses on Homework (N = 14)	# Student Responses on Exam (N = 11)
Correct	6	5
Calculus I problems	7	3
Calculus III problems	6	5

Student performance on this question is presented in Table 1. Slightly less than half of the students answered the question correctly, both on the homework assignment and on the exam. This poor performance is somewhat surprising, given that the exam question was nearly identical to the homework question, which had been graded and returned to students along with a detailed answer key depicting the solution.

Several noteworthy aspects of student reasoning were observed. A sizeable proportion of the students displayed one or more difficulties with the process of differentiation in their responses, referred to as *Calculus I problems* in Table 1. Several students had

[1] There is a more elegant way to solve this problem; simply by recognizing that $-\kappa$ and β are the coefficients of the total differential of the logarithm of the system volume. Thus, by Clairaut's theorem, their complementary partials must be equal.

difficulties in applying the product rule in their differentiation. A common approach in student strategy was to simply factor the $1/V$ term out of the derivatives of both β and κ, as if volume were not a function of either pressure or temperature. A related problem involved incorrect or missing differentiation of either the $1/V$ term or of the partial derivative terms themselves.

Other students had specific difficulties in applying of the chain rule. When differentiating the $1/V$ term with respect to pressure or temperature, some students simply wrote down $-1/V^2$, neglecting to further differentiate V with respect to P or T. Consequently, the first terms in equations (1) and (2) were not identical opposites in these students' derivations.

Fully half of the students made one or more of these mistakes on the homework assignment. This inability to correctly differentiate a relatively simple expression calls into question many of the skills that most physics professors assume their incoming upper-level students possess. It is important to note, however, that this homework assignment was only the second assignment of ten total. Fewer students made one of these mistakes on the exam, which occurred after the homework and several weeks later in the semester. While these problems were much more prevalent on the homework question than on the exam, these responses are still troubling in terms of prerequisite knowledge and skills that students are expected to bring into an upper level physics course.

Another specific flaw in student reasoning was observed, labeled *Calculus III problems* in Table 1. This type of response indicated a higher order mathematical difficulty than simple differentiation problems, namely the role of fixed variables in partial differentiation. Consider this typical student response: "If κ and β are defined as such, then $\left(\frac{\partial\beta}{\partial P}\right)_T = \left(\frac{\partial\kappa}{\partial T}\right)_P = 0$ since P has already been held constant for β and T has already been held constant for κ." The student is saying that, since the variable T has been held constant in the first derivative of volume within the definition of κ, then any subsequent differentiation with respect to that variable will yield zero as simply the derivative of a constant. This specific difficulty was as prevalent in student responses as correct answers, and a few otherwise correct answers relied on this reasoning to arrive at the correct result on the homework problem.

Such responses seem to indicate confusion between the terms "constant" and "fixed;" one implying a permanent constraint and the other a temporary one. Typical treatments of partial differentiation tend to be less than precise when introducing the terminology, often giving the verbal definition of $\left(\frac{\partial z}{\partial x}\right)_y$ as the partial derivative of z with respect to x *at constant y*, rather than specifically stating that the subscript variable is to be *fixed at a particular value only during the differentiation.* This casual use of terminology can be confusing to students. The notation itself could also be confusing, as few disciplines other than thermal physics make explicit reference to those variables being held fixed, i.e. $\frac{\partial}{\partial P}\left(\left(\frac{\partial V}{\partial T}\right)_P\right)_T$ compared with $\frac{\partial^2 V}{\partial P \partial T}$.

DISCUSSION

It seems that students' desire to set the mixed second partials identically equal to zero is a persistent and strongly held difficulty. Approximately the same proportion of students held these ideas on the exam question as had them in their homework assignment, despite receiving an answer key and a brief explanation by the instructor. Additionally, based on classroom observation data, students in the 2004 course were noticeably animated about this question when their exams were handed back to them. Several of them expressed disbelief that the mixed second partial could be anything but zero due to the variable being held constant.

We believe, however, that this tendency arises chiefly from mathematical errors, and not from any student ideas about the physics. In order to explicate this claim, a brief digression into the graphical interpretation of Clairaut's Theorem is illustrative. Just what does it mean for the mixed second partials of a function to be equal, and yet not equal to zero? In particular, what does the equality of mixed second partials tell us about the state function of volume?

The first partials tell us how the function varies along one axis, i.e., the tangent slope. Further differentiation by the complementary variable tells us how *that slope* changes with respect to an orthogonal independent variable. This rate of change of the two orthogonal tangent slopes (in the limit) must therefore be equal by Clairaut's Theorem. In the case of an ideal gas (Figure 2(a)), both slopes decrease at the same rate as pressure and temperature increase.

For a function with zero mixed second partials, as in Figure 2(b), there would be no change in the slope along either axis as we move along the other axis. Such a situation is even less interesting than the ideal gas, and is constrained quite artificially. That constraint need not be as severe as the tilted plane shown in Figure 2(b), but must be limited along one of the independent axes such that all slopes with respect to that variable can change along that variable axis, but must be constant (parallel tangents) at all locations while moving along the orthogonal axis.

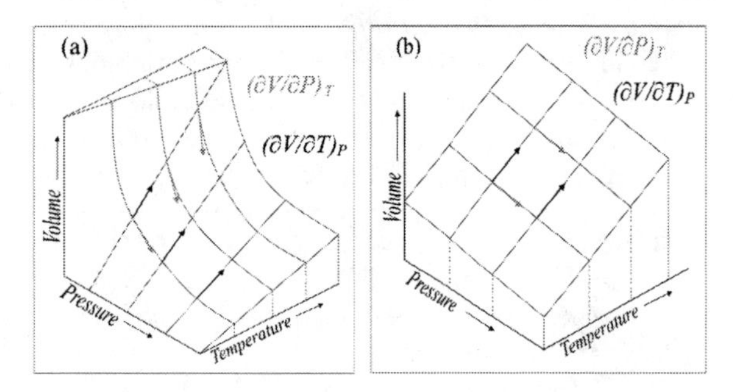

FIGURE 2. *P-V-T* diagram for (a) an ideal gas, and (b) a substance with zero mixed second partials of volume.

Thinking about the physical situation should help students in realizing that the situation is unlikely at best. Consequently, we have developed a question designed to see if students are aware of the physical implications of forcing the derivatives of β and κ to be zero: "The thermal expansivity of mercury is 37.5 K^{-1} at 1 atm and a given temperature. Do you expect the expansivity to increase, decrease, or remain the same if the sample pressure were 1000 atm instead at the same temperature? Please explain your reasoning." This question, along with the graphical reasoning depicted above, will be incorporated into a tutorial designed to improve student understanding of mixed second partials. We expect that students should then be able to use this mathematical reasoning to verify any physical intuitions about changes in these material properties.

Comparison with the Maxwell Relations

We have documented related difficulties with mixed second partials used in the Maxwell relations, which are applications of Clairaut's Theorem to the so-called thermodynamic potentials (e.g., *U, H, F, G*) [5]. In particular, students seemed to have difficulties with the physical interpretations of the equated partials. Even those students who could derive the Maxwell relations often lacked any ability to apply them in a physical context, or even to interpret their meaning.

In contrast to the results in the β–κ question, *no* students indicated that the partial derivatives generating the Maxwell relations were identically equal to zero. This suggests that students may not consider the mathematical significance of the Maxwell relations, i.e., that they are mixed second partial derivatives. We believe two factors support this idea. First, the Maxwell relations are typically used as relationships between first partials of various thermodynamic functions, e.g., entropy, volume, pressure, temperature, rather than as second partials of the thermodynamic potentials, while β and κ are defined using first partials of volume, so that their derivatives clearly include second partials. That is, for the Maxwell relations, students are interpreting the functions as *variables* rather than *coefficients* (of a total differential), allowing a nonzero mixed differentiation. Second, it may be that students consider β and κ as physical constants rather than functions, leading to derivative values of zero.

Summary

In particular, we see evidence that students misinterpret the meaning of "holding a variable constant" during partial differentiation, considering the fixed variable to remain constant after differentiation rather than being fixed only during the process.

Our results also suggest that students do not see the Maxwell relations as relating mixed second partial derivatives, implying a disconnect between the mathematical and physical meanings of these relations.

The results presented here, in conjunction with prior work, indicate that students often enter upper-level physics courses lacking the necessary (and assumed) prerequisite mathematics knowledge and/or the ability to apply it productively in a physical context. Students avoid using physical reasoning to verify their mathematical results. Taken as a whole, these results point to difficulties among advanced students in incorporating mathematics and physics into a coherent framework.

We are currently developing curricular materials aimed at addressing some of these issues in the context of state functions and material properties.

ACKNOWLEDGMENT

Supported in part by NSF Grant PHY-0406764.

REFERENCES

1. M.E. Loverude, P.R.L. Heron and C.H. Kautz, *Am. J. Phys.* **70**, 137-148 (2002).
2. D.E. Meltzer, *Am. J. Phys.*, **72**, 1432-1446 (2004).
3. D.E. Meltzer, *2004 Phys. Educ. Res. Conf.*, edited by J. Marx *et al.*, AIP Conf. Proceedings **790**, 31-34 (2005).
4. B.R. Bucy, J.R. Thompson, and D.B. Mountcastle, *2005 Phys. Educ. Res. Conf.*, edited by P. Heron *et al.*, AIP Conf. Proceedings **818**, 81-84 (2006).
5. J.R. Thompson, B.R. Bucy, and D.B. Mountcastle, *ibid.*, 77-80 (2006).
6. A. H. Carter, *Classical and Statistical Thermodynamics*, Upper Saddle River, NJ: Prentice-Hall, Inc., 2001.

Comparison of Teaching Methods for Energy Conservation

M. L. Horner and Monwhea Jeng* and Rebecca Lindell

Dept. of Physics, Southern Illinois University Edwardsville, Edwardsville, IL 62026

Abstract. Three sections, taught by different instructors, of Conceptual Physics were taught energy conservation using three different techniques: traditional - no visualization, energy bar charts, and energy bars. Performance of the groups of students on final exam questions is compared and contrasted.

Keywords: Energy Conservation, Student Understanding
PACS: 01.40.Fk, 01.40.G-, 01.40.gb

INTRODUCTION

By now the idea that students learn via different modes has become part of the standard teaching lore.[1,2] In a related but distinct vein, there is evidence in physics that expert problem-solvers employ multiple paths in solving a single problem more often than novice problem-solvers do.[3-5] These strands come together in the command "Draw a picture" found in problem-solving rubrics over at least the last two decades and sometimes now divided into "picture" and "physical representation".[6,7]

Work and energy are standards of the introductory physics curriculum. Students struggle to apply these concepts in varied contexts and to apply concepts without a numerical check in qualitative problems.[8] Until 2001 there was no common (i.e. shared in the literature) physical representation system for work-energy problems. Van Heuvelen and Zou defined energy bar charts as a physical representation of work-energy problems and analyzed the effects of their use on Ohio State University students in the calculus-based introductory sequence. They found a positive effect on qualitative problem-solving success among students taught to use the energy bar charts though only some students spontaneously drew the charts on free-response questions when it was appropriate to do so.[6]

One of us [MLH] created energy bars as a means of making energy conservation easier to see. Horizontal energy bars are divided into sections for various types of energy. The lengths of sections change over time as energy is moved from one system to another or changes form within the system. In the following sections, we discuss a comparison, based on common exam questions, among three sections of the same course using respectively no physical representation, energy bar charts derived from those of Van Heuvelen and Zou, and energy bars.[6]

DESCRIPTION OF COURSE

We drew our comparison sections from a one-semester, conceptual physics course as part of the introductory component of the Southern Illinois University Edwardsville (SIUE) general education curriculum. The course is intended for freshmen and sophomores although, in practice, the course has a substantial population of juniors and seniors. This course is not a prerequisite for anything else in the curriculum, and as such there is considerable diversity in its teaching.

We used three regularly-existing Fall 2005 sections for this study, taught by three different professors of similar age and experience. Each instructor had taught this course recently at least once before the term in question and no instructor made substantial changes from these previous semester. All of the instructors volunteered to participate in the study and had already planned to spend approximately the same amount of time on the topic of energy. Students chose sections without any indication of teaching method. The sections were all of typical size for this class at SIUE (~50 students).

* Currently at Dept. of Physics, Syracuse University, Syracuse, NY 13244

CP883, *2006 Physics Education Research Conference*, edited by L. McCullough, L. Hsu, and P. Heron

Overview of Instructional Techniques

Each section spent approximately a week covering work and energy. The "traditional" section used lecture and examples. No system of physical representation was employed, although pictures were used in explanations and problems.[MJ] The energy bar charts section used interactive methods and a group activity involving bar charts based on those of Van Heuvelen and Zou.[6][RL] The energy bar section used lecture, examples, and group problem-solving all employing energy bars.[MLH] Figure 1 is an example of energy bar and work-energy bar chart physical representations of problems. The energy bars were taught using the simpler form shown in figure 1a.

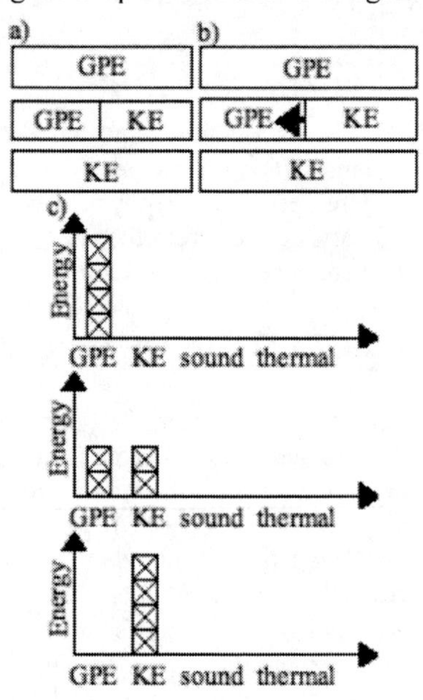

FIGURE 1. Sample Problem illustrating energy bars and energy bar charts: A ball is dropped from the top of a 10m tall building. How fast is it falling when it has fallen half-way to the ground if air resistance can be ignored? Physical representations in (a) the simple version of energy bars used in this study, (b) a more elaborate version of energy bars with an arrow indicating that GPE is increasing at the expense of KE as time passes, and (c) energy bar charts are shown. GPE=gravitational potential energy. KE=kinetic energy.

Research Questions

In this study we attempted to answer the following research questions. What effect do energy visualizations have on overall understanding of energy conservation? Do students using the energy bars conserve energy more frequently than students using energy bar charts?

Evaluation

We used common final exam problems addressing content common to all sections and within the typical styles of questions our students could expect on exams. These problems included multiple choice, numerical, physical representation (not for the traditional section), and free response qualitative and quantitative types. We will focus on the first three types here.

Common Exam Questions

MC1: A pendulum is released at an angle, and then swings back and forth for some time. After a while, it eventually comes to rest. Which of the following is a correct statement about the pendulum's energy? (a) The total energy of the pendulum is conserved, because any object's energy is always conserved. (b) The total energy of the pendulum is conserved, because it is on the Earth. (c) The total energy of the pendulum has increased. (d) The total energy of the pendulum has decreased.

MC2: A bullet is sitting in a gun. I point the gun in the air, and pull the trigger. The bullet comes out, and flies far up into the air, reaching a maximum height, high above my head. How, if at all, has the bullet's total energy changed, from the time just before I pulled the trigger, to the time it reached its maximum height in the air? (a) The total energy of the bullet has increased. (b) The total energy of the bullet has not changed. (c) The total energy of the bullet has decreased. (d) There is not enough information to choose between the above.

MC3: When a pendulum, with no friction or air resistance, is swinging, when is the total energy of the pendulum a maximum? (a) Never. Energy is conserved. (b) At the bottom, when the pendulum is moving the fastest, is when the energy is a maximum. (c) At the side, when the pendulum is at its maximum height, is when the energy is a maximum. (d) Somewhere just before the pendulum is at the bottom of its swing is where the kinetic energy and gravitational potential energy add to a maximum.

MC4 (bar charts and energy bars sections only): At the beginning of the test a popper toy was demonstrated. At the instant it was released on the table its energy could be described as below. [graphic of all initial energy as SPE] Which of the figures below best describes the energy of the popper toy at the top of its flight? (a) [graphic of sound and GPE – conserved] (b) [graphic of sound, GPE, and KE – less energy] (c) [graphic of KE and GPE – discrepant versions: conserved in bar charts, less energy in bars] (d/e) [graphic of SPE – conserved] (e/f) [graphic of sound and GPE – less energy] (f/d) [graphic of sound, GPE and KE – conserved]

WP1: You launch a ball directly up into the air. It is initially on the ground, with a kinetic energy of 60J (shown for bar charts and energy bars sections). Ignore air resistance. (A) Consider the ball when it has reached its maximum height. (i) What is the ball's kinetic energy? (ii) What is the ball's gravitational potential energy? (iii) (bar chart and energy bar sections only) Draw the energy bar (chart) for the ball at this point. (B) Repeat (A) for when the ball is moving up and halfway to its maximum height.

Data Analysis

Primary Trait Analysis[11] was utilized to determine the overall effectiveness of the three different instructional techniques by comparing the results on all common problems. The following coding scheme was utilized to compare energy bars to energy bar charts on problems referring directly to the representation technique.

Conservation of Energy	Energy Types	Relative Sizes
C - Conserved	C – Correct Energy Types	C – Correct
NC – Not Conserved	I – Incorrect Energy Types	I – Incorrect
B - Blank	B - Blank	B - Blank

Table 1: Coding scheme for analyzing energy bar / energy bar charts

RESULTS AND DISCUSSION

We addressed our first research question by the primary trait analysis discussed above. As shown in Table 2 and Figure 2, the results run counter to the leading belief that interactive engagement methods produce higher conceptual gains. In fact, the traditional section scored the highest.

Instructional Technique	N	Average Percent Correct (Standard Deviation)
Energy Bar Charts	46	55% (21%)
Energy Bars	55	59% (25 %)
Traditional	61	75% (22%)

Table 2: Primary trait analysis results for three instructional types. Results were calculated out of a possible seven points. Standard deviations shown in parenthesis. Using ANOVA results are significant at $p \leq 0.0001$.

We addressed our second research question by applying the coding scheme in Table 1 to student responses to MC4 and part (iii) of WP1. While Table 3 shows no significant difference between visualization techniques, Table 4 shows significant differences in particular sub-categories of applying the techniques.

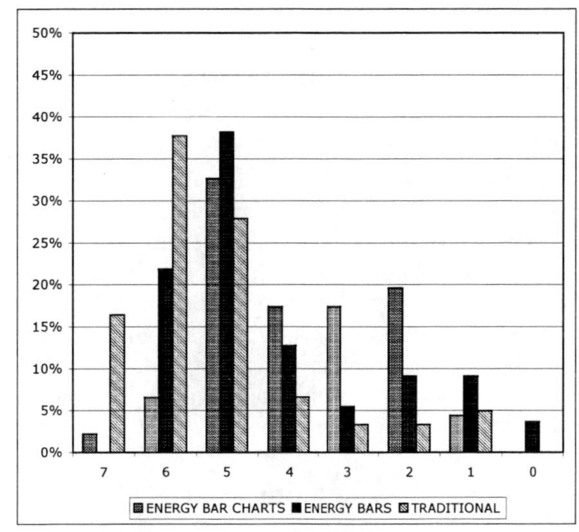

Figure 2: Distribution of student scores for each instructional technique. Results were calculated out of possible seven points. Results are significant at $p \leq 0.0001$.

Instructional Technique	N	Average Percent Correct (Standard Deviation)
Energy Bar Charts	46	71% (31%)
Energy Bars	55	74% (30 %)

Table 3: Comparison results for visualization questions. Results were calculated out of a possible seven points. Standard deviations are shown in parenthesis. Using a t-test, results not significant at the 0.05 level.

Instructional Technique	Energy Bar Charts	Energy Bars N = 55
Average Percent Correct (Standard Deviation)	N = 46	
Energy Conservation $p \leq 0.0003$	84% (33%)	53% (48%)
Energy Types $p \leq 0.007$	63% (48 %)	80% (38 %)
Relative Sizes Not Significant at the 0.05 level	73% (39 %)	85% (30 %)

Table 4: Comparison results for visualization questions separated by content evaluated. Results were calculated out of a possible two points. Standard deviations are shown in parenthesis.

Students using energy bar charts were more likely to maintain energy conservation, while students using the energy bars are more likely to correctly characterize the types of energy present in the various situations. The latter results may be due to asking the students to ignore air resistance, which was specifically included in all examples utilized in that section. The former difference may be due to the ease with which energy conservation can be drawn using energy bars (drawing same-sized rectangles versus adding up the number of squares in various columns) so that the expected reinforcement is not occurring.

CONCLUSIONS & IMPLICATIONS

We observed several statistically significant results. According to these results, traditional instruction seems to be the superior technique. Energy bar charts and energy bars seem to be similar; energy bar charts seem to encourage the use of energy conservation, while energy bars produce a more correct analysis of energy types present. Although these results are statistically significant, they may not be educationally significant as there are numerous confounding factors that need to be taken into account.

Confounding factors

Differences between test versions: There were drawing errors between the two versions of MC4. Having one answer conserve energy on one exam and not on the other means the distribution of answers with respect to energy conservation and energy type was not the same for the two exams. It is possible that having the answers in a different order affected the outcome as well. In addition, one of us [MLH] wrote two versions of the exam as a deterrent to copying and altered the order of answers to the other multiple choice questions for approximately half of that section.

Inherent problems with questions: There may be an author effect occurring with the questions. Instead of using either standard questions or co-authoring problems, we used problems suggested or written by one person (most often MJ) with editing and approval from the other two. As the students in the traditional section were accustomed to questions written by this author it may have inadvertently biased the sample.

Lack of Randomness: While students self-assorted into sections, we do not have background information on the students to assess the randomness of this assortment.

Small Sample Size: Due to our class sizes, our sample size was small.

Student-Teacher Dynamics: Because three different people taught the sections, there are a host of possible teaching-method and student-teacher dynamics that we did not control for. On the other hand, each instructor used the methods he or she preferred, so there was no bias resulting from one instructor preferring or being more comfortable with one method.

FUTURE WORK

In addition to the questions discussed here, we had two free-response questions on a mid-term exam. We hope that analysis of these responses will illuminate causes of the trends discussed here.

Further studies of this type could be carried out, with either single or multiple instructors, in either the course used here, or other courses with multiple sections

ACKNOWLEDGMENTS

We would like to acknowledge both the SIUE Department of Physics and the College of Arts and Sciences for financial support for this project. We would also like to acknowledge Dr. Andrew Neath for statistics consultation.

REFERENCES

1. Travers, Robert M. W., *Research and Theory Related to Audiovisual Information Transmission. Revised Edition*, Salt Lake City, UT: Utah University, 1967, ED081245.
2. Bruininks, Robert H. Clark, Charlotte., *Auditory and Visual Learning in First-, Third-, and Fifth-Grade Children. Research Report14*, Minneapolis, MN: Minnesota University, 1970, ED080181.
3. Larkin, Jill H., "Understanding, Problem Representations, and Skill in Physics," in *Thinking and Learning Skills, v2: Research and Open Questions,* edited by Susan F. Chipman, Judith W. Segal, and Robert Glaser, Hillsdale, NJ: Lawrence Erlbaum Assoc., 1985, pp. 141-159.
4. Larkin, Jill H., "Teaching Problem Solving in Physics: The Psychological Laboratory and the Practical Classroom," in *Problem Solving and Education: Issues in Teaching and Research,* edited by David T. Tuma and Frederick Reif, Hillsdale, NJ: Lawrence Erlbaum Assoc., 1980, pp. 111-125.
5. Pretz, Jean E, Naples, Adam J. and Sternberg, Robert J., "Recognizing, Defining, and Representing Problems," in *The Psychology of Problem Solving,* edited by Janet E. Davidson and Robert J. Sternberg, Cambridge, UK: Cambridge U. Press, 2003, pp. 3-30.
6. Van Heuvelen, Alan, and Zou, Xueli, *Am. J. Phys* **69**, 184-194 (2001).
7. Davidson, Janet, "Insights about Insightful Problem Solving," in *The Psychology of Problem Solving,* edited by Janet E. Davidson and Robert J. Sternberg, Cambridge, UK: Cambridge U. Press, 2003, pp. 149-175.
8. Singh, Chandralekha and Rosengrant, David, *Am. J. Phys* **71**, 607-617 (2003).
11. Walvood B, and McCarthy, L. P., *Thinking and Writing in College.* Urbana, IL: National Council of Teachers of English, 1990.

Addressing Students' Difficulties in Understanding Two Different Expressions of Gravitational Potential Energy (I) : mgh & -GMm/r

Gyoungho Lee* and Jinseog Yi

Department of Physics Education, Seoul National University, San56-1, Sillim-dong, Gwanak-gu, Seoul, South Korea
*ghlee@snu.ac.kr** *colbert0@snu.ac.kr*

Abstract. During our investigation of students' understanding of gravitational potential energy, we found some difficulties that students have with this topic. Many students who took upper-level mechanics courses had difficulties in understanding why there are two different expressions of gravitational potential energy. These students said they had some difficulties in understanding why there should be two different signs (+ & —) and two different forms (g & 1/r) even though these expressions were considered as representing the same gravitational potential energy. To gain understanding of the sources of student difficulties, we used weekly reports and individual interviews. We analyzed student difficulties in terms of conceptual knowledge, procedural knowledge, and contextual knowledge. The results of these research have guided the development of teaching material that addresses students' difficulties in understanding gravitational potential energy. We will show the development process and contents of the material in the second paper on this topic [1].

Introduction

Although gravitational force and gravitational potential energy are familiar topics to students from secondary schools, many students have difficulties in understanding these topics [2]. In particular, one of the frequent difficulties stems from the two different expressions of Gravitational Potential Energy (G.P.E): mgh & -GMm/r. Students sometimes pose the question, "Why should there be two different signs (+ & —) and two different forms (g & 1/r) even though we are considering the same gravitational potential energy." In this study, we have tried to understand the sources of the students' difficulties, from three kinds of knowledge perspectives [3]: conceptual knowledge, procedural knowledge, and contextual knowledge.

Research Context

This study took place at Seoul National University, a public university. The student population studied was an upper-level mechanics class of 68 students (2005, 2006 years). The course under investigation was a two-semester course taken by physics education majors who are preservice teachers.

The following data was collected and analyzed during the course of this study:

CP883, *2006 Physics Education Research Conference*, edited by L. McCullough, L. Hsu, and P. Heron

- weekly report

In 2005~2006, we collected weekly reports from students in the course. In this study, we revised the original weekly report [4]. These weekly reports revealed what students had learned and their difficulties in learning mechanics. We used the following questions in the weekly reports:

1. What do you learn in this week lectures? How could you learn them?

2. What kinds of difficulties do you have in this week lectures? (precisely)

3. Do you have experienced conflicts between your previous knowledge and new knowledge that you have learned in this week.

- Individual interviews

We also conducted individual interviews with three students. The interview questions were based on the weekly reports. It took about 30 minutes with a student, respectively.

Results

Table 1 shows the results of the students' difficulties in learning two different expressions of Gravitational Potential Energy (G.P.E): mgh & -GMm/r. As can be seen, during the 2005-2006 study period, 42.6% students experienced difficulties in understanding the two different expressions of gravitational potential energy (mgh & -GMm/r). There are four sources causing the difficulties: +/- signs, reference point, proportion to r & 1/r, G.P.E graphs.

From the perspective of knowledge structure [3], the sources of difficulties could be divided into

two: conceptual knowledge and procedural knowledge. The 'reference point' is related to conceptual knowledge. On the other hand, '+/- signs, proportion to r & 1/r, G.P.E graphs' are related to procedural knowledge.

Among the 29 students who had difficulties with the topic, the source of difficulties for 41.4% students was related to reference points. They could not understand why we should consider separate reference points for mgh & -GMm/r.

We also found that many students experienced the +/- signs as a source of difficulty. They believe that the sign of energy should be +, if there is energy. Many students are reluctant to admit expressions of energy with zero(0) or minus(-) because they consider existent energy as an ability of working or pushing an object. In addition, students have long been used to representing with a plus(+) as kinetic energy ($\frac{1}{2}mv^2$) or potential energy (mgh) since having learned physics early in their school careers. In other words, students do not understand that the methods of energy expression can be various; for instance, that existent energy can be expressed

TABLE 1. Students' difficulties[29/68 : 42.6%](2005. 3 & 2006. 3)

Difficulties	Sources of difficulties	Examples	Frequency	Percent(%)
mgh vs. $-\frac{GMm}{r}$	+ (Absolute) vs − (relative)	I believe that energy is existent(positive value). How can we use the negative value?	9	31.0
	reference point	How can I decide the sign and the datum point for the algebraic expression?	12	41.4
	r vs $\frac{1}{r}$	Why is the same object of gravitational potential energy represented differently? (proportion or inverse proportion r)	3	10.3
	Different graphs	Are the two graphs different or the same?	5	17.3

with a plus(+) or a minus(-). In an interview, student K said,

"Why can energy expressed with minus sign? If I were a teacher, how would I explain this?......my concepts......Since I have been considering this (topic), I have been confused."

Here, the minus sign(-) of universal gravitational potential energy($-\frac{GMm}{r}$) gave rise to difficulty (in other words, cognitive conflict) with the student's belief that, "Energy should be expressed with a plus sign(+)".

In addition, students also had confusion over using absolute zero and relative zero. Generally, we use zero with two concepts. Absolute zero stands for "no existence", meaning "not to be". In contrast, relative zero stands for "comparative reference point": it is the same as a scale of thermometer pointing 0℃. Relative zero depends on the reference point of the relative scale rather than the absolute value of energy existence. For instance, a temperature pointing to 0 does not mean that there is no internal energy. Likewise, potential energy pointing to 0 does not mean that there is no potential energy. Thus, even though potential energy has a minus(-) sign, we can say that it exists. The students' belief that energy should be expressed with a plus(+) obstructs their vision in distinguishing between absolute zero(0) and relative zero(0), providing another reason to be reluctant to accept the minus(-) expression for gravitational potential energy.

The third source of students' difficulties was the different expression of G.P.E as a 'r' function. They thought that the proportion to r is in controversy with the proportion to 1/r. Thus, they could not match the two different expressions of G.P.E. In an interview, student K said,

"Although there is the same gravitational field, why is gravitational potential energy in the surface of th earth not represented with $-\frac{GMm}{r}$ *but with* mgr. *why must we use the expression ' mgr ' in the surface of the earth."*

The last source of students' difficulties was the graphs of G.P.E. Intuitively, it is difficulty to consider them as the graphs that express that G.P.E originated from the same earth.

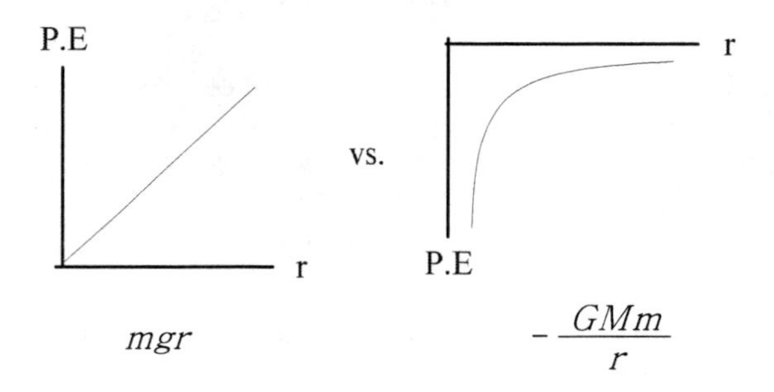

Figure 1. The graphs of mgr and $-\frac{GMm}{r}$

Summary & Future Study

In this study, we found that students' difficulties in understanding two different expressions of Gravitational Potential Energy (mgh & -GMm/r) were the major difficulty in their learning mechanics: 42.6% of students had difficulties with this topic during the two year study period. We also found the sources of difficulties in understanding the two different expressions of Gravitational Potential Energy. These sources were the +/- signs, reference points, the proportion to r & 1/r, and G.P.E graphs. From the perspective of knowledge structure, the 'reference point' is related to conceptual knowledge. On the other hand, '+/- signs, proportion to r & 1/r, and G.P.E graphs' are related to procedural knowledge.

In the future, we intend to probe the resources for addressing the students' difficulties concerning this topic in more detail. Even though this is a preliminary study, the results of this study have guided the development of some new teaching material for resolving students' difficulties in understanding G.P.E. We will report upon the development process and major contents of these materials in our second paper [1] which has the same title submitted to the PERC proceedings.

Acknowledgments

This work was supported by the Brain Korea 21 Project in 2006.

References

1. Jinseog Yi and Gyoungho Lee, "Addressing Students' Difficulties In Understanding Two Different Expressions of Gravitational Potential Energy(Ⅱ)," *Submitted to 2006 PERC.*
2. M. E. Loverude, "Student Understanding Of Gravitational Potential Energy And The Motion Of Bodies In A Gravitational Field," *PERC* 77-80 (2004).
   Alan Van Heuvelen and Xueli Zou, *Am. J. Phys.* 69, 184-194 (2001).
   Arnold B. Arons, *Am. J. Phys.* 67, 1063-1069 (1999).
3. Gyoungho Lee, *AAPT Announcer,* 33(4), 105 (2003).
4. Etkina, E. "Weekly Reports: A two-way feedback tool," *Sci. Ed.* 84, 594-605 (2000).

Strategy Levels for Guiding Discussion to Promote Explanatory Model Construction in Circuit Electricity

E. Grant Williams and John J. Clement

School of Education and Scientific Reasoning Research Institute, University of Massachusetts – Amherst
428 Lederle Graduate Research Tower, Amherst, MA 01003, USA

Abstract: A framework for describing and tracking the whole-class discussion-based teaching strategies used by a teacher to support students' construction and development of explanatory models for concepts in circuit electricity is described. A new type of diagram developed to portray teacher-student discourse patterns facilitated the identification of two distinct types, or levels, of teaching strategies: 1) those that support *dialogical* or conversational elements of classroom interaction; and 2) those that support cognitive *model construction* processes. The latter include the higher-level goals of promoting a cycle of Observation, model Generation, model Evaluation, and model Modification. While previous studies have focused primarily on the dialogical strategies that are essential for fostering communication as an enabling condition, the cognitive strategies identified herein are aimed at fostering conceptual model construction.

Keywords: Teaching strategies, model-based learning, whole-class discussion, high school physics.
PACS: 01.40.Fk, 01.40.gb, 01.40.sk

INTRODUCTION

As a long-term goal, we have been engaged in identifying *cognitive strategies for fostering model construction* that could complement known *dialogical strategies for encouraging conversations in large group discussions*. This paper introduces a distinction between cognitive and dialogical strategies using examples from high school electricity teaching and describes a diagramming system for tracking both kinds of strategies.

THEORETICAL FRAMEWORK

Discussion-Based Teaching

There is considerable agreement that classroom discussions, both student-to-student and between students and teacher, can serve as important means for facilitating the construction of scientific knowledge [1]. In particular, the idea that student reasoning abilities can be supported through whole-class discursive interactions with others draws on the early work of Vygotsky [2]. Many contemporary researchers [3,4] also believe that discussion plays an important part in conceptual change and that teaching with a focus on discussion can improve scientific reasoning ability.

Model-Based Learning

In the context of this research, a model is considered to be a simplified representation of a system, which concentrates attention on specific aspects of the system [5]. An *explanatory* model for a system is a hypothesized, theoretical, qualitative mental model (such as molecules, waves, and fields) that provides a (usually causal) description of a hidden, non-observable mechanism that explains how the system works. These can enable aspects of the system, i.e., objects, events, or ideas that are either complex, abstract, or on a different scale to that which is normally perceived, to be rendered more readily visible [6] and are the focus here. Model-centered or model-based learning is grounded in the theory that humans construct "mental models," internal cognitive representations that support reasoning and understanding by simulating the behavior of systems in the real world [7]. In model-based learning, it is assumed that learners construct explanatory mental models of phenomena in response to particular learning tasks, by integrating pieces of information about structure, function/behavior, and causal mechanisms. Learners then use and continuously re-evaluate their models, discarding or revising them as needed.

CP883, *2006 Physics Education Research Conference*, edited by L. McCullough, L. Hsu, and P. Heron

A Model-Based Approach to Learning about Electricity

In an approach to learning electricity via explanatory models, students are encouraged to focus first on the causes behind what is happening in the wires. The mathematical quantification of voltage, current, etc., is usually left until later in instruction, serving to verify and support the students' working mental models. Clement and Steinberg [8] discuss an approach to teaching complex models of electricity based on a *model construction cycle* of Generation, Evaluation, and Modification (referred to as a GEM cycle). Through this process of *model evolution*, or incremental growth in sophistication of a model, students are led to reassess and revise their model many times. Their learning approach is centered around Steinberg's [9] CASTLE (Capacitor Aided System for Teaching and Learning Electricity) curriculum, which utilizes the introduction of large, non-polar capacitors into basic electric circuits as a way to focus students' attention on the transient states of potential differences that exist throughout the circuit. By using the analogy of voltage as a type of "pressure" that exists in the "compressible electric fluid" of a circuit, students are encouraged to generate images of dynamic pressure changes occurring throughout the circuit as these capacitors go through their charging and discharging cycles.

Studies [10, 11] found that students in these model-based learning classes recorded significantly greater gains in electric circuit problem solving and reasoning abilities than their counterparts who learned the concepts of electricity through more traditional, lecture and equation-based means. In this paper, we describe a method of tracking specific types and levels of whole-class-discussion-based teaching strategies that one teacher used to foster such students gains.

STUDY CONTEXT & SETTING

This grounded theory study was conducted with a ninth grade science class that was studying an instructional unit on the fundamental concepts of circuit electricity. The 18 students (9 male and 9 female) at a small private suburban high school in New England spent approximately six weeks working with the CASTLE curriculum on activities of incremental model building. Their male teacher was a 30-year veteran of high school physics and was very familiar with model-based instruction. The structure of the classroom sessions had students working in pairs, alternating between assembling and testing circuit experiments, completing readings and responses in their student workbooks, and participating in full class discussions moderated by the teacher.

DATA COLLECTION & ANALYSIS

This work builds on a research methodology and analytical techniques [12] in which videotaped segments of whole-class discussions were used in the identification of model-based teaching strategies. Several five- to six-minute segments, in which the teacher and students appeared to be engaged in the co-construction of explanatory models, were selected for analysis. Once detailed transcripts of these segments were prepared, interpretive descriptions of the teacher's strategies or moves were generated. A diagrammatic representation of the teacher-student discourse patterns was developed in an attempt to: a) present the spoken contributions of teachers and students, b) describe the functions of these utterances, c) categorize the levels of various teaching strategies, d) track the evolution of the explanatory models as revealed through the discourse Analysis utilized a constant comparative method leading to open coding, followed by refinement of categories in joint coding with the second author.

RESULTS

The following is a description of the diagrams that resulted from the analysis outlined above. Because it typically requires a diagram 2-4 ft. wide to portray 5-6 minutes of model-building activity, an abridged diagram of 15 seconds of discussion is presented here to illustrate the technique. Figure 1 shows time running from left to right. The horizontal strip across the middle of the diagram contains icons (generalized images) of the evolving model. These are based on the sequentially numbered student and teacher statements, above and below the icons. The icons represent our hypotheses about the students' collective model, as revealed through their discourse, at given points in the discussion. In Fig. 1, the first icon portrays a slight compass deflection and an unlit bulb, observed in earlier class experiments. (A circuit containing too many resistors for a bulb to light still caused a compass to deflect). The second icon portrays one student's hypothesis: resistors are absorbers of charge, leaving enough charge in the wires to deflect a compass but not enough to light the bulb.

One layer further away from the icon strip, in each direction, are brief descriptions of the function of each student and teacher contribution. For the teacher, instructional strategies are separated into two distinct categories or levels: 1) those that support the *dialogical*, or conversational, elements of classroom interaction that are essential to effective two-way communication and sharing of ideas, 2) those that appear directly to influence the cognitive *model construction* processes.

Dialogical teacher strategies are generally observed to be conversational in nature; occur within a very short timeframe; support dialogical interaction; encourage increased student participation and ownership in the discussion; and foster a classroom culture that promotes and encourages student input, values opinions, and considers alternative conceptions and viewpoints. Previous studies [13,14] have made valuable contributions to the identification and description of these types of teacher strategies. Examples identified in this teacher's repertoire that are *not* portrayed in the abridged diagram include:

* Repeating student answers for emphasis
* Praising students for contributing
* Allowing scientifically incorrect statements
* Eliciting student voting
* Using diagrams to clarify
* Paraphrasing student responses for emphasis/clarity

Model construction teacher moves can generally be observed to utilize cognitive strategies for fostering model construction and evolution through questions and comments that focus on students' pre-conceptions, patterns in the data, and the processes of reasoning about the scientific concepts at hand. Generally, these moves appeared to influence the direction of discussion for longer periods than the conversational moves described above. Examples in addition to those shown in Fig. 1 include:

* Initiating Retroactive Discrepant Events – asking students to recall the results of a previous experiment or demonstration to generate cognitive dissonance
* Helping students focus on patterns in the data
* Asking students to differentiate or integrate aspects of the model
* Requesting that experimental observations be explained
* Initiating Discrepant Thought Experiments – asking students to run a mental model and to make predictions from it to compare with experimental data

The levels on the extreme top and bottom of the diagram outline the progression of the whole-class-discussion sequence through the various phases of a higher-order strategy for driving the co-construction of models. A brief indication of two of the phases is shown. The complete cycle is Observation, Generation, Evaluation, and Modification, with multiple instances of each phase occurring within the conversation excerpted here. Previous studies [15-16] have contributed precursory diagramming techniques and have identified the last three phases of this cycle as they occurred during teacher- student co-construction of explanatory models in science classes, and during model construction activities of expert scientists [17]. The model construction cycle, as previously described, was: Generation (G), Evaluation (E), and Modification (M). Because of the nature of the CASTLE curriculum with its frequent and integrated use of student experiments and teacher demonstrations, recent studies [12] added an additional *Observation* (O) phase to the process, resulting in the OGEM cycle discussed here.

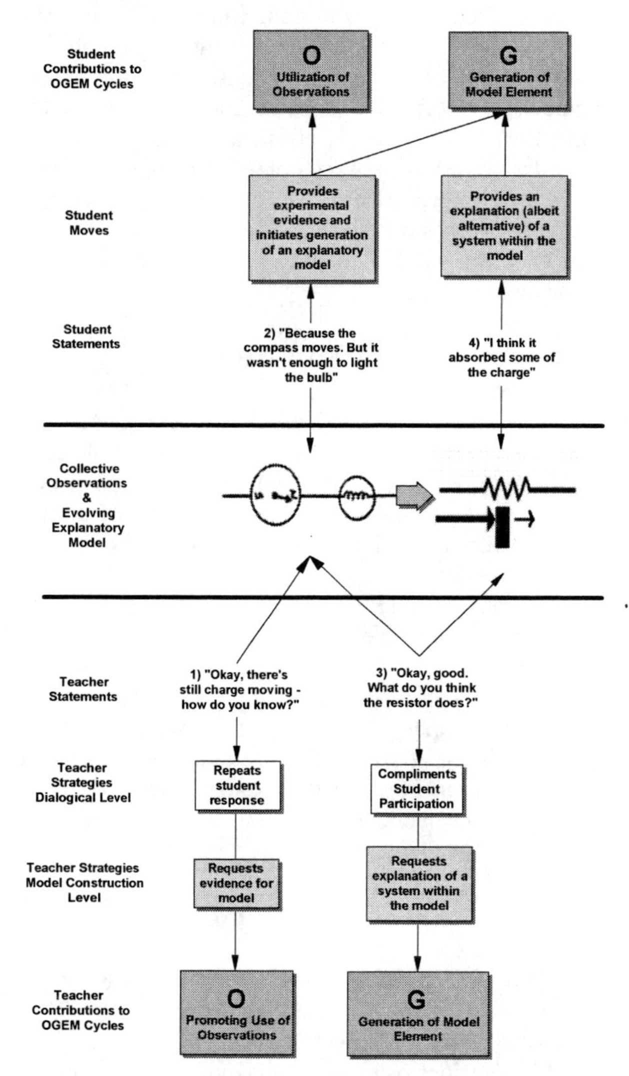

FIGURE 1. Model Construction through Discussion

The teacher strategies portrayed in the second-to-bottom level of the diagram are shown to contribute to higher-order goals shown in the bottom level. The two strategies shown here contribute to the higher-order goals *promoting the use of observations* (O) and *generating model elements* (G). Other examples of teaching strategies, not included in the above diagram: asking a discrepant question to promote the evaluation of a misconception (*promotes model evaluation*, E), and asking for alternative descriptions or explanations (*promotes model modification*, M).

The diagram allows us to portray teacher-student co-construction in detail. It was noted that this teacher frequently implemented dialogical and model construction moves together, in single utterances, in the service of keeping his students engaged in the conversation yet continuously constructing their models. The diagram also allows us to tie the hypothesized strategies to transcript observations as a way to constrain our theorizing about the influence of these strategies on students' construction of mental models. We see this work as complementing other important work [18, 19] on meta-cognitive strategies and the tradeoffs between content and process goals. In our next phase of work, we plan to triangulate from teacher interviews about these video segments to enhance the validity of our diagrammatic models of teaching strategies.

CONCLUSION

This paper has outlined the development of a methodology for describing and tracking whole-class discussion-based teaching strategies. Through a new diagramming technique, two distinct levels of teacher moves were posited: *dialogical* teacher strategies that foster students' active participation in science discussions, and cognitively-focused *model construction* strategies that the teacher used to support students' construction of explanatory mental models for circuit electricity concepts. Of the various teaching strategies found to exist at the model construction level, it is hypothesized that each can also contribute to the higher-order goals of promoting a cycle of Observation, model Generation, model Evaluation, and model Modification.

ACKNOWLEDGMENTS

This material is based upon work supported by the National Science Foundation under Grant REC-0231808, John J. Clement, PI. Any opinions, findings, and conclusions or recommendations expressed in this paper are those of the author(s) and do not necessarily reflect the views of the National Science Foundation.

REFERENCES

1. J. Solomon, The Rise and Fall of Constructivism. *Studies in Science Education*, 23, 1-19, (1994)
2. L. S. Vygotsky, *Thought and Language*. Cambridge, MA: MIT Press, 1962.
3. T. Sprod, I Can Change Your Opinion on That: Social Constructivist Whole Class Discussions and Their Effect on Scientific Reasoning. *Research in Science Education.* Vol. 28 (4), pp. 463-480, (1998).
4. J. Roschelle, Learning by Collaborating: Convergent Conceptual Change. *The Journal of the Learning Sciences*, 3, 265-283, (1992).
5. A. M. Ingham & J. K. Gilbert, The Use of Analogue Models by Students of Chemistry at Higher Education Levels. *International Journal of Science Education*, 13, pp. 193-202, (1991).
6. J. Gilbert, The Role of Models and Modeling in Some Narratives in Science Learning. *Paper Presented at the Annual Meeting of the American Educational Research Association*, Apr. 18-22, 1995. San Fransisco, CA.
7. D. L. Schwartz, and J. B. Black, Analog Imagery in Mental Model Reasoning: Depictive Models. *Cognitive Psychology*, 30, 154-219. (1996).
8. J. Clement & M. S. Steinberg, Step-Wise Evolution of Mental Models of Electric Circuits: A "Learning-Aloud" Case Study. *The Journal of the Learning Sciences*, 11(4), 1pp. 389-452. (2002).
9. M. S. Steinberg, *The CASTLE Project Student Manual.* PASCO Scientific. (2004).
10. D.E. Brown, Teaching Electricity with Capacitors and Causal Models: Preliminary Results from Diagnostic and Tutoring Study data Examining the CASTLE Project. (1992).
11. E. G. Williams & J. J. Clement, Model-Based Learning of Electricity: Improving Student Reasoning and Confidence. *Paper Presented at the AAPT Annual Conference*, Syracuse, NY, July 2006.
12. E. G. Williams, Teacher Moves during Large-Group Discussions of Electricity Concepts: Identifying Supports for Model-Based Learning. *Proceedings of the NARST Annual Conference* – San Francisco, CA, April 2006.
13. E. van Zee & J. Minstrell, Reflective Discourse: Developing Shared Understandings in a Physics Classroom. *International Journal of Science Education*, 19, pp. 209-228, (1997).
14. D. Hammer, Student Inquiry in a Physics Class Discussion. *Cognition and Instruction.* 13(3), 401-430 (1995).
15. M. A. Rea-Ramirez& M.C. Nunez-Oviedo, Discrepant Questioning as a Tool to Build Complex Mental Models of Respiration. In: *Proceedings of the Annual International Conference of the Association for the Education of Teachers in Science* (Charlotte, NC, Jan 10-13, 2002).
16. M. C. Nunez-Oviedo & J. Clement, Modes of Co-Construction During Large Group Discussion for Model-Based Learning in Science. (to appear)
17. J. Clement, " Learning via Model Construction and Criticism: Protocol Evidence on Sources of Creativity in Science" in *Handbook of Creativity: Assessment, Theory and Research, edited by J.* Glover, R. Ronning, and C. Reynolds, NY: Plenum, 341-381, 1989.
18. D. Hammer, Discovery Learning and Discovery Teaching. *Cognition and Instruction*, 15(4), 485-529, 1997.
19. Duschl, R. A., & Osborne, J. (2002). Supporting and promoting argumentation discourse in science education. *Studies in Science Education, 38,* 39-72.

Students' Use of Symmetry with Gauss's Law

Adrienne L. Traxler,[1] Katrina E. Black,[2] John R. Thompson[1,2]

[1]*Center for Science and Mathematics Education Research and* [2]*Department of Physics and Astronomy*
The University of Maine, Orono, ME

Abstract. To study introductory student difficulties with electrostatics, we compared student techniques when finding the electric field for spherically symmetric and non-spherically symmetric charged conductors. We used short interviews to design a free-response and multiple-choice-multiple-response survey that was administered to students in introductory calculus-based courses. We present the survey results and discuss them in light of Singh's results for Gauss's Law, Collins and Ferguson's epistemic forms and games, and Tuminaro's extension of games and frames.

Keywords: physics education research, electrostatics, symmetry, Gauss's Law, epistemic frames, epistemic games.
PACS: 01.40.-d, 01.40.Fk, 03.50.De, 41.20.Cv.

INTRODUCTION

Gauss's Law relates the charge enclosed by a closed surface to the total flux through that surface. In particular symmetry cases (spherical, cylindrical, planar), the flux integral can be simplified and rearranged to find an expression for the electric field. Anecdotal evidence suggests that many introductory physics students use Gauss's Law to find electric fields in a rote way, without concern for the symmetry conditions required to do so. Following initial interviews, we developed a written survey that was distributed in two semesters of the introductory calculus-based electricity and magnetism course. Our intent was to see whether students could recognize the non-applicability of Gauss's Law to a problem situation lacking any of the standard symmetries.

Additionally, our interview results and other research in the field [1] suggest that many students are unnecessarily restrictive when choosing Gaussian surfaces to find the electric flux, while others are not restrictive enough when choosing Gaussian surfaces to find the electric field.

SURVEY DESIGN

We asked interview and survey subjects about both a spherical charged conductor and a "football-shaped" (prolate spheroid) charged conductor (see Figure 1). Students were asked to sketch the electric field lines and to find an expression for the field both inside and outside the conductor surface, and to explain their reasoning for each step, or, if the field could not be determine, to describe why not. For the written survey, we also provided students with a list of Gaussian surfaces (sphere, cylinder, cube, football, other) and asked the students to select those surfaces that would be useful for finding the electric field in the given situation.

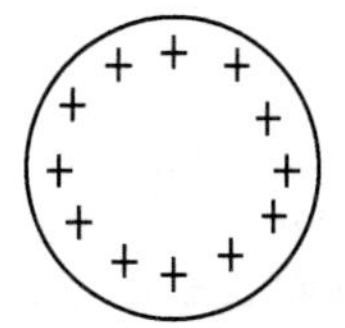
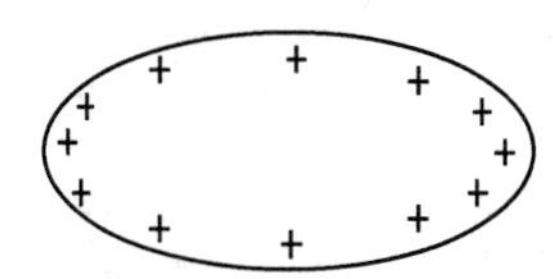

FIGURE 1. Charged, conducting objects for which students were asked to provide field lines and an electric field expression. Students were given that the conductors had total charge Q and surface areas A (circle) and B (football).

Based on our interviews, we expected students to draw reasonably correct field lines for both the sphere and football and either recall expressions for the sphere's electric field or derive them using Gauss's Law. We also suspected that many of the students would use the same electric field method for both the sphere and football, inappropriately generalizing from the spherical case. Although we did not ask about choosing Gaussian surfaces in our interviews, based on Singh's result, we expected some of the students to choose Gaussian surfaces from the list that would not be useful for finding the field.

In the fall of 2005, the survey was offered for nominal extra credit immediately following an exam that covered magnetism ideas. Forty-seven students took the survey, with 45 completing the sphere question and 36 completing the football question. In the spring of 2006 the survey was administered in two

CP883, *2006 Physics Education Research Conference*, edited by L. McCullough, L. Hsu, and P. Heron

out of 8 recitation sections; the TAs for these sections offered to administer the survey during class time. The survey was distributed following an exam that covered Gauss's Law, but no extra credit was offered. Sixteen students completed the survey. All students completed both questions. (The groups will be referred to by year, i.e., 2005 and 2006.)

RESULTS AND DISCUSSION

Although we were primarily interested in how students developed a mathematical expression for the electric field, many students had difficulty with the supporting ideas required to begin such a development, including the differences between electric and magnetic ideas and the construction of field lines.

A few students in 2005 seemed to use magnetic ideas in tackling the problem. One student drew X's around the conductor and drew perpendicular force and velocity vectors, although she referred to these as "electric field lines." We interpret the X's as indicating a magnetic field directed into the page. Other students wrote expressions such as $\oint_E \vec{E} \cdot d\vec{l} = \frac{\varepsilon_0 I_{induced}}{2\pi}$, apparently confusing Gauss's Law with Ampère's Law. These students, however, were in the minority.

A number of students showed trouble with one or more ideas from electrostatics. For example, even in the simpler case of the sphere about 40% of students in 2005 (n = 45) drew field lines with some kind of error (curved rather than straight, inside the conductor, or pointing toward the conductor). A third of the students gave a nonzero field expression for the inside of the conductor and 18% gave an expression for the field outside the conductor with no dependence on distance. Surprisingly, 18% also drew or described negative charges attracted by the positive charge on the conductor. These negative charges sometimes resided on the Gaussian surface, indicating an incomplete understanding of the purpose of the Gaussian surface. All of these difficulties— confusion with magnetic ideas, trouble drawing field lines, and qualitatively incorrect field behaviors inside and outside conductors—would make appropriate use of Gauss's Law difficult.

In 2005, only 7 of 45 students wrote a correct expression for the electric field outside the spherical conductor, and of these students 4 included some form of correct explanation. In 2006, 10 (of 16) gave the correct expression; of these, 3 accompanied it with a correct explanation.

The football question was even more difficult for students. In 2005, 5 of 36 students gave what we considered a correct answer for the field outside the football (that is, the electric field cannot be found), but none provided a correct explanation of why this was the case. One student, however, seemed to be on the right track with the explanation "the electric field is not evenly dispersed so without the dimensions the electric field cannot be found." Students in 2006 fared no better, with 3 of 16 stating that the electric field could not be found. Two of these students explained their response in a similar manner to the student in 2005.

TABLE 1. Percentage of correct electric field expressions for the two conductors (with or without correct explanation).

Field Location	2005 semester		2006 semester	
	Sphere (n=45)	Football (n=36)	Sphere (n=16)	Football (n=16)
Inside	29%	36%	44%	56%
Outside	16%	14%	63%	19%

Overall, fewer than half of the students explained the reasoning behind their responses on the written survey in 2005, either with words or with a step-by-step mathematical solution, for either correct or incorrect responses. In 2006, we strengthened language in the directions explicitly asking for a written explanation, but still, only 9 of the 16 students provided any sort of explanation. Even from the scant information supplied, however, some interesting patterns emerge.

The rarity of step-by-step problem solving was a more widespread issue. Very few students provided any mathematical steps leading to their final electric field expressions. This suggests that in the case of the sphere they simply remembered and wrote the kq/r^2 form for a spherical charge distribution.

Although the data were somewhat "noisier" than we expected, the predicted pattern did emerge. In particular, of the students who gave electric field expressions for both problems (34 students in 2005 and 16 in 2006), 66% in 2005 and 57% in 2006 used essentially the same method for both (i.e., the same steps if a derivation was used, or the same functional form if only an equation was given).

The results of the multiple-choice-multiple-response Gaussian surface question, summarized in Table 2, were as expected. For the sphere, 40% (2005) and 75% (2006) of students chose only a spherical Gaussian surface, but a significant minority (31% and 13%, respectively) included some, but not all, other listed surfaces in addition to the sphere. Most frequently, the additional surface was the cylinder. Additionally, 24% and 12% of students chose either all the listed surfaces or explicitly wrote that any closed surface would be a useful Gaussian surface.

For the football, 25% (2005) and 75% (2006) of students chose only a football-shaped Gaussian surface. Fewer students (8% and 13%) chose the football along with some other surfaces, and 25% and 12% of students chose all the listed surfaces or any closed surface. Interestingly, although only 2% of students in 2005 (and none in 2006!) did not include the spherical Gaussian surface when choosing appropriate surfaces for the sphere, 17% and 6% of students did not include the football-shaped surface in the football situation.

TABLE 2. Shapes of Gaussian surfaces chosen for both shapes of conductor

Surface Chosen	2005 semester		2006 semester	
	Sphere (n=45)	Football (n=36)	Sphere (n=16)	Football (n=16)
Shape Only	40%	25%	75%	75%
Shape + others	31%	8%	13%	13%
Other than shape	2%	17%	0%	6%
Any or all	24%	25%	12%	12%

The results of the Gaussian surface question indicate several difficulties with the interpretation of Gaussian surfaces as a (mathematical) tool.

Interpretation of Results Using Epistemic Frames and Games

Epistemic frames [2,3] and games [3,4] as discussed in earlier work and then applied to physics problem solving [3], provide an interesting lens through which to view our data. Briefly, an epistemic frame can be described as a student's perception and expectations surrounding the situation at hand (for example, being in a lecture setting, or doing physics homework in a dorm room). An epistemic game is the series of moves followed by the student (e.g., writing notes and staying quiet in a lecture, or working through the solution to the physics problem and asking friends for help) to generate some target structure, known as the epistemic form [4]. (In a lecture setting, the epistemic form associated with the *note-taking* game is a set of notes that reflect the lecture content.) A student's current frame influences which games are seen as useful in order to achieve the perceived target structure.

We interpreted our results using this framework, in the hopes of gaining additional insights into student responses. Tuminaro identifies three distinct frames students use when solving physics problems: *rote equation-chasing*, *qualitative sense-making*, and *quantitative sense-making*.

To successfully navigate both survey tasks, students must use the *qualitative* (for the field line sketch) and *quantitative* (for the field expression) *sense-making* frames. For the sphere, *rote equation-chasing* could potentially lead to a correct response. Since using any memorized general form of Gauss's Law (including restricted forms such as $EA = q/\varepsilon_0$) leads to the correct solution, students do not need to make sense of Gauss's Law or consider if it applies in order use it correctly. For the football-shaped conductor, however, the electric field is stronger near the "ends" of the football. Because of this lack of symmetry, it is essentially impossible to find a Gaussian surface for which the magnitude of $\boldsymbol{E}$ is constant without already having an expression for the field, rendering Gauss's Law impotent in this situation. Students in the *rote equation-chasing* frame might simply extend their solution from the sphere task, while students in a sense-making frame might realize that the change in the problem situation prohibited a straightforward use of Gauss's Law to find the field.

In general, we expected many students to be in the *rote equation-chasing* frame, remembering a series of steps to use Gauss's Law for the sphere and applying those same steps to the football. We found that many students were even more rote in their use of the process than we expected, to the point of some refusing to answer without numbers to substitute for the values of the charge on the conductors. Many students seemed to process a kind of "if-check": if the electric field equation was remembered, then write it down; otherwise, don't worry about deriving it. Figure 2 summarizes a frames view of the survey tasks and results.

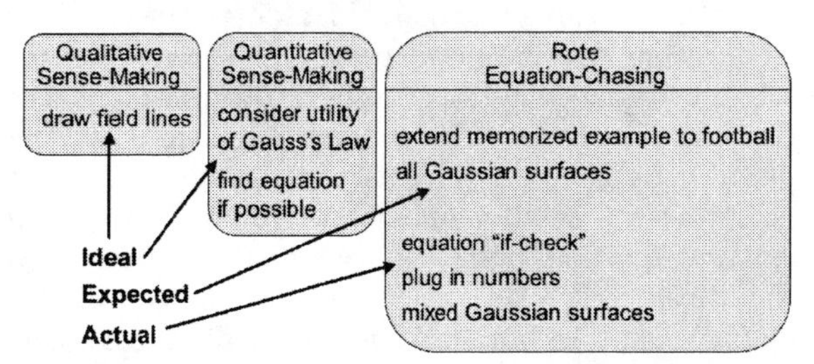

FIGURE 2. Epistemic frames and associated epistemic games associated with the task of determining the electric field for the objects in Figure 1.

The idea of epistemic forms also provides one possible explanation for the lack of student explanations when completing the survey. If students perceived the expression for $\boldsymbol{E}$ to be the target structure, then detailed steps would not necessarily be worth writing (especially if the final equation could simply be recalled). Since the grading of homework and exam questions in our traditional courses typically emphasize the final result (even if the problem asks for explanation), it is not unreasonable to suspect that students considered providing reasoning as unimportant, rather than being unable to provide reasoning. For the researchers, on the other hand, the process was as interesting as the final expression, so the complete derivation was the hoped-for epistemic form. If further surveys were done in the future on this topic, it might be helpful to rewrite the language to de-emphasize the importance of the final field expression.

Complicating Factors

Our choice of survey instruments, when examined in the light of epistemic frames, contains a dilemma: is the use of both "working" and "non-working" cases of Gauss's Law an unfair trick to play on students? If they recall how Gauss's Law is used in the case of the spherical conductor, that first task could cause students to overgeneralize this method to the football conductor without careful thought. In addition, in traditional physics classes (such as those from which we drew our subjects) the correct answer is rarely "you can't do this problem," further encouraging the application of the solution for the sphere to the football.

On the other hand, if students are prompted into using Gauss's Law inappropriately, it indicates that they may use the law in an algorithmic way rather than as a sometimes-useful tool. The prevalence of this behavior was exactly what interested us; thus potentially priming students with the sphere was an acceptable possibility.

One other pattern of interest that emerged from the data was the generally better performance of the students in 2006. For example, students in 2006 did not reason using magnetic ideas or write E-field expressions that did not depend on distance from the conductor, and they used the idea of induced charge far less frequently. One possible explanation is that there were far fewer students participating in 2006 as in 2005; the smaller number of respondents might have magnified quirks in the data. Alternately, significant ability differences between the two classes might exist. In our view, however, the most likely cause of the improvement in 2006 was timing. The 2005 students were given the survey late in the course, after an exam covering magnetism, while the 2006 students received the survey after an exam that included Gauss's Law problems. Not only was the material fresher in the minds of students in 2006, they had no instruction on magnetism to confuse with electrostatic topics.

CONCLUSIONS

Student performance in deciding how to determine the electric field of spherical and prolate spheroidal charged conductors suggests that Gauss's Law and the underlying procedures and concepts, especially the restrictive applicability of Gauss's Law, are not well understood by many introductory students.

Although the sample size was small, we noticed a qualitative difference in the types of responses between semesters. Some responses in one group seemed influenced by course content introduced between Gauss's Law and the administration of the survey. This suggests that while students may understand some of the consequences of Gauss's Law at the time that it is taught, newer electric and magnetic concepts can be confused with or replace the earlier content. While confusion between electric and magnetic ideas has been documented previously [5], it has not been seen with Gauss's Law.

The overall pattern of matched responses was in line with our expectations in type, if not in degree: most students who attempted both problems used the same method for both, indicating that they do not consider the shape of and charge distribution on a conductor to be an important cue. However, very few students made an attempt to derive an expression for the electric field if they could not recall it, as we had hoped they might. One possible explanation for this lack of written reasoning is a mismatch of epistemic frames between students and researchers, with few of the students considering documentation of the problem-solving process as an important part of the solution.

Finally, we partially replicated Singh's result: many students are too permissive when choosing appropriate Gaussian surfaces to find electric fields.

Gauss's Law can be a powerful tool, but it requires a deep understanding of both physical and mathematical principles in order to use it effectively. Students who lack this understanding are unlikely to use (or not use) Gauss's Law appropriately in unfamiliar situations.

ACKNOWLEDGMENT

We thank Michael C. Wittmann for helpful discussions and input, especially with regard to epistemic frames and games.

REFERENCES

1. C. Singh, *2004 Physics Education Research Conference* AIP Conf. Proc. **790** 65-68 (2005).
2. D. Tannen, *Framing in Discourse* (New York: Oxford University Press, 1993).
3. J. Tuminaro, "A Cognitive Framework for Analyzing and Describing Introductory Students' Use and Understanding of Mathematics in Physics", unpublished Ph.D. dissertation, University of Maryland (2004).
4. A. Collins and W. Ferguson, *Educational Psychologist,* **28(1)**, 25 (1993).
5. R. Driver, A. Squires, P. Rushworth and V. Wood-Robinson, *Making sense of secondary science* (New York: Routledge, 1994), 126-127.

Analysis Of Shifts In Students' Reasoning Regarding Electric Field And Potential Concepts

David E. Meltzer

Department of Physics, University of Washington, Seattle, WA 98195, USA

Abstract. Students' reasoning regarding the relationships among electric fields, forces, and equipotential line patterns was explored using pre- and post-test responses to selected multiple-choice questions on the Conceptual Survey of Electricity and Magnetism. Students' written explanations of their reasoning, provided both pre- and post-instruction, allowed additional assessment of the changes in their thinking. In particular, the data indicate that although students largely abandon an initial tendency to associate stronger fields with wider equipotential line spacing, many of them persist in incorrectly associating electric field magnitude at a point with the electric potential at that point.

Keywords: physics education, electricity and magnetism.
PACS: 01.30.Cc; 01.40.Fk

INTRODUCTION

There has been extensive discussion in recent years regarding the desirability of extracting, from multiple-choice test results, information regarding student thinking that is more detailed than that provided simply by net scores of correct and incorrect responses [1, 2]. With that objective in mind, I have carried out a detailed analysis of student responses to several related items from the Conceptual Survey of Electricity and Magnetism (CSEM) [3] that were administered in an algebra-based physics course taught with interactive-engagement methods at Iowa State University from 1998-2002 [4]. The 1998-2001 sample consists of four separate classes (N_{total} = 299); students responded to the test both on a first-day pretest and on a final-day posttest. In 2002, students wrote explanations of their answers on selected test items; this sample consists of an unmatched set (pretest, N = 72; posttest, N = 68).

The CSEM test items I will discuss here are (1) Item #18, which asks students to compare electric-field magnitudes at three points shown on equipotential-line diagrams, and (2) Item #20, which asks students to compare electric force magnitude and direction at two different points on an equipotential-line diagram, for a proton located at those points. (See Fig. 1.) The correct answer on Item #18 is *D*; response *E* is that all field magnitudes at the selected points are equal (note that V = 40V at all three points), while response *C* corresponds to *lower* field magnitude where line spacing is tightest. Both responses *B* and *D* on Item #20 are consistent with the correct *D* response on Item #18 (that is, larger force vectors where line spacing is tightest). Responses *A* or *C* on Item #20 could both be seen as consistent with *both C* and *E* responses on Item #18: with *C*, because force is larger where line spacing is wider, and with *E* because force is larger where potential is larger. However, although these responses may be consistent in the manner described, the items do *not* test exactly the same concepts; in one case field magnitude is at issue, and in the other case it is force magnitude.

ANALYSIS OF RESPONSE PATTERNS

Consistency Comparison For Correct And Incorrect Responses On Item #18

The proportion of total responses corresponding to the correct answer *D* on Item #18 increases from 46% (pre-) to 75% (post-instruction), reflecting a significant increase in correct responses. The *D* responses may be further broken down into "consistent" and "inconsistent" responses, where "consistent" corresponds to the proportion of those students who answered *D* on Item #18 who *also* answered either *B* or *D* on Item #20. For the pre-instruction results, the proportion of *D* responses that are consistent (45%) is almost that which would correspond with random guessing on Item #20, i.e., 40%. Similarly, of students who an-

CP883, *2006 Physics Education Research Conference*, edited by L. McCullough, L. Hsu, and P. Heron

swered Item #18 *incorrectly* pre-instruction, 40% also gave a *B* or *D* response on Item #20. Thus, before instruction, students who answered Item #18 correctly were not significantly more likely ($p > 0.38$) to give the preferred *B* or *D* responses on Item #20 than were students who answered Item #18 incorrectly.

Post-instruction, however, the situation is different. The proportion of *D* answers that are accompanied by *B* or *D* responses on Item #20 rises to 83%, far above the random-guessing rate. This implies that a large majority of those students who gave the correct *D* response on Item #18 after instruction were able to answer Item #20 in a manner consistent with their answer on Item #18, whereas before instruction that was not the case. The rise in consistency from 45% pre-instruction to 83% post-instruction is statistically significant ($p = 0.001$ by a paired two-sample *t*-test).

By comparison, of those students who had *incorrect* answers on Item #18 after instruction, only 57% gave *B* or *D* responses on Item #20. Thus, students who had correct answers on Item #18 were significantly more likely ($p < 0.001$) to give *B* or *D* responses on Item #20 than students who had incorrect answers on Item #18, but *only* after instruction. Before instruction, correct answers on Item #18 implied *no* increased probability of the favored *B* or *D* response on Item #20, despite the similarities between the two items. The implication is that *D* responses on the posttest corresponded to significantly better student understanding of the targeted concept than did *D* responses on the pretest, even though both would ordinarily be scored simply as "correct" responses.

Pre- To Post-Instruction Consistency Shift For "E" Responses On Item #18

The proportion of total responses on Item #18 corresponding to answer *E* barely changes from pre- to post-instruction. Before instruction, 18% of all students give answer *E*, while after instruction that rate is virtually unchanged at 20%; the difference is negligible and not statistically significant. Evidently (and disappointingly), the net popularity of this incorrect answer was essentially unaltered by instruction. (It is interesting that the data reported by Maloney et al. [3] also show increases in the popularity of this response on the posttest, compared to the pretest.)

In order to explore the consistency of thinking reflected by *E* responses on Item #18, we examine how frequently students who gave that response also gave the related responses *A* or *C* on Item #20. Before instruction, we find that 51% of students who answered *E* on Item #18 also gave an *A* or *C* response on Item #20. Although this is slightly above the random-guessing rate of 40%, the difference is not statistically

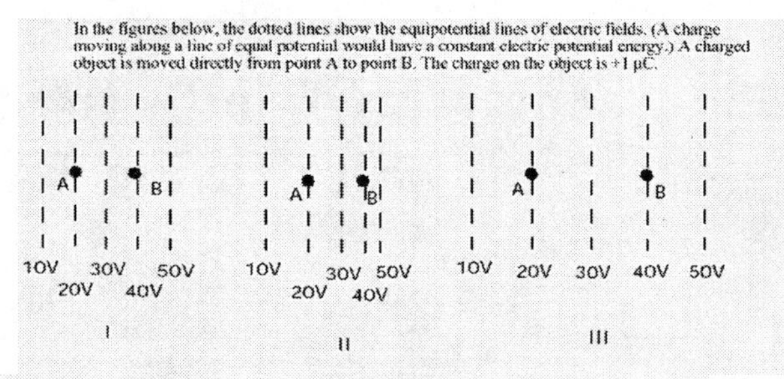

How does the magnitude of the electric field at B compare for these three cases?
(a) I > III > II
(b) I > II > III
(c) III > I > II
(d) II > I > III
(e) I = II = III

Item #18

A positively-charged proton is first placed at rest at position I and then later at position II in a region whose electric potential (voltage) is described by the equipotential lines. Which set of arrows on the left below best describes the relative magnitudes and directions of the electric force exerted on the proton when at position I or II?

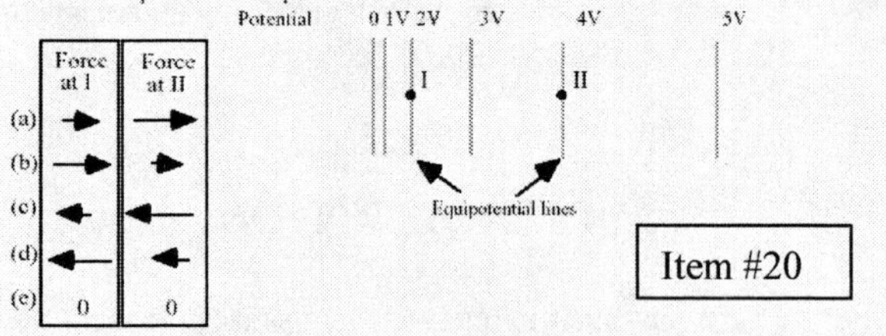

FIGURE 1. CSEM Items #18 and #20 discussed in text.

significant; it is virtually identical to the 50% rate found among students who did *not* answer *E* on Item #20. That is, an *E* pretest response on Item #18 was associated with a marginally greater-than-random probability of giving an *A* or *C* response on Item #20, but no more so than shown by students who did *not* answer *E* on Item #18. After instruction, two changes in these results are noted. First, the proportion of students giving answer *E* on Item #18 who answered *A* or *C* on Item #20 is down to 38% (compared to 51% pre-instruction); this drop is statistically significant ($p = 0.01$) according to a paired two-sample (pre- and post-instruction) *t*-test where the four independent subsamples correspond to the years 1998-2001. This contrasts sharply with the *increase* in consistency that was clearly shown for the correct response *D* (as discussed in the section immediately above). This implies that in this case, the incorrect response *E* when given on the posttest is less closely associated with *consistent* incorrect thinking, than it is when given on the pretest.

The second change noted was among students who did *not* answer *E* on Item #18; the proportion of these students giving an *A* or *C* response on Item #20 is now significantly lower (at 18%) than it was before instruction (i.e., 50%). The difference in *A*/*C* response rate on Item #20 for the *E*-responders on Item #18

(38%) compared to the non-*E*-responders (18%) is now statistically significant ($p < 0.03$), and so even though the *A* and *C* responses on Item #20 become less popular with *E*-responders than before, they are still more often given by this group than by those who do *not* answer *E* on Item #18.

Pre- To Post-Instruction Shift For "C" Responses On Item #18

Besides choices *D* and *E*, the one other answer option on Item #18 to garner a substantial proportion of pretest responses was answer *C*. Before instruction, 26% of all students gave response *C*. On the posttest, however, the response rate for choice *C* dropped sharply to only 3%; that is, support for this option essentially disappeared. Apparently, instruction was effective in eliminating the attractiveness of this answer.

Response *C* on Item #18 can be considered consistent with responses *A* and *C* on Item #20. Of all students who gave the *C* response on Item #18 pre-instruction, 53% also gave an *A* or *C* response on Item #20, not significantly different than random guessing ($p > 0.10$). This suggests relatively low consistency in students' thinking reflected by this *C* response.

ANALYSIS OF EXPLANATIONS

Students' pre- and post-instruction written explanations provided in 2002 offer considerable additional insight into student thinking.

Response "D": Correct Answer, Incorrect Explanation

Nearly half of all students—46%—gave the correct response *D* for Item #18 on the pretest in 1998-2001. In the 2002 offering of the same course, students were specifically asked whether or not they had previously studied electricity and magnetism, and in particular whether they had studied the material represented on the questions dealt with here. Of the students who gave the correct answer *D* (53% of all students in 2002), almost 60% said that they had not previously studied electricity and/or magnetism, and less than 20% said that they had studied the material represented on these questions. How, then, did so many of them get the correct answer?

Their explanations indicate that many students either based their correct answer on some "intuitive" reasoning, or they simply guessed. A sample of some of the explanations offered for answer *D* bears this out:

"Chose them in the order of closest lines"
"Magnitude decreases with increasing distance"
"Greatest because 50 [V] is so close"
"More force where fields are closest"
"Because charges are closer together"
"Guessed"

Although it is difficult to categorize many of the responses on the pretest, it is clear that many of the students who chose answer *D* justified their answer with vague or inconsistent arguments based on the tighter spacing of equipotential lines. However, for most of this group, inconsistent answers on Item #20 (i.e., choosing *A*, *C*, or *E* on that item), ambiguous explanations of what "closer spacing" implied, or explicit admissions that they were guessing, all implied that most of students' correct answers on this pretest item were not indicative of a clear understanding of the conceptual principles involved.

On the posttest, by contrast, most students (83%) who answered *D* on Item #18 gave consistent answers on Item #20 (*B* or *D;* see section above), and most (63%) of the group with this *D/B-or-D* answer pattern gave explanations on *both* items that were consistent with an adequate understanding of the concept involved.

Response "C": Wider Spacing ⇒ Stronger Field

On the pretest, the second most popular response for Item #18 was *C*, consistent with wider equipotential line spacing being associated with larger-magnitude electric fields. Similarly, in 2002, 24% of pretest responses on this item were *C*. About half of the explanations offered made explicit reference to the line spacing, although few of these specifically related wider spacing to larger field magnitude. Most of the explanations in this category indicated that the students were unsure of their reasoning, or indeed that they were making outright guesses.

Here is a sampling of students' pretest explanations for response *C*:

"III is the farthest apart, then 1 and then 2"
"The equipotential lines are farther apart so a greater magnitude is needed to maintain an electrical field"
"I guessed"

On the posttest, as discussed above, *C* responses on Item #18 fell to an insignificant 3% of all responses.

Response "E": Field Magnitudes All Equal

As discussed above, this response is consistent with the idea that electric-field magnitude scales with the value of the electric potential at a particular point. About half of the explanations offered for this response, both pre- and post-instruction, explicitly argued along these lines. On the posttest, another popular explanation was based on the (incorrect) claim that the diagram portrayed a uniform electric field, thus ensuring equal field magnitudes at all points. This particular explanation—which ignores the line spacing issue—was not present on the pretest. Otherwise, pre- and posttest explanations were similar. For example,

Pretest explanations:

"They are all at the same voltage"
"The magnitude is 40 V on all three examples"
"The voltage is the same for all 3 at B"
"The change in voltage is equal in all three cases"

Posttest explanations:

"The potential at B is the same for all three cases"
"They are all from 20 V–40 V"
"The equipotential lines all give 40 V"
"They all have the same potentials"

As discussed in the sections above and below, response *E* retained its popularity after instruction as it accounted for 20% of all posttest responses on this item, compared to 18% on the pretest. It seems that the addition of new justifications for this response (e.g., uniform electric field) helped sustain its attractiveness even after instruction.

AN ALTERNATIVE PERSPECTIVE ON CONSISTENCY SHIFTS

Another way to analyze pretest-to-posttest shifts in consistency is to probe for possible changes in the popularity of various student "models," in a sense analogous to that of Bao and Redish [1]. Here a model will be defined as a specific set of answers to two or more related questions. Certainly, using responses on only two questions to attempt to specify a student model is far from optimal, if only because the probability that random guessing could produce the specified response patterns is relatively high. Moreover, as discussed above, the questions on this test were not designed to probe exactly identical ideas regarding field and potential. For these reasons, the present discussion regarding student models is meant to be illustrative only.

We may for instance presume that one student model corresponds to the idea that "*electric-field magnitude is larger at a point where electric potential is larger*," and define this model as corresponding to an *E* response on Item #18 when accompanied by an *A* or *C* response on Item #20. With that definition we find that 9% of all responses (1998-2001) on the pretest were consistent with this model, along with 8% of responses on the posttest. Analogously, a model corresponding to "*electric-field magnitude is larger at a point where equipotential lines have wider separation*" could correspond to a *C* response on Item #18, and an *A* or *C* response on Item #20. This model dropped from 14% of the responses on the pretest to only 2% on the posttest. Meanwhile, the "correct" model (*D* on Item #18, *B* or *D* on Item #20) rose from 20% on the pretest to 63% on the posttest.

Although this method of analysis is potentially useful and informative, it is worth probing in more detail to try and determine the extent to which the specified models actually correspond to distinct patterns of student thinking. We can look at the students' 2002 written explanations to shed light on this issue. For example, on the posttest, we find 6% of all responses following the *E*/*A-or-C* pattern, while only half of those (3%) offered explanations consistent with the idea that "*electric-field magnitude is larger at a point where electric potential is larger*." On the pretest, the corresponding figures were 7% of all responses, with 3% offering consistent explanations for this model. Similarly, although 17% of pretest responses corresponded to the pattern *C*/*A-or-C*, only one-quarter of those (4% overall) provided explanations consistent with "*electric-field magnitude is larger at a point where equipotential lines have wider separation.*" For the "correct" model (*D*/*B-or-D*) the proportion of consistent explanations was higher, but still only about two-thirds of pretest explanations accompanying that pattern were consistent with a correct explanation.

Although these results are illustrative rather than definitive, they do suggest the need for caution in interpreting patterns of multiple-choice responses as corresponding to well-defined student models.

ACKNOWLEDGMENTS

The original motivation for this work was a series of discussions with Lei Bao regarding his work on Concentration Analysis. This work was supported in part by NSF REC-#0206683 (T. J. Greenbowe, Co-Principal Investigator), DUE-#0243258, and DUE-#0311450.

REFERENCES

1. L. Bao and E. F. Redish, *Am. J. Phys.* **69**, S49-S53 (2001).
2. R. J. Dufresne, W. J. Leonard, and W. J. Gerace, *Phys. Teach.* **40**, 174-180 (2002).
3. D. P. Maloney, T. L. O'Kuma, C. J. Hieggelke, and A. Van Heuvelen, *Am. J. Phys.* **69**, S12-S23 (2001).
4. D. E. Meltzer and K. Manivannan, *Am. J. Phys.* **70**, 639-654 (2002).

Improving Student Understanding of Coulomb's Law and Gauss's Law

Zeynep Isvan and Chandralekha Singh

Department of Physics and Astronomy, University of Pittsburgh, Pittsburgh, PA, 15260, USA

Abstract. We discuss the development and evaluation of five research-based tutorials on Coulomb's law, superposition, symmetry and Gauss's Law to help students in the calculus-based introductory physics courses learn these concepts. We compare the performance of students on the pre-/post-tests given before and after the tutorials in three calculus-based introductory physics courses. We also compare the performance of students who used the tutorials and those who did not use it on a multiple-choice test which employs concepts covered in the tutorials.

INTRODUCTION

Electrostatics is an important topic in most calculus-based introductory physics courses. Although Coulomb's law, superposition principle, and Gauss's law are taught in most of these courses, investigations have shown that these concepts are challenging for students [1, 2, 3]. Despite the fact that students may have learned the superposition principle in the context of forces in introductory mechanics, this learning does not automatically transfer to the abstract context of electrostatics and students get distracted by the very different surface features of the electrostatics problems. Effective application of Gauss's law implicitly requires understanding the principle of superposition for electric fields and the symmetry that ensues from a given charge distribution. Helping students learn these concepts can improve their reasoning and meta-cognitive skills and can help them build a more coherent knowledge structure. Here, we discuss the development and evaluation of research-based tutorials and the corresponding pre-/post-tests to help students develop a functional understanding of these concepts.

TUTORIAL DEVELOPMENT AND ADMINISTRATION

Before the development of the tutorials, we conducted investigation of student difficulties with these concepts [3] by administering free-response and multiple-choice questions and by interviewing individual students. We found that many students have difficulty distinguishing between the electric charge, field and force. Students also have difficulty with the principle of superposition and in recognizing whether sufficient symmetry exists for a particular charge distribution to calculate the electric field using Gauss's law. Choosing appropriate Gaussian surfaces to calculate the electric field using Gauss's law when sufficient symmetry exists is also challenging for students. Distinguishing between electric field and flux was often difficult.

We then developed the preliminary version of five tutorials and the corresponding pre-/post-tests based upon the findings of the difficulties elicited in previous research and a theoretical task analysis of the underlying concepts. Such analysis involves making a fine-grained flow chart of the concepts involved in solving specific class of problems and can help identify stumbling blocks where students may have difficulty. The first two tutorials were developed to help students learn about Coulomb's law, superposition principle and symmetry in the context of discrete and continuous charge distributions (conceptually), the third tutorial focused on distinguishing between electric flux and field, and the fourth and fifth tutorials dealt with symmetry and Gauss's law and on revisiting superposition principle after Gauss's law. Some tutorials on related topics have been developed by the University of Washington group. Those tutorials are complementary to the ones we have developed which focus on achieving competency with symmetry ideas in Coulomb's law, superposition, and Gauss's Law. We administered each pre-test, tutorial and post-test to 5 students individually who were asked to talk aloud while working on them. After each administration, we modified the tutorials based upon the feedback obtained from student interviews. These individual administrations helped fine-tune the tutorials and improve their organization and flow. Then, the tutorials were administered to four different calculus-based introductory physics classes with four lecture hours and one recita-

CP883, *2006 Physics Education Research Conference*, edited by L. McCullough, L. Hsu, and P. Heron

TABLE 1. Average percentage scores obtained on individual questions on the pre-/post-tests (matched unless only the post-test was given) for each of the five tutorials (I-V). The pre-tests were administered after traditional instruction but before the tutorial. As shown in the table, additional questions were included either in the pre-test or post-test and the pre-tests for tutorials II and V were not administered in some of the classes. The symblo n refers to the matched number of students in a given class for a given pre-/post-tests and *Total* refers to the total average percentage score including all questions on a pre-test or post-test administered to a given class for a particular tutorial. For tutorial II, the relative weights for the three pre-test questions for class 2 were 30%, 30% and 40% respectively. For tutorial IV, the relative weights for the pre-test and post-test questions for classes 1 and 2 were 10%, 10%, 20%, 20%, 20%, 20% and 20%, 10%, 20%, 10%, 20%, 20% respectively while the relative weights for the pre-test and post-test questions for class 3 were 20%, 20%, 30%, 30% and 30%, 20%, 30%, 20% respectively. For tutorial V, the relative weights for both the pre-test and post-test questions for class 2 were 30%, 40% and 30% and the weights for the post-test questions for the other two groups were 10%, 20%, 20%, 20%, 10%, 20% respectively. For all other cases, the same weight is assigned to each pre-test or post-test question.

Tutorial	Class	n	PRETEST							POSTTEST						
			1	2	3	4	5	6	Pre-Total	1	2	3	4	5	6	Post-Total
I	1	82	64	57	46	45	—	—	53	92	86	93	—	—	—	90
	2	60	52	58	38	47	—	—	48	96	96	95	—	—	—	96
	3	52	44	29	45	—	—	—	39	85	77	88	—	—	—	83
II	1	84	—	—	—	—	—	—	—	68	84	68	72	90	—	76
	2	63	56	6	41	—	—	—	35	90	96	87	87	98	—	92
	3	63	—	—	—	—	—	—	—	75	84	77	77	92	—	81
III	1	78	42	—	—	—	—	—	42	77	85	72	92	81	—	81
	2	55	44	—	—	—	—	—	44	74	88	82	95	—	—	85
	3	49	40	—	—	—	—	—	40	81	78	84	96	87	—	85
IV	1	65	28	22	58	58	41	17	40	83	91	95	91	91	93	91
	2	62	39	19	51	52	42	6	36	85	84	93	96	95	84	89
	3	49	45	6	38	28	—	—	30	87	90	88	81	—	—	87
V	1	85	—	—	—	—	—	—	—	71	61	75	70	96	64	71
	2	57	21	26	35	—	—	—	27	82	76	84	—	—	—	80
	3	64	—	—	—	—	—	—	—	92	81	89	89	95	90	88

tion hour per week. Students worked on each tutorial in groups of two or three either during the lecture section of the class or in the recitation depending upon what was most convenient for an instructor. Table 1 shows the pre-/post-test data on each question from three of the classes in which the tutorials were administered. The details of each question will be discussed elsewhere. In the fourth class, the post-tests were returned without photocopying them and we only have data on student performance on the cumulative test administered after all tutorials. As shown in Table 1, for some tutorials, additional questions were included in the pre-test and/or post-test after the previous administration and analysis of data. The pre-/post-tests were not identical but focused on the same topics covered in a tutorial.

All pre-tests and tutorials were administered after traditional instruction in relevant concepts. Instructors often preferred to alternate between lectures and tutorials during the class and give an additional tutorial during the recitation. This way all of the five tutorials from Coulomb's law to Gauss's law were administered within two weeks. For the tutorials administered in lecture section of the class, pre-tests were given to students right before they worked on the tutorials in groups. Since not all students completed a tutorial during the class, they were asked to complete them as part of their homework assignment. At the beginning of the next class, students were given an opportunity to ask for clarification on any issue related to the part of the tutorial they completed at home and then they were administered the corresponding post-test before the lecture began. Each pre-/post-test counted as a quiz and students were given a full quiz grade for taking each of the pre-test regardless of students' actual performance. The pre-tests were not returned but the post-tests were returned after grading. When a tutorial was administered in the recitation (tutorials II and V which were shorter), the teaching assistant (TA) was given specific instruction on how to conduct the group work effectively during the tutorial. Moreover, since the TA had to give the post-test corresponding to the tutorial during the same recitation class in which students worked on the tutorials (unlike the lecture administration in which the post-tests were in the following class), the pre-tests were skipped for some of these tutorials due to

TABLE 2. Percentage average pre-/post-test scores (matched pairs) for each of the five tutorials (I-V), divided into three groups according to the pre-test performance. N denotes the total number of students who worked through a tutorial and took both the pre-/post-tests, and n_i (i=1,2,3) denote the number of students in a particular class. For tutorials II and V, only one of the classes took both the pre-/post-tests. For students in the high pre-test range, sometimes there are very few students for a meaningful statistical interpretation.

N	Tutorial	Range (%)	n1 (class 1)	pre	post	n2 (class 2)	pre	post	n3 (class 3)	pre	post
194	I	All	82	53	90	60	48	96	52	39	83
		0-33	24	19	77	21	20	92	29	22	76
		34-67	33	55	92	18	43	97	21	58	92
		68-100	25	83	99	21	80	99	2	100	100
63	II	All				63	35	92			
		0-33				30	17	89			
		34-67				32	50	94			
		68-100				1	90	90			
182	III	All	78	42	81	55	44	85	49	40	85
		0-33	31	18	76	22	20	81	22	15	85
		34-67	38	52	84	26	55	87	19	53	85
		68-100	9	84	88	7	79	88	8	78	85
176	IV	All	65	40	91	62	36	89	49	30	87
		0-33	26	15	89	31	17	85	29	16	83
		34-67	32	50	91	27	51	94	17	47	92
		68-100	7	83	96	4	78	94	3	70	93
57	V	All				57	27	80			
		0-33				42	14	77			
		34-67				10	47	90			
		68-100				5	96	92			

a lack of time. Sometimes, the instructors gave the pre-tests in the lecture section of the class for a tutorial that was administered in the recitation.

In all of the classes in which the tutorials were used, 2-2.5 weeks were sufficient to cover all topics from Coulomb's law to Gauss's law. This time line is not significantly different from what the instructors in other courses allocated to this material. The main difference between the tutorial and the non-tutorial courses is that fewer solved examples were presented in the tutorial classes (students worked on many problems themselves in the tutorials). We note that since many of the tutorials were administered during the lecture section of the class, sometimes two instructors (e.g., the instructor and the TA) were present during these "large" tutorial sessions to ensure smooth facilitation. In such cases, students working in groups of three were asked to raise their hands for questions and clarifications. Once the instructor knew that a group of students was making good progress, that group was invited to help other groups in the vicinity which had similar questions. Thus, students not only worked in small groups discussing issues with each other, some of them also got an opportunity to help those in the other groups.

DISCUSSION

Out of the five tutorials, the first two focused on Coulomb's law, superposition and symmetry. The first tutorial started with the electric field due to a single point charge in the surrounding region and then extended this discussion to two or more point charges. The second tutorial further continued the conceptual discussion that started in the first tutorial (which was mainly about discrete charges) to continuous charge distributions. The tutorials guided students to understand the vector nature of the electric field, learn the superposition principle and recognize the symmetry of the charge distribution. Students worked on examples in which the symmetry of the charge distribution (and hence the electric field) was the same but the charges were embedded on objects of different shapes (e.g., four equidistant charges on a plastic ring vs. a plastic square). Common misconceptions were explicitly elicited often by having two students discuss an issue in a particular context. Students were asked to identify the student with whom they agreed.

The third tutorial was designed to help students learn to distinguish between the electric field and flux. The tutorial tried to help students learn that the electric field

TABLE 3. The average percentage of correct responses to the "Superposition, Symmetry and Gauss's Law Test" for different groups of students. N refers to the total number of students for a given group. In all undergraduate courses, the test was administered after instruction on these concepts in that course except in the upper-level $E\&M$ course in which it was given both as a pre-test and post-test (since students had instruction in these concepts at the introductory level). The "without tutorial" group and the "Honors students" group are from the same student population (mainly physical science and engineering freshmen but some sophomores) as the tutorial group from the same institution. The second row of the table gives the p value for t-tests (which performed a pair-wise comparison of the performance of the tutorial group with each of the other groups before rounding off the numbers).

	Without tutorial but otherwise same type of courses (2 classes)	Honors Students (2 classes)	Upper-level Undergrads		With Tutorial (4 classes)	First Year Grads (2 classes)
			Pre-test	Post-test		
	N=135	N=182	N=33	N=28	N=278	N=33
Average	38	42	44	49	59	75
p value	1.34E-04	7.85E-04	4.33E-03	5.29E-02		1.37E-03

is a vector while the electric flux is a scalar. Also, electric field is defined at various *points* in space surrounding a charge distribution while the electric flux is always through an *area*. Students learn about Gauss's law and how to relate the flux through a closed surface to the net charge enclosed. Rather than emphasizing the symmetry considerations, this tutorial focused on helping students use Gauss's law to find the net flux through a closed surface given the net charge enclosed and vice versa.

The fourth tutorial was designed to help students learn to exploit Gauss's law to calculate the electric field at a point due to a given charge distribution if a high symmetry exists. Students were helped to draw upon the superposition and symmetry ideas they learned in the first two tutorials to evaluate whether sufficient symmetry exists to exploit Gauss's law to calculate the electric field. Then, students learn to choose the appropriate Gaussian surfaces that would aid in using Gauss's law to find the electric field. Finally, they use Gauss's law to calculate the electric field in these cases. The last tutorial revisits the superposition principle after students have learned to exploit Gauss's law to calculate the electric field. For example, students learn to find the electric field at a point due to two non-concentric uniform spheres of charge or due to a point charge and an infinitely long uniform cylinder of charge.

The pre-tests and post-tests were graded by two individuals, and the inter-rater reliability was good. The average pre-/post-test scores on matched pairs for a particular class graded by them differ only by a few percent. Table 1 shows the student performance (on each question and also overall) on the pre-test and post-test in each of the five tutorials (I-V) in percentage for each class. The classes utilizing each tutorial may differ either because additional pre-/post-test questions were added or the pre-tests for tutorial II and V were not administered to some of the classes. The differences in the performance of different classes may also be due to the differences in student samples, instructor/TA differences or the manner in which the tutorials were administered. Table 2 shows the performance of students on the pre-/post-tests for each tutorial partitioned into three separate groups based upon the pre-test performance (see the Range column). As can be seen from Table 2, tutorials generally helped all students including those who performed poorly on the pre-test. The post-test scores are unusually high and can be attributed to a variety of reasons including Hawthorne effect. Another possibility is that the tutorials were "teaching to the test" and immediately after working on a tutorial, the concepts were fresh in students' minds. We therefore administered a cumulative test which includes concepts from all of the tutorials [3]. Table 3 shows the average percentage scores from the cumulative test administered to different student populations. Although the performance of the tutorial group is not as impressive on the cumulative test as on the pre-/post-tests administered with the tutorials, Table 3 shows that students who worked through the tutorials significantly outperformed both Honors students and those in upper-level undergraduate courses, but not first year physics graduate students.

CONCLUSION

We developed and evaluated tutorials to help calculus-based introductory students learn Coulomb's law, superposition, symmetry and Gauss's law. Pre-/post-tests for each tutorial and a test that includes content on all of the tutorials show that the tutorials can be effective in improving student understanding of these concepts.

ACKNOWLEDGMENTS

We are grateful to the NSF for award DUE-0442087.

REFERENCES

1. D. Maloney, T. O'Kuma, C. Hieggelke, A. V. Heuvelen, Am. J. Phys. **69**, S12 (2001).
2. S. Rainson, G. Transtromer, L. Viennot, Am. J. Phys. **62** (11), 1026 (1994).
3. C. Singh, Am. J. Phys., **74**(10), 923-236, (2006).

Student Difficulties with Quantum Mechanics Formalism

Chandralekha Singh

Department of Physics and Astronomy, University of Pittsburgh, Pittsburgh, PA, 15260, USA

Abstract. We discuss student difficulties in distinguishing between the physical space and Hilbert space and difficulties related to the Time-independent Schroedinger equation and measurements in quantum mechanics. These difficulties were identified by administering written surveys and by conducting individual interviews with students.

INTRODUCTION

Here, we describe the difficulties with the formalism of quantum mechanics identified by administering written surveys to eighty-nine advanced undergraduates and more than two hundred graduate students from seven different universities. In the written surveys, students were asked to explain their reasoning. We also conducted individual interviews using a think aloud protocol with a subset of students to better understand the rationale for their responses. During the semi-structured interviews, students were asked to verbalize their thought processes while they answered qualitative questions posed to them. Students were not interrupted unless they remained quiet for a while. In the end, we asked them for clarifications of the issues they had not made clear earlier. Below we discuss some of the difficulties with the formalism.

Difficulty Distinguishing between the Physical Space and Hilbert Space

In quantum theory, one must interpret the outcome of real experiments performed in three dimensional (3D) space by making connection with an abstract Hilbert space (state space) in which the wavefunction lies. The physical observables that one measures in the laboratory correspond to Hermitian operators in the Hilbert space whose eigenstates span the space. Knowing the initial wavefunction and the Hamiltonian of the system allows one to determine the time-evolution of the wavefunction unambiguously and the measurement postulate can be used to determine the possible outcomes of individual measurements and ensemble averages (expectation values).

It is difficult for many students to distinguish between vectors in the 3D laboratory space and Hilbert space. For example, S_x, S_y and S_z denote the orthogonal components of the spin angular momentum vector of an electron in the 3D space, each of which is a physical observable that can be measured in the laboratory. However, the Hilbert space corresponding to the spin degree of freedom for a spin-1/2 particle is two dimensional (2D). In this Hilbert space, $\hat{S}_x$, $\hat{S}_y$ and $\hat{S}_z$ are operators whose eigenstates span the 2D space. The eigenstates of $\hat{S}_x$ are vectors which span the 2D space and are orthogonal to each other (but not orthogonal to the eigenstate of $\hat{S}_y$ or $\hat{S}_z$). Also, $\hat{S}_x$, $\hat{S}_y$ and $\hat{S}_z$ are operators and *not* orthogonal components of a vector in the 2D space. If the electron is in a magnetic field with the gradient in the z direction in the laboratory (3D space) as in a Stern-Gerlach experiment, the magnetic field is a vector in 3D space but not in the 2D space. It does not make sense to compare vectors in the 3D space with vectors in the 2D space as in statements such as "the magnetic field gradient is perpendicular to the eigenstates of $\hat{S}_x$". Unfortunately, these distinctions are difficult for students to make and such difficulties were frequently observed in response to the survey questions and during individual interviews. This difficulty is discussed below in the context of a two part question related to the Stern-Gerlach experiment:

Question: Notation: $|\uparrow_z\rangle$ and $|\downarrow_z\rangle$ represent the orthonormal eigenstates of $\hat{S}_z$ (z component of the spin of the electron). SGA is an abbreviation for a Stern-Gerlach apparatus. The electron is in the SGA for an infinitesimal time. Ignore the Lorentz force on the electron.

(a) A beam of electrons propagating along the y direction (into the page) in spin state $(|\uparrow_z\rangle + |\downarrow_z\rangle)/\sqrt{2}$ is sent through an SGA with a vertical magnetic field gradient in the $-z$ direction. Sketch the electron cloud pattern that you expect to see on a distant phosphor screen in the x-z plane. Explain your reasoning.

(b) A beam of electrons propagating along the y direction (into the page) in spin state $|\uparrow_z\rangle$ is sent through an SGA with a horizontal magnetic field gradient in the $-x$ direction. Sketch the electron cloud pattern that you expect to see on a distant phosphor screen in the x-z plane. Explain your reasoning.

CP883, *2006 Physics Education Research Conference*, edited by L. McCullough, L. Hsu, and P. Heron

In question (a), students have to realize that the magnetic field gradient in the -z direction would exert a force on the electron due to its spin angular momentum and one should observe two spots on the phosphor screen due to the splitting of the beam along the z direction corresponding to electron spin component in $|\uparrow_z\rangle$ and $|\downarrow_z\rangle$ states. All responses in which students noted that there will be a splitting along the z direction were considered correct even if they did not explain their reasoning. Only 41% of the students provided the correct response. Many students thought that there will only be a single spot on the phosphor screen as in these typical survey responses:

- *SGA will pick up the electrons with spin down since the gradient is in -z direction. The screen will show electron cloud only in $-z$ part.*
- *All of the electrons that come out of the SGA will be spin down with expectation value $-\hbar/2$ because the field gradient is in $-z$ direction.*
- *Magnetic field is going to align the spin in that direction so most of the electrons will align along -z direction. We may still have a few in the +z direction but the probability will be very small.*

In the interviews, students were often confused about the origin of the force on the particles and whether there should be a force on the particles at all as they pass through the SGA.

Question (b) is more challenging than (a) because students have to realize that the eigenstate of $\hat{S}_z$, $|\uparrow_z\rangle$ can be written as a linear superposition of the eigenstates of $\hat{S}_x$, i.e., $|\uparrow_z\rangle = (|\uparrow_x\rangle + |\downarrow_x\rangle)/\sqrt{2}$. Therefore, the magnetic field gradient in the $-x$ direction will split the beam along the x direction corresponding to the electron spin components in $|\uparrow_x\rangle$ and $|\downarrow_x\rangle$ states and cause two spots on the phosphor screen. Only 23% of the students provided the correct response. The most common difficulty was assuming that since the spin state is $|\uparrow_z\rangle$, there should not be any splitting as in the examples below:

- *Magnetic field gradient cannot affect the electron because it is perpendicular to the wavefunction.*
- *Electrons are undeflected or rather the beam is not split since $\vec{B}$ is perpendicular to spin state.*
- *The direction of the spin state of the beam of electrons is y, and the magnetic field gradient is in the $-x$ direction. The two directions have an angle 90^0, so the magnetic field gradient gives no force to electrons.*
- *With the electrons in only one measurable state, they will experience a force only in one direction upon interaction with $\vec{B}$.*

Thus, many students explained their reasoning by claiming that since the magnetic field gradient is in the $-x$ direction but the spin state is along the z direction, they are orthogonal to each other, and therefore, there cannot be any splitting of the beam. Student responses suggest that they were incorrectly connecting the direction of magnetic field in the 3D space with the "direction" of state vectors in the Hilbert space. Several students in (b) drew a monotonically increasing function. One interviewed student drew a diagram of a molecular orbital with four lobes and said "this question asks about the electron cloud pattern due to spin...I am wondering what the spin part of the wavefunction looks like." Then he added, "I am totally blanking on what the plot of $|\uparrow_z\rangle$ looks like otherwise I would have done better on this question". From responses such as these it appears that the abstract nature of spin poses special problems in teaching quantum physics.

Compared to question (a), many more students in (b) thought that there will be only one spot on the screen, but there was no consensus on the direction of deflection despite the fact that students were asked to ignore the Lorentz force. Some students drew the spot at the origin, some showed deflections along the positive or negative x direction, some along the positive or negative z direction. They often provided interesting reasons for their choices. Some students who drew two spots were confused about the direction in which the magnetic field gradient will cause the splitting of the beam. Thirteen percent of the students (including questions (a) and (b)) drew the splitting of the beam in the wrong direction (along the x axis in (a) and along the z axis in (b)). One interviewed student who drew it in the wrong direction said, "I remember doing this recently and I know there is some splitting but I don't remember in which direction it will be."

Students were posed another question involving $\vec{S}\cdot\hat{n}$ where $\hat{n}$ is a general unit vector pointing in an arbitrary direction in the physical three dimensional space. This dot product is a scalar product between two vectors in the physical space. Students were given that for spin 1/2, a state χ goes to χ' via $\chi' = e^{i(\vec{S}\cdot\hat{n})\phi/\hbar}\chi$ where $e^{i(\vec{S}\cdot\hat{n})\phi/\hbar}$ effects a rotation through angle ϕ about the axis $\hat{n}$. They were asked to construct a 2×2 matrix representing rotation by $\phi = \pi$ about the x axis and show that it converts χ_+ to χ_- ($\chi_\pm$ are the eigenstates of $\hat{s}_x$ with eigenvalues $\pm\hbar/2$). Written responses and interviews suggest that one major difficulty was that many students were confused about whether $\vec{S}\cdot\hat{n}$ (which can be written as $s_x n_x + s_y n_y + s_z n_z$) is a dot product in the physical space or Hilbert space. Students were not clear about the fact that in a Hilbert space, the possible states of the system are the vectors and the inner products of these states are the scalar products. Similar confusion between physical space and Hilbert space were observed in the context of questions posed in surveys and interviews about a one-dimensional infinite square well. The physical space for this problem is one-dimensional (e.g., in a quantum wire) but the Hilbert space is infinite dimensional.

Difficulties Related to Time-independent Schroedinger Equation and Measurement

One of the questions on the survey asks students to consider the following statement: "By definition, the Hamiltonian acting on any allowed state of the system ψ will give the same state back, i.e., $H\psi = E\psi$, where E is the energy of the system." Students were asked to explain why they agree or disagree with this statement. We wanted students to disagree with the statement and note that it is only true if ψ is a stationary state. In general, $\psi = \sum_{n=1}^{\infty} C_n\phi_n$, where ϕ_n are the stationary states and $C_n = \langle\phi_n|\psi\rangle$. Then, $\hat{H}\psi = \sum_{n=1}^{\infty} C_n E_n \phi_n \neq E\psi$. For this question, just writing down "disagree" was not enough for the response to be counted correct. Students had to provide the correct reasoning. Only 29% of the students provided the correct response with correct reasoning. Thirty-nine percent of students wrote (incorrectly) that the statement is unconditionally correct. Typically, these students were reasonably confident of their responses as can be seen from these examples:

- *Agree, this is a fundamental postulate of quantum mechanics which is proved to be highly exact until present.*
- *Agree. Hamiltonian does not alter the state of the system.*
- *Agree. Hamiltonians give back physical observables energy. It is an observable and real.*

In response to this question, 10% of students agreed with the statement as long as the Hamiltonian is not time-dependent. They often claimed (incorrectly) that if $\hat{H}$ is not time-dependent, the energy for the system is conserved so $\hat{H}\psi = E\psi$ must be true. The following are typical examples:

- *Agree, if the potential energy does not depend on time.*
- *Agree but only if the energy is conserved for this system.*
- *Agree because energy is a constant of motion.*
- *Agree if it is a closed system since H is a linear operator and gives the same state back multiplied by the energy.*

While the energy is conserved if the Hamiltonian is time-independent, $\hat{H}\psi = E\psi$ need not be true. For example, if the system is in a linear superposition of stationary states, $\hat{H}\psi \neq E\psi$ although the energy is conserved.

Eleven percent (11%) of the students answering this question believed (incorrectly) that any statement involving a Hamiltonian operator acting on a state is a statement about the measurement of energy. Some of these students who (incorrectly) claimed that $\hat{H}\psi = E\psi$ is a statement about energy measurement agreed with the statement while others disagreed. Those who disagreed often claimed that $\hat{H}\psi = E_n\phi_n$ because as soon as $\hat{H}$ acts on ψ, the wavefunction will collapse into one of the stationary states ϕ_n and the corresponding energy E_n will be obtained. The examples below are typical of students with this misconception:

- *Agree. If you make a measurement of energy by applying H to a state of an electron in hydrogen atom you will get the energy.*
- *Disagree. Hamiltonian acting on a state (measurement of energy) will return an energy eigenstate.*
- *Disagree. Quantum measurements will perturb the system so that it jumps into an eigenstate after measurement.*
- *Disagree. If it is a mixed state, the measurement of energy will force it to end up with some base state.*
- *Disagree. When Ψ is a superposition state and $\hat{H}$ acts on Ψ, then Ψ evolutes to one of the Ψ_n so we have $\hat{H}\Psi = E_n\Psi_n$.*

Interviews and written reasonings suggest that these students believed that the measurement of a physical observable in a particular state is achieved by acting with the corresponding operator on the state. The incorrect notions expressed above are over-generalizations of the fact that *after* the measurement of energy, the system is in a stationary state so $\hat{H}\phi_n = E_n\phi_n$.

Individual interviews related to this question suggest that some students believed that whenever an operator $\hat{Q}$ corresponding to a physical observable Q acts on *any* state ψ, it will yield the corresponding eigenvalue λ and the same state back, i.e., $\hat{Q}\psi = \lambda\psi$. Some of these students were over-generalizing their "$\hat{H}\psi = E\psi$" reasoning and attributing $\hat{Q}\psi = \lambda\psi$ to the measurement of an observable Q. Before over-generalizing to any physical observables, these students often agreed with the $\hat{H}\psi = E\psi$ statement with arguments such as "the Hamiltonian is the quantum mechanical operator which corresponds to the physical observable energy" or "if H did not give back the same state it would not be a hermitian operator and therefore would not correspond to an observable". Of course, $\hat{Q}\psi \neq \lambda\psi$ unless ψ is an eigenstate of $\hat{Q}$ and in general $\psi = \sum_{n=1}^{\infty} D_n\psi_n$, where ψ_n are the eigenstates of $\hat{Q}$ and $D_n = \langle\psi_n|\psi\rangle$. Then, $\hat{Q}\psi = \sum_{n=1}^{\infty} D_n\lambda_n\psi_n$ (for observable with a discrete eigenvalue spectrum).

Many students believed that even when answering questions related to the probability of different possible outcomes for the measurement of an observable other than energy, the wavefunction should be expanded in terms of the energy eigenfunctions and the absolute square of the expansion coefficients will give the probability of measuring different values of that observable. In contrast, the wave function should be expanded in terms of the eigenfunctions of the operator corresponding to the physical observable to be measured and the absolute square of the expansion coefficients then will give the probability of measuring different possible values of that observable. Student's belief that the energy eigenfunctions are always the "preferred" basis vectors is not surprising because quantum mechanics courses often exclusively focus on solving time-independent Schroedinger equation to find the energy eigenfunctions and eigenvalues. Also, for questions related to the time-development of the wavefunction one must expand the wavefunction in terms of the energy eigenfunctions.

Moreover, students often believed that successive measurements of continuous variables, *e.g.*, position, produce "somewhat" deterministic outcomes whereas successive measurements of discrete variables, *e.g.*, spin, can produce very different outcomes. In an interview, one student began with a correct statement: "*if you measure (an observable) Q, the system will collapse into an eigenstate of that operator. Then, if you wait for a while the measurement will be different*". But then he added incorrectly: "*if Q has a continuous spectrum then the system would gently evolve and the next measurement won't be very different from the first one. But if the spectrum of eigenvalues is discrete then you will get very different answers even if you did the next measurement after a very short time*". When the student was asked to elaborate, he said: "*For example, imagine measuring the position of an electron. It is a continuous function so the time dependence is gentle and after a few seconds you can only go from A to its neighboring point. [Pointing to an x vs. t graph that he sketches on the paper]...you cannot go from this place to this without going through this intermediate space*". When asked to elaborate on the discrete spectrum case, he said: "*...think of discrete variables like spin...they can give you very different values in a short time because the system must flip from up to down. I find it a little strange that such [large] changes can happen almost instantaneously. But that's what quantum mechanics predicts...*" This type of response may also be due to the difficulty in reconciling classical and quantum mechanical ideas; in classical mechanics the position of a particle is deterministic and can be unambiguously predicted for all times from the knowledge of the initial conditions and potential energy.

Confusion between the Probability of Measuring Position and $\langle x\rangle$

Born's probabilistic interpretation of the wavefunction can also be confusing for students. In one question, students were told that immediately after the measurement of energy which yields the first excited state, a measurement of the electron position is performed. They were asked to describe qualitatively the possible values of position they could measure and the probability of measuring them. We hoped that students would note that one can measure position values between $x = 0$ and $x = a$ (except at $x = 0, a/2, a$ where the wavefunction is zero), and according to Born's interpretation, $|\phi_2(x)|^2 dx$ gives the probability of finding the particle in a narrow range between x and $x + dx$. Only 38% of the students provided the correct response. Partial responses were considered correct for tallying purposes if students wrote anything that was correct related to the above wavefunction, *e.g.*, "The probability of finding the electron is highest at $a/4$ and $3a/4$.", "The probability of finding the electron is non-zero only in the well", etc.

Eleven percent of the students tried to find the expectation value of position $\langle x\rangle$ instead of the probability of finding the electron at a given position. They wrote the expectation value of position in terms of an integral involving the wavefunction. Many of them explicitly wrote that $Probability = (2/a)\int_0^a x\ sin^2(2\pi x/a)dx$ and believed that instead of $\langle x\rangle$ they were calculating the probability of measuring the position of electron. During the interview, one student said (and wrote on paper) that the probability is $\int x\,|\Psi|^2 dx$. When the interviewer asked why $|\Psi|^2$ should be multiplied with x and if there is any significance of $|\Psi|^2 dx$ alone without multiplying it by x, the student said, "$|\Psi|^2$ gives the probability of the wavefunction being at a given position and if you multiply it by x you get the probability of *measuring* (student's emphasis) the position x". When the student was asked questions about the meaning of the "wavefunction being at a given position", and the purpose of the integral and its limits, the student was unsure. He said that the reason he wrote the integral is because $x\,|\Psi|^2 dx$ without an integral looked strange to him. Similar confusion about probability in classical physics situations have been found [1].

CONCLUSION

Instructional strategies that focus on improving student understanding of these concepts should take into account these difficulties [2, 3, 4]. We are currently developing and assessing Quantum Interactive Learning Tutorials (QuILT) suitable for use in advanced undergraduate quantum mechanics courses [5].

ACKNOWLEDGMENTS

We are grateful to the NSF for award PHY-0244708.

REFERENCES

1. M. Wittmann, J. Morgan, R. Feeley, Phys. Rev. ST. PER **2**, 020104-1-8, (2006).
2. D. Zollman, N. Rebello, and K. Hogg Am. J. Phys., **70(3)**, 252-259, (2002).
3. Wittmann, Steinberg, Redish, Am. J. Phys. **70**, 218, (2002).
4. C. Singh, Am. J. Phys., **69(8)**, 885-895, (2001).
5. C. Singh, M. Belloni, W. Christian, Physics Today, **8**, 43-49, August (2006).

Use of Physical Models to Facilitate Transfer of Physics Learning to Understand Positron Emission Tomography*

Bijaya Aryal, Dean Zollman and N. Sanjay Rebello

Department of Physics, Kansas State University, Manhattan, KS, 66506-2601

Abstract. In this paper we describe a qualitative study of the role of the physical models in transferring physics ideas to understanding positron emission tomography technology. Sixteen students enrolled in an introductory level physics class individually participated in two sessions of a teaching experiment. In this study we noted that many students used reasoning from prior experiences in inappropriate ways. A result from this study is that physical models are effective in triggering appropriate transfer provided that the activities using the models are introduced in the right sequence. Given the appropriate sequencing of the activities, we find that the transfer of abstract ideas is facilitated through interactive learning with the aid of physical models. Three different types of non-scaffolded transfer have been identified: spontaneous, semi-spontaneous and non-spontaneous transfer.

Keywords: Physical models, positron emission tomography, transfer of learning
PACS: 01.40Fk

INTRODUCTION

This study investigates the transfer of learning from ideas in physics to positron emission tomography (PET), a medical imaging device. Students cover some of the physics concepts relevant to PET such as momentum conservation and distance in introductory level physics. Thus, we investigate how students use those physics ideas in the context of PET. Rather than beginning with PET we first examine students' models of some of the physics ideas relevant to PET. Students learn from simple classical models and later apply those ideas to the electron-positron annihilation process in order to understand the image construction process in PET. For this purpose we conduct teaching interviews [1] to investigate not only students' models but also the ways in which we can facilitate student learning of the correct model.

Our aim in this study was to answer the following research questions.

1. What cognitive resources do introductory college students bring to bear when interacting with physical models?
2. How does sequencing of different physical models affect activation of these resources?
3. How do students transfer what they have learned in physics from physical models to their understanding of PET?

LITERATURE REVIEW

From the encoding specificity [2] perspective in cognitive psychology, transfer of learning is deemed easy if the problem structures are superficially similar in learning stage and transfer stage. Contrary to this view, some researchers [3] have reported that increasing concreteness of the problem does not promote transfer of abstract ideas.

We have developed activities to facilitate transfer of abstract ideas from one context to another. The activities in the learning and transfer contexts are therefore intended to be different at the superficial level but similar at an abstract level. For the learning session we designed structurally simple hands-on activities with underlying abstract physics concepts. On the other hand for the transfer session we introduced the problems that were similar to problems in the learning session at an abstract level but in different contexts and modes of presentation.

In this paper we mainly discuss the influence of students' prior ideas while learning PET's relevant physics principles using our interactive activities. Another part of discussion of this paper will be the transfer of learning from the models of the activities to the PET image construction process.

*Supported by the National Science Foundation under grant # 04-26745.

CP883, *2006 Physics Education Research Conference*, edited by L. McCullough, L. Hsu, and P. Heron

METHODOLOGY

Sixteen students, eight females and eight males, participated in this study. The students were enrolled in an introductory algebra-based physics course. The study was done during the second half of spring 2006. We choose that time of the semester because students would be familiar with some basic kinematics terms by then.

The teaching interview sequence consisted of a fixed protocol with scaffolding activities introduced dependent upon students' responses. Each student participated in two different sessions, both one-hour in length, that were about a week apart. The sessions were conducted by one of the authors (BA) and some were observed by the other authors.

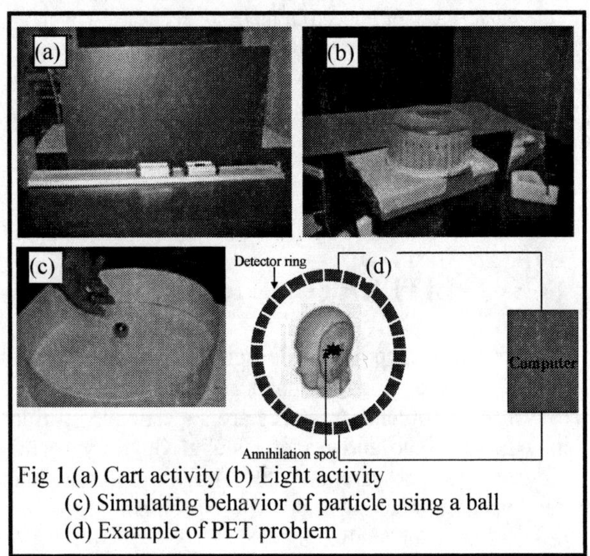

Fig 1.(a) Cart activity (b) Light activity
(c) Simulating behavior of particle using a ball
(d) Example of PET problem

We introduced two activities in the first session. In both of the activities students were asked to figure out the locations of events that were not visible directly, but whose effects were visible. In the 'cart activity' in Fig. 1(a) two magnetic carts were brought together on a hidden track and then released. Students could see the carts only at the ends of the track. Given the velocity of the carts students were asked to determine the location on the track from where the carts were released. The 'light activity' in Fig. 1(b) simulated an explosion inside a cylindrical enclosure. The result was two pulses of light which were visible on the surface of the cylinder. Students could see two light pulses on the wall of barrier, but not the simulated explosion. To facilitate students' understanding of the particles that caused the simulated explosion in the 'light activity' a model of balls on a whiteboard was used (Fig 1 (c)). In this model, students were asked to predict the directions of two smaller balls resulting from the explosion of a bigger ball. Through these activities, students had to figure out the locations of such events and pinpoint the common region of such events.

The second session began with a discussion of general ideas about positron emission tomography (PET). We then moved on to problems related to the PET technique (See, Fig. 1(d)). Students were asked to build a model of how the location of the exact position of electron-positron annihilation in the brain could be ascertained. Finally they were asked to complete activities which enabled them to find the region of a tumor in which the annihilations were occurring.

The teaching activities of both sessions were video recorded and later transcribed. The field notes of the teaching interview during and after the activities and students' worksheets served as valuable data sources.

RESULTS AND DISCUSSION

A phenomenographic [4] analysis was conducted. We examined the activity transcripts, field notes and student worksheets to find recurrent categories. Different categories were labeled and their criteria were defined. An inter-rater reliability ranging from 72% to 84% was established among five researchers.

Reasoning Resources

The following themes emerged from the analysis:

Central Tendency:

Most of the students (87%) located the events in both the cart and light activities at the center of a circle or line. For instance, in the light activity, students explicitly mentioned that if two lights appeared at two points on the circumference the source producing these lights must be at the center of that circle. When asked to explain their reasoning most students appeared to have arrived at this conclusion based on their intuition.

During the teaching interview we challenged this idea. Figure 2 shows a representative sketch of how students' ideas progressed with scaffolding.

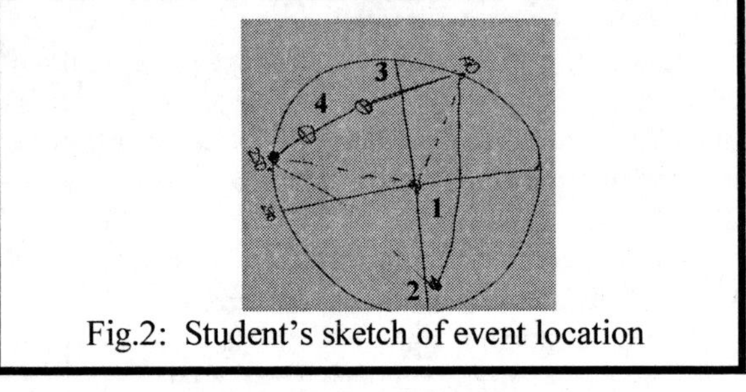

Fig.2: Student's sketch of event location

The stages are labeled as 1, 2, 3 and 4 in the sketch. Before receiving any hints the student thought that the event should be at the center (1) and later decided that it could be anywhere inside the circle (2).

After the student was told that two fragments from an explosion produce light, he said that the event must be at the center of a line joining two lights (3). Finally he realized that it could be anywhere along that line (4). It is reasonable to deduce that a symmetry argument might have led them to have the central tendency.

Factors to Predict Event Location Along a Line:

About 40% of the students who participated in this study relied on intensity of light and another 20% relied on size of light to locate the events. The appropriate factor to be considered was time, which was considered by 40% of the students. It is interesting to note that most of the students who considered 'time' were engaged in the cart activity before the light activity as discussed further in the next section.

Predictions of Number and Direction of Gamma Rays:

Most of the students used classical analogies and everyday experiences to explain the number and direction of gamma rays produced by annihilation (i.e. collision of automobiles and balls rolling on a table). A huge majority (87%) of students stated that only an even number of gamma rays could cancel momentum to conserve it. The following statement by a student is a typical example of this line.

There have to be... almost... there have to be an even number... that way... so that for each one produced it may... it will have opposite one it will cancel out so that it doesn't have any movement anywhere...

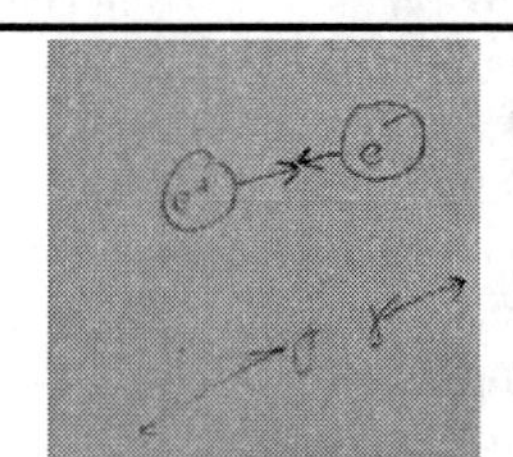

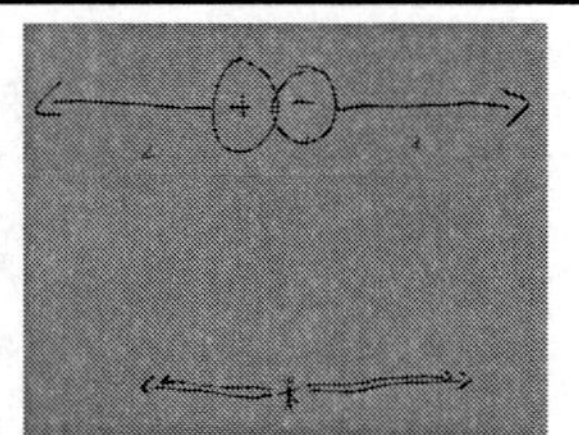

Fig 3(a) and (b) Student sketches of gamma ray direction as a result of annihilation

Figures 3(a) and 3(b) illustrate an idea held by a majority of the interviewees (73%) regarding the direction of gamma rays. Many of them explicitly mentioned that a conservation law should be applied in direction also. They said that the conservation of the line should hold so that the gamma rays should travel back along the same line that the electron and positron traveled before they met to annihilate. Upon probing we found that this was a result of the influence of the 1-dimensional activities that they had seen in lecture demos and other prior activities.

Sequencing Activities

The two activities used in first session were introduced in two different sequences. For the first group (LC) of 11 students the light activity was introduced first followed by the cart activity. For the second group (CL) of five students the cart activity was introduced before the light activity. The second group had fewer participants because students were unavailable.

In response to a question on how to locate the event producing light in the light activity seven out of 11 from the LC group mentioned that light intensity should be the determining factor whereas two students mentioned the apparent size of the light. The remaining two students considered that the time for the light to reach the cylinder would be the determining factor. In the second (CL) group on the other hand, all five students considered 'time' to be the relevant factor to locate the event that produces light.

We speculate that most of the students in the second group (CL) activated their resource of 'time' as being associated with the 'location' of the event. This activation was apparently based on the cart activity. Shortly afterward they were introduced to the light activity. Apparently the associations they made between 'time' and 'location' with the cart activity were so strong that the light activity which followed was unable to suppress the activation of the 'time' resource and displace it with the 'intensity' or 'size' resource. But, for first group of students they didn't have an opportunity to make an association of 'time' with 'location' with the aid of cart activity before doing the light activity. So, they activated their resources from their everyday life experiences where they usually locate lights based on their intensity or size. This group of student appeared to have activated the p-prim 'closer is brighter' and 'closer is bigger' [5] to conclude that the location should be determined by the intensity or size of the lights. This p-prim-based association was apparently so strong that it could not be displaced by the association between 'time' and 'location' when the cart activity was introduced. In other words, the sequence of the activities CL vs. LC made a difference in the activation of knowledge resources by these students as they explained this activity.

Transfer of Learning

The main objective of this study was to see if and how students could transfer their learning from physical models to understand PET. Thus, in the first session we involved students with classical hands-on activities using physical models. The purpose of the second session was to help them learn the physics of PET. The problem structures in both sessions were the same in level of abstraction. We observed that students' responses indicated a transfer of ideas from the first to the second session. We classified the responses into four categories.

Response indicates Spontaneous Transfer (ST)
If students immediately related PET with the activities of the first session we labeled this as spontaneous transfer (ST). For example

Interviewer: How will the PET machine be able to determine the exact location here?
Student: By the process that we went through last time... knowing the difference in time...knowing which gamma ray reach the sensor first...so if the gamma ray reaches this sensor first and the computer can figure out which point it is in between the two sensors...

Response indicates Semi-Spontaneous Transfer (SST)
If students related PET back to the activities of the first session upon being asked the reason for their answer we called this semi spontaneous transfer (SST). For example:

Interviewer: What caused you to answer in that way?
Student: It is like the cars last week...where the event took place ...since you can't time whenever this event took place (refers to the annihilation)...then you could say whichever detector goes first and time it when it goes off and then the time to reach ...

Response indicates Non-Spontaneous Transfer (NST)
The third category is non-spontaneous transfer (NST). We categorized these students if they related PET with the first session only after being asked if they had seen a situation similar to PET somewhere before. E.g:

Interviewer: Any prior learning prompted at this point?
Student: Last week when we did the cake exercise (light activity) trying to figure out the source of light that kinda helped too...

Response indicates No Transfer (NT)
There were some instances in which very few of the students did not transfer at all from the first session to the second session. Here is an example.

Interviewer: (after introducing the picture of coincidence detection) How does the machine get the exact location of annihilation?
Student: I can tell it ...can tell about here (detector).... can't tell how far from here ...I can't tell how to get the exact location... because I never saw this machine and don't know how it works...

Almost one half of the participants (47%) in this study were found to be in the SST category. The second largest population (27%) exhibited ST transfer. Only 13% of students demonstrated NST transfer and another 13% of students' responses were in the NT category. Unlike many of the earlier studies [6] we found significant transfer from learning context (first session) to transfer context (second session). It could, of course, be argued that the students already had some ideas about the transfer context due to hints that they were in a two-part study. But it is encouraging that they made an association with the activities of the first session with those of the second session without any hints in spite of the different problem structure in the two sessions. The students were not told about the activities in the transfer session by those in the learning session but they themselves built or changed their ideas while interacting with the physical models. This result suggests that the exercise helped them construct ideas by active learning and eventually led them to apply in the transfer task.

CONCLUSIONS

The results of this study show that the introductory physics students who participated in this study rely on everyday experiences even when dealing with complex physics problems. They also appear to transfer their learning from familiar physics experiments to new situations. Despite the usefulness of such prior ideas and reasoning, in some cases these transfer processes may have an adverse effect on learning.

The analysis of the activities of two groups of students showed the importance of sequencing different activities. Based upon these results, we can suggest that the sequence of the activities has an important role in activating different conceptual resources. This result has important implications in designing teaching materials. Depending upon the ideas we want students to apply in a new situation, we can decide where and when an activity should be introduced to facilitate spontaneous transfer for a majority of students.

REFERENCES

1. P. V. Engelhardt, E. G. Corpuz, D. J. Ozimek et al., presented at the Physics Education Research Conference, 2003, Madison, WI, 2003 (unpublished).
2. R. H. Bruning, *Cognitive Psychology and Instruction.* (Pearson, Upper Saddle River, NJ, 2004).
3. R.L. Goldstone and Y. Sakamoto, Cognitive Psychology **46** (414-446) (2003).
4. F. Marton, Journal of Thought **21**, 29 (1986).
5. A. DiSessa, in *Constructivism in the computer age*, edited by G. Forman and P.B. Pufall (Lawrence Erlbaum Associates, Hillsdale, NJ, 1988), pp. 49.
6. M. Wertheimer, *Productive Thinking.* (Harper & Row, New York, NY, 1959).

Investigating Students' Ideas About X-rays While Developing Teaching Materials for a Medical Physics Course*

Spartak Kalita and Dean Zollman

Department of Physics, Kansas State University, Manhattan, KS, 66506-2601

Abstract. The goal of the Modern Miracle Medical Machines project is to promote pre-med students' interest in physics by using the context of contemporary medical imaging. The X-ray medical imaging learning module will be a central part of this effort. To investigate students' transfer of learning in this context we have conducted a series of clinical and teaching interviews. In the latter interview, some of the proposed learning materials were used. The students brought to our discussion pieces of knowledge transferred from very different sources such as their own X-ray experiences, previous learning and the mass media. This transfer seems to result in more or less firm mental models which often are not always internally consistent or coherent.

Keywords: physics education, medical physics, transfer of learning, mental models, X-rays

INTRODUCTION

Pre-med students often complain that physics classes lack relevance to their future profession and the traditional required physics curriculum rarely makes any effort to relate the content to this vast (and mainly diligent) student population. Some of these issues have been addressed in Medical Physics courses designed and implemented around the country in recent years, [1,2] but these are optional courses that are designed to follow compulsory physics classes and do not replace them. Thus, a more systematic effort here is urgent.

For this purpose the Modern Miracle Medical Machines (MMMM) project has been undertaken. Its main goal is to conduct research on the reasoning and models that students use as they transfer basic physics knowledge in the application of physics to contemporary medicine. [3]

We utilize the general framework for dynamic transfer of learning that was developed by the KSU Physics Education Research Group [4]. We seek evidence of transfer from physics and other science classes, students' personal experiences and any other sources that students may find relevant.

The X-ray module will be a central one in our MMMM instruction. However, we have not found in the literature any explorations of students' understanding of X-rays. This lack of effort can be easily explained – the concept of X-rays is not so universal and omnipresent in our life (as, for instance, light that is somewhat connected to it) and probably would not require our special attention if it were not for the purpose of making the algebra-based physics course more appealing and relevant to pre-med students.

For other proposed modules in the MMMM project, such as Positron Emission Tomography or Magnetic Resonance Imaging, the investigation of students' pre-conceptions probably would not be useful as these topics are not covered in regular courses even in passing and students know them in the best case only by their names.

For the purpose of the development of the X-rays module our situation is fortunately a bit better. Almost all of the students either have undergone some X-ray procedure in their lives or know of someone who has. They already have some preconceived ideas about how it might work or may be strongly inclined to build models right on the spot when asked to do it even if they haven't thought about it before (as was confirmed by our research).

* Supported by NSF grant # 04-26745

CP883, *2006 Physics Education Research Conference*, edited by L. McCullough, L. Hsu, and P. Heron

METHODOLOGY

Phase 1 – Fall 2004

At the very beginning, a series of preliminary unstructured interviews (with some semi-structured elements modeled after Piaget [6]) was conducted. The protocol format allowed follow-up questions and thus questions that come later in the prepared list were able to be modified or even omitted if a student had already answered them in one of the follow-up series of questions.

In this preliminary study we had not yet narrowed our attention on the pre-med student population and we were looking at students' ideas about X-rays in general. Thus, we decided to interview students with various backgrounds and very different levels of preparation. Among 16 of these students eight were from a conceptual physics class - 4 females studying elementary education and 4 males who were non-science majors. 8 were from a calculus-based physics class - all male - with engineering majors - electrical, mechanical or civil engineering.

All of the students were in either their sophomore or junior years, and conceptual physics or calculus-based physics were the only physics courses that they had all taken in college. All but one of them had taken physics classes in high school. Half of the students were motivated by extra credit and half were attracted by a small cash payment. Each interview lasted for about 30-40 minutes.

In the beginning of each interview students were presented with four X-ray pictures - three medical ones - the hand, the skull and the breast and one non-medical - an image of a bag screened by an airport security camera. They were also shown three or four other medical images that resulted from ultrasound, MRI and CAT scans and were asked what they could tell about them. Then our discussion went through various physics concepts that students eventually brought up in the conversation - light, waves, particles, spectrum, etc. The students were asked to compare X-rays to ultrasound and other imaging techniques, prompted to recall details from their personal experiences with X-rays and encouraged to use any information from other sources that they found relevant.

Phase 2 – Fall 2005

In phase 2 of our study we interviewed 10 junior and senior pre-med or other health-related majors who were currently taking algebra-based physics. We used a rigid, semi-structured (but otherwise very similar to Phase 1) protocol, that also included a general self-reflective discussion where students were free to express any opinion about the topic of X-rays and medical imaging, their relevance in the pre-med Physics curriculum and their views on how they should be taught. We also added the question "How would you explain X-rays to a 12-year old child?" giving the students another chance to express their views in more simple and clear if not scientific terms. This last question allowed us to double-check their mental models about the phenomena.

Phase 3 – Spring 2006

Having accumulated extensive information on what to expect from our targeted pre-med audience, during Phase 3, we focused more on the fact that the final results of our research would be the development of teaching materials. Therefore, we extended the interview process into two stages - one clinical and one teaching interview with each student. The first stage remained basically unchanged from Phase 2 (since it proved to be comprehensive enough and allowed comparison for reliability purposes).

Addressing the issue of the electromagnetic (light) nature of X-rays, their different ways of interaction with the materials of different properties (and also addressing geometrical issues that may arise during CAT scan image processing) we designed a small individual lab that used LEGO bricks.

We also have taken into account the learning cycle paradigm [7] - so the exploration, introduction of a concept and the application of a concept stages could be more or less clearly identified. This sequence was built around the "most convenient" concept - the blocking ability of a material. (Strictly speaking, this blocking ability should be separated into reflection, refraction and absorption; however for our image-processing purposes this elaboration was not necessary).

The students were presented with a "black box" which was built from semi-transparent LEGO bricks and were told that an object of an unknown shape was inside and that it is made out of the same semi-transparent bricks as the walls. (There was no non-destructive way that they could see it directly.). Students were asked how they could determine the shape of this object. If students were unable to answer the question, a couple of scaffolding steps were provided, which included showing students the source of light (red LED, light-emitting diode) and the light detector (photovoltmeter).

To facilitate the task and the discussion another similar box with an object of a different shape inside was shown to students. In this case they were allowed to open it before making assumptions (the exploration stage finished here).

Subsequently the students were asked how they thought the readings of the photovoltmeter depended on the number of bricks through which the light had gone through. They were prompted to make a prediction about what would happen if we were to add more bricks one by one and measure the light that had gone through that sequence. This concluded the concept introduction stage.

Then the students made their final prediction about the result – what is inside the first box (the final application of the concept). They again engaged in a general discussion about the activities and their order and relevance in the pre-med labs followed.

Overall, five pre-med students, five other health-related majors and two engineering students were interviewed. One student who participated in Phase 2 took part only in the teaching interview (stage 2) this time.

RESULTS & DISCUSSION

Many students felt confused when they were asked what else they could say about X-rays as they had almost never thought about the subject. They even declared that they did not know anything - although they often actually knew enough to answer our main questions. They just needed some encouragement, patience and scaffolding from the interviewer to invoke the transfer. Calculus-based physics students were more knowledgeable about the general topics of physics but that did not help them much to build up a coherent model of X-rays.

The pre-med students who were clinically interviewed during Phase 2 and 3 were much more interested in the dialog, felt that they should have known about it, and even very knowledgeable and assiduous students expressed this "constructive frustration" during the early stages of the interview. Then they openly and extensively talked about it in the self-reflective part of the interviews.

All of the pre-med students were enthusiastic about the Phase 3 teaching interview. Engineering majors were more reluctant while other pre-health-professionals could be described as having been moderately willing.

One of the most eager pre-med students responded: "*I think it's really cool... interesting and... I mean... it's one of the most interesting physics kind of labs... kinds of things I ever done. It really gives the idea of what's going on...*"

There was only one exception where a student thought that the proposed lab was somewhat "irrelevant" but even in that case the student liked the routine; she just did not believe that a physics lab could be likable: "*I don't know whether it would be any better... I mean I liked this... you could really see more visually... we did really do stuff like this with X-rays.*"

Essentially our study revealed the following main themes:

1) Pre-med (and some other) students' ideas about X-rays can be described as models although these models are rarely consistent. Even for the students with greater knowledge (like pre-engineers) and greater interest (like pre-meds) these models are incoherent. Although in the interview process, through Socratic dialogue and careful leading, they often successfully tried to put together the pieces of knowledge transferred from their physics classes and combine them with other pieces of information. These models typically include the following components:

a) Almost all the interviewed students associate X-ray visibility of different objects with their density (12 out of 13 for the last phase) although not necessarily regular mass density, sometimes they mean something different - like concentration. However they sometimes mention other possible options: "*They* (darker regions on an X-ray picture) *are dense... Or may be just because of the structure of it? Permeability, I guess... But I think it has more with the density.*" So these strong and stable intuitive associations can still be characterized as basic elements of reasoning (how things "simply happen") so called phenomenological primitives (p-primes) [8]

b) Those students who successfully invoked and transferred their knowledge that X-rays are a part of "the spectrum" (12 out of 13) usually cannot recall whether they belong to the longer or shorter part of the wavelength spectrum (7 out of 13); they even tend to put them not randomly but rather mistakenly in the longer part of the spectrum, apparently making the association "longer – bigger – stronger.", here we observe the whole combination or chain of p-primes

The characteristic of wavelength comes into their mind much more easily and quickly than frequency and this fact affects their further conclusions a lot.

c) This previous association is coupled together with another important one - that X-rays are more damaging than most of the others so they have to be "stronger" and "bigger" in some relation.

d) When prompted to think about other wave characteristics of X-rays - like frequency - those students who chose longer wavelength for X-rays tended to change their opinion - now higher frequency is already associated with stronger, more dangerous waves including X-rays.

2) Students also easily recognize ultrasound (sonogram) pictures. They successfully transfer almost all of the sound properties to ultrasound, although how

exactly ultrasound pictures are produced appears to be a mystery to them. They easily assume that taking ultrasound pictures isn't hazardous since it's used for looking at delicate unborn babies without any safety measures: "*With X-rays I have to take a lot of precautions and you want to limit the exposure... With ultrasound I've never heard that... so I am thinking that it's OK*". Students understand that sound is more like a "vibration" and light is something different. They express this distinction using different terms such as radiation, photons, particles, perpendicular magnetic and electrical and transfer different "signature" features and concepts. Of course the usual particle-wave duality difficulties arise here and we often pursued them although this was not the main purpose of our research.

3) Students who can recall more details from the X-ray procedures tend to associate X-rays with shadow-like processes since we are recording the light that passes (or doesn't pass) through our bones instead of the light that is reflected from them as we do in regular photography (their transfer of learning from personal experience overcomes others sources here): "*Photography... is basically... when you are flashing... and getting... and it coming back to your camera... while this is still on the other side of the hand... that is going through... it's X-rays that make it through... and hit the film... and may it turn white.*"

However, when asked how they would explain x-rays to a 12-year old they often just said that it was making a picture of you with rays that could penetrate through the skin – so again they appear to retreat to the photography analogy: "*I would explain X-rays take a picture, but the flash on the camera can show through the skin... except for the bones... so it's like taking picture of the bones.*" Thus, their judgment appears to be very context-dependent.

CONCLUSION AND FUTURE WORK

During our study we found evidence that students transfer pieces of knowledge from very different sources such as their own X-ray experiences, previous Physics and other science courses and the mass media. This transfer results in mental models that are not necessary stable, consistent or coherent. However, these models are popular and persistent among pre-med students and should be taken into account when designing the X-ray module for our Medical Physics curriculum and we have to build on them when planning new instructional materials. Students liked the proposed teaching activities, enthusiastically learned from them and we will continue develop them, taking into account the issues that arose in pre-med physics education in general and in our interviews series particularly.

REFERENCES

1. Nelson Christensen, *Eur. J. Phys.* **22** 421-427, Medical physics: the perfect intermediate level physics class (2001)
2. Suzanne Amador, *Biophysical Journal* Vol. 66 June 2217-2221, Teaching Medical Physics to General Audiences (1994)
3. Dean Zollman, Modern Miracle Medical Machines: a course in contemporary physics for future physicians *Proceedings of Groupe International de Recherche sur l'Enseignement de la Physique (GIREP)*, Lund, Sweden (2002)
4. N. Sanjay Rebello, Dean A. Zollman, Alicia R. Allbaugh, Paula V. Engelhardt, Kara E. Gray, Zdeslav Hrepic, Salomon F. Itza-Ortiz, Dynamic Transfer: A perspective from Physics Education Research in J. Mestre (Ed.), *Transfer of learning: Research and perspectives.* Greenwich, CT: Information Age, 2002
5. Dean Zollman, NSF grant proposal: "Technology & Model-Based Conceptual Assessment," 1999
6. Jean Piaget. *The Principles of Genetic Epistemology; Basic Books*, New York, 1972
7. Robert Karplus, *Science Curriculum Improvement Study: Teachers Handbook.* Lawrence Hall of Science, Berkeley, CA, 1974
8. Andrea diSessa. *Cognition and Instruction*, **10**, 105-225, Toward an epistemology of physics (1993)

AUTHOR INDEX

A

B

C

D

E

F

G

H

I

J

K

L

M

O

P

R

S

T

V

W

Y

Z